TEXTBOOK OF CLASSICAL MECHANICS

TEXTBOOK OF CLASSICAL MECHANICS

By
Shashi Kant Yadav

DISCOVERY PUBLISHING HOUSE PVT. LTD.
NEW DELHI-110 002

First Published – 2011

Reprinted – 2024

ISBN: 978-81-8356-834-0

Textbook of Classical Mechanics

Published by:

DISCOVERY PUBLISHING HOUSE PVT. LTD.

4383/4B, Ansari Road, Darya Ganj

New Delhi-110 002 (India)

Phone: +91-11-23279245; 23253475; 43596065

Mobile: +91 9811179893 / +91 9871656464

E-mail: discoverypublishinghouse@gmail.com

orderdphbooks@gmail.com

web: www.discoverypublishinggroup.com

Printed at:

Infinity Imaging Systems

Delhi

Preface

The purpose of this book "Classical mechanics" is to give an introductory account of the basic principles of Lagrangian formulation and hamiltonian formulation of mechanics. It is meant to serve as a textbook for the B.Sc., Physics students of Indian Universities. Care has been taken to dealing with the subject with modern outlook. A large number of questions and problems have been given at the end of each chapter. A better way to understand the various aspect of classical mechanics is to solve many problems. Keeping this in view many solved and unsolved problems have been included in this book.

Suggestion for the further improvement of this book will be gratefully acknowledge.

Author

Contents

1

Analysis of Strain

NTRODUCTION

In this chapter the state of strain at a point will be analysed. In elementary strength of materials two types of strains were introduced: (i) the extensional strain (in x or y direction) and (ii) the shear strain in the xy plane. Fig. 1 illustrates these two simple cases of strain. In each case, the initial or undeformed position of the element is indicated by full lines and the changed position by dotted lines. These are two–dimensional strains.

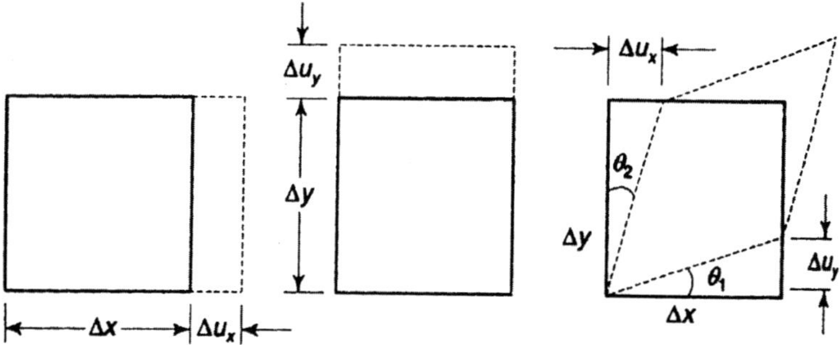

Fig 1

In Fig. 1(a), the element undergoes an extension Δu_x in x direction. The extensional or linear strain is defined as the change in length per unit initial length. If ε_x denotes the linear strain in x direction, then

$$\varepsilon_x = \frac{\Delta u_x}{\Delta x} \tag{1}$$

Similarly, the linear strain in y direction [Fig. 1(b)] is

$$\varepsilon_y = \frac{\Delta u_y}{\Delta y} \tag{2}$$

Fig. 1(c) shows the shear strain γ_{xy} in the *xy* plane. Shear strain γ_{xy} is defined as the change in the initial right angle between two line elements originally parallel to the *x* and *y axes*. In the figure, the total change in the angle is $\theta_1 + \theta_2$. If θ_1 and θ_2 are very small, then one can put

$$\theta_1 \text{ (in radians)} + \theta_2 \text{ (in radians)} = \tan\theta_1 + \tan\theta_2$$

From Fig. 1(c)

$$\tan\theta_1 = \frac{\Delta u_y}{\Delta x}, \qquad \tan\theta_2 = \frac{\Delta u_x}{\Delta y} \tag{3}$$

Therefore, the shear strain γ_{xy} is

$$\gamma_{xy} = \theta_1 + \theta_2 = \frac{\Delta u_y}{\Delta x} + \frac{\Delta u_x}{\Delta y} \tag{4}$$

Reduction in the initial right angle is considered to be a positive shear strain, since positive shear stress components τ_{xy} and τ_{yx} cause a decrease in the right angle.

In addition to these two types of strains, a third type of strain, called the volumetric strain, was also introduced in elementary strength of materials. This is change in volume per unit original volume. In this chapter, we will study strains in three dimensions and we will begin with the study of deformations.

DEFORMATIONS

In order to study deformation or change in the shape of a body, we compare the positions of material points before and after deformation.

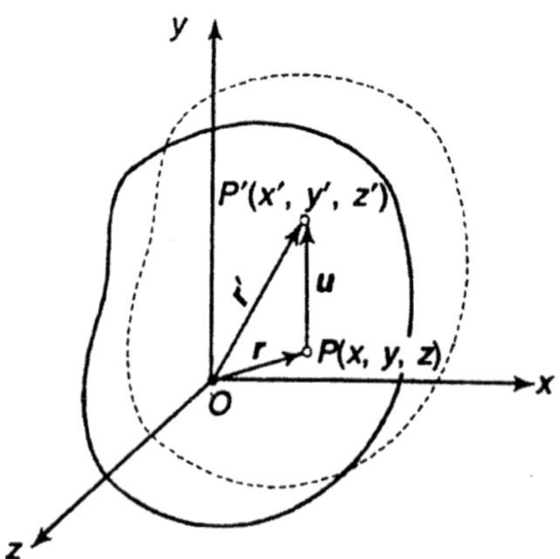

Fig 2

Let a point P belonging to the body and having coordinates (x, y, z) be displaced after deformations to P' with coordinates (x', y', z') (Fig. 2). Since P is displaced to P', the vector segment PP' is called the displacement vector and is denoted by u.

The displacement vector u has components u_x, u_y and u_z along the x, y and z axes respectively, and one can write

$$u = iu_x + ju_y + Ku_z \tag{5}$$

The displacement undergone by any point is a function of its initial coordinates. We assume that the displacement is defined throughout the volume of the body, i.e. the displacement vector u (both in magnitude and direction) of any point P belonging to the body is known once its coordinates are known. Then we can say that a displacement vector field has been defined throughout the volume of the body. If r is the position vector of point P, and r' that of point P', then

$$r' = r + u \tag{6}$$

$$u = r' - r$$

Example 1:

The displacement field for a body is given by

$$u = (x^2 + y)i + (3 + z)j + (x^2 + 2y)K$$

What is the deformed position of a point originally at (3, 1, – 2)?

Solution:

The displacement vector u at (3, 1, – 2) is

$$u = (3^2 + 1)\, i + (3 - 2)j + (3^2 + 2)K$$

$$= 10i + J + 11K$$

The initial position vector r of point P' is

$$r = 3i + J - 2K$$

The final position vector r' of point P' is

$$r' = r + u\ 13i + 2j + 9K$$

Example 2:

Two points P and Q in the undeformed body have coordinates (0, 0, 1) and (2, 0, – 1) respectively. Assuming that the displacement field given in Example 1 has been imposed on the body, what is the distance between points P and Q after deformation?

Solution:

The displacement vector at point P is

$$u(P) = (0 + 0)i + (3 + 1)j + (0 + 0)K = 4j$$

The displacement components at P are $u_x = 0$, $u_y = 4$, $u_z = 0$. Hence, the final coordinates of P after deformation are

$$P' : x + u_x = 0 + 0 = 0$$

$$y + u_y = 0 + 4 = 4$$

$$z + u_z = 1 + 0 = 1$$

or P' : (0, 4, 1)

Similarly, the displacement components at point Q are,

$$u_x = 4, \quad u_y = 2, \quad u_z = 4$$

and the coordinates of Q' are (6, 2, 3).

The distance $P'\,Q'$ is therefore

$$d' = (6^2 + 2^2 + 2^2)^{1/2} = 2\sqrt{11}$$

DEFORMATION IN THE NEIGHBOURHOOD OF A POINT

Let P be a point in the body with coordinates (x, y, z). Consider a small region surrounding the point P. Let Q be a point in this region with coordinates $(x + \Delta x, y + \Delta y, z + \Delta z)$. When the body undergoes deformation, the points P and Q move to P' and Q'. Let the displacement vector u at P have components (u_x, u_y, u_z) (Fig. 3).

The coordinates of P, P' and Q are

P: (x, y, z)

P': $(x + u_x, y + u_y, z + u_z)$

Q: $(x + \Delta x, y + \Delta y, z + \Delta z)$

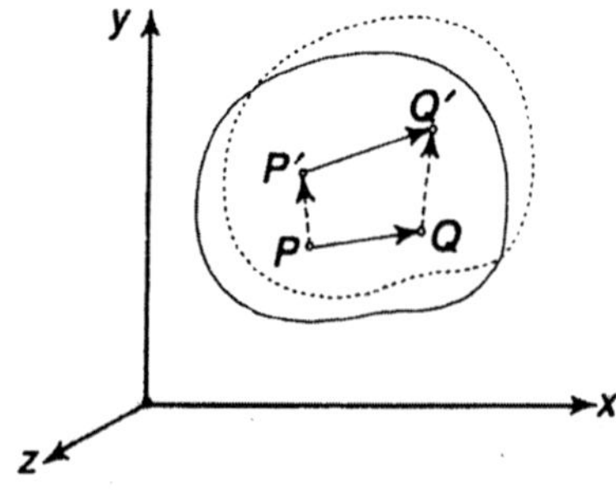

Fig 3

The displacement components at Q differ slightly from those at P since Q is away from P by Δx, Δy and Δz. Consequently, the displacements at Q are,

$$u + \Delta u_x,\ u_y + \Delta u_y,\ u_z + \Delta u_z.$$

If Q is very close to P, then to first–order approximation

$$\Delta u_x = \frac{\partial u_x}{\partial x}\Delta x + \frac{\partial u_x}{\partial y}\Delta y + \frac{\partial u_x}{\partial z}\Delta z \qquad (7a)$$

The first term on the right–hand side is the rate of increase of increase of u_x in x direction multiplied by the distance traversed, Δx. The second term is the rate of increase of u_x in y direction multiplied by the distance traversed in y direction, i.e. Δy. Similarly, we can also interpret the third term. For Δu_y and Δu_z too, we have

$$\Delta u_y = \frac{\partial u_y}{\partial x}\Delta x + \frac{\partial u_y}{\partial y}\Delta y + \frac{\partial u_y}{\partial z}\Delta z \qquad (7b)$$

$$\Delta u_z = \frac{\partial u_z}{\partial x}\Delta x + \frac{\partial u_z}{\partial y}\Delta y + \frac{\partial u_z}{\partial z}\Delta z \qquad (7c)$$

Therefore, the coordinates of Q' are,

$$Q' = (x + \Delta x + u_x + \Delta u_x,\ y + \Delta y + u_y + \Delta u_y,\ z + \Delta z + u_z + \Delta u_z) \qquad (8)$$

Before deformation, the segment PQ had components Δx, Δy and Δz along the three axes. After deformation, the segment $P'Q'$ has components $\Delta x + \Delta u_x, + \Delta y + \Delta u_y$, $\Delta z + \Delta u_z$ along the three axes. Terms like,

$$\frac{\partial u_x}{\partial x}, \frac{\partial u_x}{\partial y}, \frac{\partial u_x}{\partial z}, \text{ etc.}$$

are important in the analysis of strain. These are the gradients of the displacement components (at a point P) in x, y and z directions. One can represent these in the form of a matrix called the displacement gradient matrix as

$$\left[\frac{\partial u_i}{\partial x_j}\right] = \begin{bmatrix} \frac{\partial u_x}{\partial x} & \frac{\partial u_x}{\partial y} & \frac{\partial u_x}{\partial z} \\ \frac{\partial u_y}{\partial x} & \frac{\partial u_y}{\partial y} & \frac{\partial u_y}{\partial z} \\ \frac{\partial u_z}{\partial x} & \frac{\partial u_z}{\partial y} & \frac{\partial u_z}{\partial z} \end{bmatrix}$$

Example 3:

The following displacement field is imposed on a body

$$u = (xyi + 3x^2zj + 4K)\ 10^{-2}$$

Solution:

Consider a point P and a neighbouring point Q where PQ has the following direction cosines

$$n_x = 0.200, \quad n_y = 0.800, \quad n_z = 0.555$$

Point P has coordinates (2, 1, 3). If $PQ = \Delta s$, find the components of $P'Q'$ after deformation.

Before deformation, the components of PQ are

$$\Delta x = n_x \quad \Delta s = 0.2\ \Delta s$$

$$\Delta y = n_y \quad \Delta s = 0.8\ \Delta s$$

$$\Delta z = n_z \quad \Delta s = 0.555\ \Delta s$$

Using Eqs (7a) – (7c), the values of Δu_x, Δu_y, and Δu_z can be calculated. We are, using $P = 10^{-2}$,

$$u_x = pxy \qquad u_y = 3px^2z \qquad u_z = 4p$$

$$\frac{\partial u_x}{\partial x} = py \qquad \frac{\partial u_y}{\partial x} = 6pxz \qquad \frac{\partial u_z}{\partial x} = 0$$

$$\frac{\partial u_x}{\partial y} = px \qquad \frac{\partial u_y}{\partial y} = 0 \qquad \frac{\partial u_z}{\partial y} = 0$$

$$\frac{\partial u_x}{\partial z} = 0 \qquad \frac{\partial u_y}{\partial z} = 3px^2 \qquad \frac{\partial u_z}{\partial z} = 0$$

At point $P(2, 1, 3)$ therefore,

$$\Delta u_x = (y\Delta x + x\Delta y)P = (\Delta x + 2\Delta y)p$$

$$\Delta u_y = (6xz\Delta x + 3x^2\Delta z)P = (36\Delta x + 12\Delta z)p$$

$$\Delta u_z = 0$$

Substituting for Δx, Δy and Δz, the components of $\Delta s' = P'Q'$ are

$$\Delta x + \Delta u_x = 1.01\ \Delta x + 0.02\ \Delta y = (0.202 + 0.016)\ \Delta s = 0.218$$

Δs

$$\Delta y + \Delta u_y = (0.36\ \Delta x + \Delta y + 0.12\ \Delta z) = (0.072 + 0.8 + 0.067)\Delta s$$

$$= 0.939\ \Delta s$$

$$\Delta z + \Delta u_z\ \Delta z = 0.555\ \Delta s$$

Hence, the new vector $P'Q'$ can be written as

$$P'Q' = (0.218i + 0.939j + 0.555k)\ \Delta s$$

CHANGE IN LENGTH OF A LINEAR ELEMENT

Deformation causes a point $P(x, y, z)$ in the solid body under consideration to be displaced to a new position P' with coordinates $(x + u_x, y + u_y, z + u_z)$ where u_x, u_y and u_z are the displacement components. A neighbouring point Q with coordinates $(x + \Delta x, y + \Delta y, z + \Delta z)$ gets displaced to Q' with new coordinates $(x + \Delta x, + u_x + \Delta u_x, y + \Delta y + u_y + \Delta u_y, z + \Delta z + u_z + \Delta u_z)$. Hence, it is possible to determine the change in the length of the line element PQ caused by deformation. Let Δs be the length of the line element PQ. Its components are

$$\Delta s\text{: } (\Delta x,\ \Delta y,\ \Delta z)$$

$$\therefore \qquad \Delta s^2\text{: } (PQ)^2 = \Delta x^2 + \Delta y^2 + \Delta z^2$$

Let $\Delta s'$ be the length of $P'Q'$. Its components are

$$\Delta s'\text{: } (\Delta x' = \Delta x + \Delta u_x,\ \Delta y' = \Delta y + \Delta u_y,\ \Delta z' = \Delta z + \Delta u_z)$$

$$\therefore \qquad \Delta s^2\text{: } (P'\,Q')^2 = (\Delta x + \Delta u_x)^2 + (\Delta y + \Delta u_y)^2 + (\Delta z + \Delta u_z)^2$$

Form Eqs (7a) – (7c),

$$\Delta x' = \left(1 + \frac{\partial u_x}{\partial x}\right)\Delta x + \frac{\partial u_x}{\partial y}\Delta y + \frac{\partial u_x}{\partial z}\Delta z$$

$$\Delta y' = \frac{\partial u_y}{\partial x}\Delta x + \left(1 + \frac{\partial u_y}{\partial y}\right)\Delta y + \frac{\partial u_y}{\partial z}\Delta z \qquad (9)$$

$$\Delta z' = \frac{\partial u_z}{\partial x}\Delta x + \frac{\partial u_z}{\partial y}\Delta y + \left(1 + \frac{\partial u_z}{\partial z}\right)\Delta z$$

We take the difference between $\Delta s'^2$ and Δs^2

$$\begin{aligned}(P'Q')^2 - (PQ)^2 &= \Delta s'^2 - \Delta s^2 \\ &= (\Delta x'^2 + \Delta y'^2 + \Delta z'^2) - (\Delta x^2 + \Delta y^2 + \Delta z^2) \\ &= 2\,(E_{xx}\,\Delta x^2 + E_{yy}\,\Delta y^2 + E_{zz}\,\Delta z^2 + E_{xy}\,\Delta x\,\Delta y \\ &\qquad + E_{yz}\,\Delta y\,\Delta z + E_{xz}\,\Delta x\,\Delta z)\end{aligned} \qquad (10)$$

where $$E_{xx} = \frac{\partial u_x}{\partial x} + \frac{1}{2}\left[\left(\frac{\partial u_x}{\partial x}\right)^2 + \left(\frac{\partial u_y}{\partial x}\right)^2 + \left(\frac{\partial u_z}{\partial x}\right)^2\right]$$

$$E_{yy} = \frac{\partial u_y}{\partial y} + \frac{1}{2}\left[\left(\frac{\partial u_x}{\partial y}\right)^2 + \left(\frac{\partial u_y}{\partial y}\right)^2 + \left(\frac{\partial u_z}{\partial y}\right)^2\right]$$

$$E_{zz}=\frac{\partial u_z}{\partial z}+\frac{1}{2}\left[\left(\frac{\partial u_x}{\partial z}\right)^2+\left(\frac{\partial u_y}{\partial z}\right)^2+\left(\frac{\partial u_z}{\partial z}\right)^2\right] \tag{11}$$

$$E_{xy}=\frac{\partial u_x}{\partial y}+\frac{\partial u_y}{\partial x}+\frac{\partial u_x}{\partial x}\frac{\partial u_x}{\partial y}+\frac{\partial u_y}{\partial x}\frac{\partial u_y}{\partial y}+\frac{\partial u_z}{\partial x}\frac{\partial u_z}{\partial y}$$

$$E_{yz}=\frac{\partial u_y}{\partial z}+\frac{\partial u_z}{\partial y}+\frac{\partial u_x}{\partial y}\frac{\partial u_x}{\partial z}+\frac{\partial u_y}{\partial y}\frac{\partial u_y}{\partial z}+\frac{\partial u_z}{\partial y}\frac{\partial u_z}{\partial z}$$

$$E_{xz}=\frac{\partial u_x}{\partial z}+\frac{\partial u_z}{\partial x}+\frac{\partial u_x}{\partial x}\frac{\partial u_x}{\partial z}+\frac{\partial u_y}{\partial x}\frac{\partial u_y}{\partial z}+\frac{\partial u_z}{\partial x}\frac{\partial u_z}{\partial z}$$

It is observed that

$$E_{xy}=E_{yx},\qquad E_{yz}=E_{zy},\qquad E_{xz}=E_{zx}$$

We introduce the notation

$$E_{pQ}=\frac{\Delta s'-\Delta s}{\Delta s} \tag{12}$$

E_{PQ} is the ration of the increase in distance between the points P and Q caused by the deformation to their initial distance. This quantity will be called the relative extension of point P in the direction of point Q. Now,

$$\frac{\Delta s^{2'}-\Delta s^2}{2}=\left(\frac{\Delta s'-\Delta s}{\Delta s}+\frac{(\Delta s'-\Delta s)^2}{2\Delta s^2}\right)\Delta s^2$$

$$=\left(E_{pQ}+\frac{1}{2}E_{PQ}^2\right)\Delta s^2 \tag{13}$$

$$=E_{pQ}\left(1+\frac{1}{2}E_{pQ}\right)\Delta s^2$$

Form Eq. (10), substituting for $(\Delta s'^2-\Delta s^2)$

$$E_{pQ}\left(1+\frac{1}{2}E_{pQ}\right)\Delta s^2=E_{xx}\,\Delta x^2+E_{yy}\,\Delta y^2+E_{zz}\,\Delta z^2$$

$$+\,E_{xy}\,\Delta x\,\Delta y+E_{yz}\,\Delta y\,\Delta z+E_{xz}\,\Delta x\,\Delta z$$

If n_x, n_y and n_z are the direction cosines of PQ, then

$$n_x=\frac{\Delta x}{\Delta s},\quad n_y=\frac{\Delta y}{\Delta s},\quad n_z=\frac{\Delta z}{\Delta s}$$

Substituting these in the above expression

$$E_{pQ}\left(1+\frac{1}{2}E_{pQ}\right)=E_{xx}n_x^2+E_{yy}n_y^2+E_{zz}n_z^2+E_{xy}n_xn_y$$

$$+ E_{yz} n_y n_z + E_{xz} n_x n_z \qquad (14)$$

Equation (14) gives the value of the relative extension at point P in the direction PQ with direction cosines n_x, n_y and n_z.

If the line segment PQ is parallel to the x axis before deformation, then $n_x = 1$, $n_y = n_z = 0$ and

$$E_x\left(1+\frac{1}{2}E_x\right) = E_{xx} \qquad (15)$$

Hence, $E_x = [1 + 2E_{xx}]^{1/2} - 1$ (16)

This gives the relative extension of a line segment originally parallel to the x–axis. By analogy, we get

$$E_y = [1 + 2E_{yy}]^{1/2} - 1, \qquad E_z = [1 + 2E_{zz}]^{1/2} - 1 \qquad (17)$$

CHANGE IN LENGTH OF A LINEAR ELEMENT LINEAR COMPONENTS

Equation (11) in the previous section contains linear quantities like $\partial u_x/\partial x$, $\partial u_y/\partial y$, $\partial u_x/\partial y$, . . ., etc., as well as not–linear terms, like $(\partial u_x/\partial x)^2$, $(\partial u_x/\partial x . \partial u_x/\partial y)$, ... , etc. If the deformation imposed on the body is small, the quantities like $\partial u_x/\partial x$, $\partial u_y/\partial y$, etc. are extremely small so the their squares and products can be neglected. Retaining only linear terms, the following equations are obtained

$$\varepsilon_{xx} \frac{\partial u_x}{\partial x}, \quad \varepsilon_{yy} \frac{\partial u_y}{\partial y}, \quad \varepsilon_{zz} \frac{\partial u_z}{\partial z} \qquad (18)$$

$$\gamma_{xy} \frac{\partial u_x}{\partial y}+\frac{\partial u_y}{\partial x}, \quad \gamma_{yz} \frac{\partial u_y}{\partial z}+\frac{\partial u_z}{\partial y}, \quad \gamma_{xz} \frac{\partial u_x}{\partial z}+\frac{\partial u_z}{\partial x} \qquad (19)$$

$$E_{PQ} \approx \varepsilon_{PQ} = \varepsilon_{xx} n_x^2 + \varepsilon_{yy} n_y^2 + \varepsilon_{zz} n_z^2 + \varepsilon_{xy} n_x n_y + \varepsilon_{yz} n_y n_z + \varepsilon_{xz} n_x n_z \qquad (20)$$

Equation 20 directly gives the linear strain at point P in the direction PQ with direction cosines n_x, n_y *nz. When* $n_x = 1$, $n_y = n_z = 0$, the line element PQ is parallel to the axis and the linear strain is

$$E_x \approx \varepsilon_{xx} = \frac{\partial u_x}{\partial x}$$

Similarly, $E_y \approx \varepsilon_{yy} = \dfrac{\partial u_y}{\partial y}$ and $E_z \approx \varepsilon_{zz} = \dfrac{\partial u_z}{\partial z}$

are the linear strains in y and z directions respectively. In the subsequent analyses, we will use only the linear terms in strain components and neglect squares and products of strain components. The relations expressed by Eqs (18) and (19) are known as the strain displacement relations of Cauchy.

RECTANGULAR STRAIN COMPONENTS

ε_{xx}, ε_{yy} and ε_{zz} are the linear strains at a point in x, y and z directions. It will be shown later that γ_{xy}, γ_{yz} and γ_{xz} represent shear strains in *sy, yz* and *xz* planes respectively. Analogous to rectangular stress components, these six strain components are called the rectangular strain components at a point.

THE STATE OF STRAIN AT A POINT

Knowing the six rectangular strain components at a point P, one can calculate the linear strain in any direction PQ, using Eq. (20). The totality of all linear strains in every possible direction PQ defines the state of strain at point P. This definition is similar to that of the state of stress at a point. Since all that is required to determine the state of strain are the six rectangular strain components, these six components are said to define the state of strain at a point. We can write this as

$$\left[\varepsilon_{ij}\right]=\begin{bmatrix}\varepsilon_{xx} & \gamma_{xy} & \gamma_{xz}\\ \gamma_{xy} & \varepsilon_{yy} & \gamma_{yz}\\ \gamma_{xz} & \gamma_{yz} & \varepsilon_{zz}\end{bmatrix} \tag{21}$$

To maintain consistency, we could have written

$$\varepsilon_{xy} = \gamma_{xy}, \qquad \varepsilon_{yz} = \gamma_{yz}, \qquad \varepsilon_{xz} = \gamma_{xz}$$

but as it is customary to represent the shear strain by γ, we have retained this notation. In the theory of elasticity, $1/2\gamma_{xy}$ is written as e_{xy}, i.e.

$$\frac{1}{2}\gamma_{xy} = \frac{1}{2}\left(\frac{\partial u_x}{\partial y}+\frac{\partial u_y}{\partial x}\right) = e_{xy} \tag{22}$$

If we follow the above notation and use

$$e_{xx} = \varepsilon_{xx}, \qquad e_{yy} = \varepsilon_{yy}, \qquad e_{zz} = \varepsilon_{zz}$$

then Eq. (20) can be written in a very short form as

$$\varepsilon_{PQ} = \sum_i \sum_i e_{ij} n_i n_j$$

where i and j are summed over x, y and z Note the $e_{ij} = e_{ji}$

INTERPRETATION OF $\gamma_{xy}, \gamma_{yz}, \gamma_{xz}$ AS SHEAR STRAIN COMPONENTS

It was shown in the previous section that

$$\varepsilon_{xx} = \frac{\partial u_x}{\partial x}, \qquad \varepsilon_{yy} = \frac{\partial u_y}{\partial y}, \qquad \varepsilon_{zz} = \frac{\partial u_z}{\partial z}$$

represent the linear strains of line elements parallel to the x, y and z axes respectively. It was also stated that

$$\gamma_{xy} = \frac{\partial u_x}{\partial y} + \frac{\partial u_y}{\partial x}, \quad \gamma_{yz} = \frac{\partial u_y}{\partial z} + \frac{\partial u_z}{\partial y}, \quad \gamma_{xz} = \frac{\partial u_x}{\partial z} + \frac{\partial u_z}{\partial x}$$

represent the shear strains in the xy, yz and xz planes respectively. This can be shown as follows.

Consider two line elements, PQ and PR, originally perpendicular to each other and parallel to the x and y axes respectively (Fig. 4a).

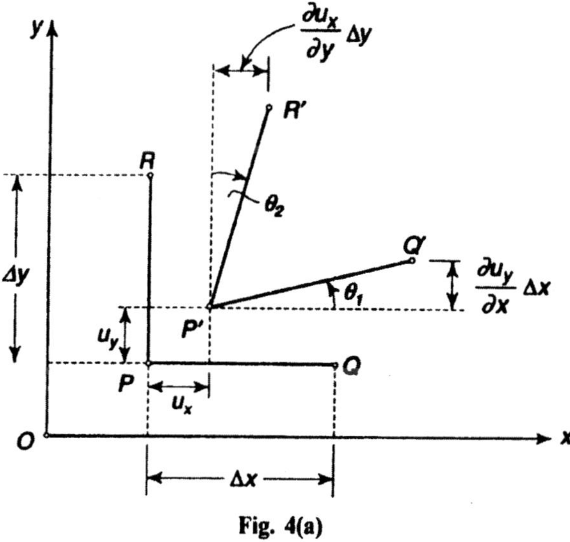

Fig. 4(a)

Let the coordinates of P be (x, y) before deformation and let the lengths of PQ and PR be Δx and Δy respectively. After deformation, point P moves to P', point Q to Q' and point R to R'.

Let u_x, u_y be the displacements of point P, so that the coordinates of P' are $(x + u_x y + u_y)$. Since point Q is Δξ δισταvχε away from P, the displacement components of Q $(x + \Delta x, y)$ are

$$u_x + \frac{\partial u_x}{\partial x}\Delta x \quad \text{and} \quad u_y + \frac{\partial u_y}{\partial x}\Delta x$$

Similarly, the displacement components of R $(x + y, \Delta y)$ are

$$u_x + \frac{\partial u_x}{\partial y}\Delta y \quad \text{and} \quad u_y + \frac{\partial u_y}{\partial y}\Delta y$$

From Fig. 4(a), it is seen that if θ_1 and θ_2 are small, then

$$\theta_1 \approx \tan\ \theta_1 = \frac{\partial u_y}{\partial x}$$

$$\theta_2 \approx \tan\ \theta_2 = \frac{\partial u_x}{\partial y} \qquad (23)$$

so that the total change in the original right angle is

$$\theta_1 + \theta_2 = \frac{\partial u_x}{\partial y} + \frac{\partial u_y}{\partial x} = \gamma_{xy} \qquad (24)$$

This is the shear strain in the xy plane at P. Similarly, the shear strains γ_{yx} and γ_{zx} can be interpreted appropriately.

If the element PQR undergoes a pure rigid body rotation through a small angular displacement, then from Fig. 4(b) we note

$$\omega_{zo} = \frac{\partial u_y}{\partial x} = -\frac{\partial u_x}{\partial y}$$

taking the counter-clockwise rotation as positive. The negative sign in $(-\partial u_x / \partial y)$ comes since a positive $\partial u_x / \partial y$ will give a movement from the y to the x axis as shown in Fig. 4(a). No strain occurs during this rigid body displacement. We define

$$\omega_z = \frac{1}{2}\left(\frac{\partial u_y}{\partial x} - \frac{\partial u_x}{\partial y}\right) = \omega_{yx} \qquad (25)$$

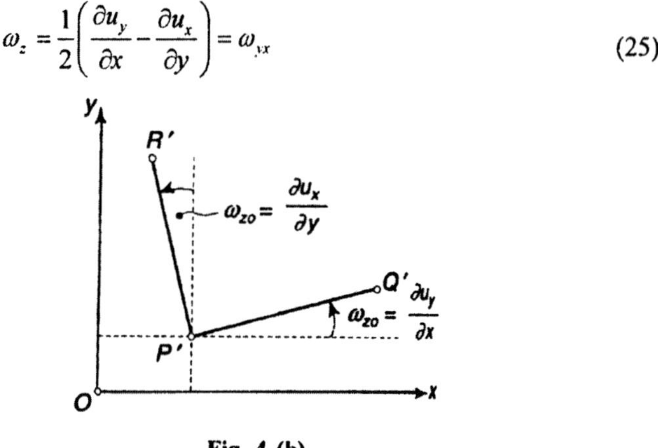

Fig. 4 (b)

This represents the average of the sum of rotations of the x and y elements and is called the rotational component. Similarly, for rotations about the x and y axes, we get

$$\omega_x = \frac{1}{2}\left(\frac{\partial u_z}{\partial y} - \frac{\partial u_y}{\partial z}\right) = \omega_{zy} \qquad (26)$$

$$\omega_y = \frac{1}{2}\left(\frac{\partial u_x}{\partial z} - \frac{\partial u_z}{\partial x}\right) = \omega_{xz} \tag{27}$$

Example 4:

Consider the displacement field

$$u = [y^2 i + 3yzj + (4 + 6x^2)k]\ 10^{-2}$$

What are the rectangular strain components at the point P(1, 0, 2)?

Solution:

Use only linear terms.

$$u_x = y^2 \cdot 10^{-2} \qquad u_y = 3yz \cdot 10^{-2} \qquad u_z = (4 + 6x^2) \cdot 10^{-2}$$

$$\frac{\partial u_x}{\partial x} = 0 \qquad \frac{\partial u_y}{\partial x} = 0 \qquad \frac{\partial u_z}{\partial x} = 12x \cdot 10^{-2}$$

$$\frac{\partial u_x}{\partial y} = 2y \cdot 10^{-2} \qquad \frac{\partial u_y}{\partial y} = 3z \cdot 10^{-2} \qquad \frac{\partial u_z}{\partial y} = 0$$

$$\frac{\partial u_x}{\partial z} = 0 \qquad \frac{\partial u_y}{\partial z} = 3y \cdot 10^{-2} \qquad \frac{\partial u_z}{\partial z} = 0$$

The linear strains at (1, 0, 2) are

$$\varepsilon_{xx} = \frac{\partial u_x}{\partial x} = 0, \qquad \varepsilon_{yy} = \frac{\partial u_y}{\partial y} = 6\times 10^{-2}, \qquad \varepsilon_{zz} = \frac{\partial u_z}{\partial z} = 0$$

The shear strains at (1, 0, 2) are

$$\gamma_{xy} = \frac{\partial u_x}{\partial y} + \frac{\partial u_y}{\partial x} = 0 + 0 = 0$$

$$\gamma_{yz} = \frac{\partial u_y}{\partial z} + \frac{\partial u_z}{\partial y} = 0 + 0 = 0$$

$$\gamma_{xz} = \frac{\partial u_x}{\partial z} + \frac{\partial u_z}{\partial x} = 0 + 12\times 10^{-2} = 12\times 10^{-2}$$

CHANGE IN DIRECTION OF A LINEAR ELEMENT

It is easy to calculate the change in the orientation of a linear element resulting from the deformation of the solid body. Let PQ be the element of length Δs, with direction cosines n_x, n_y and n_z. After deformation, the element becomes $P'Q'$ of length $\Delta s'$, with direction cosines n'_x, n'_y and n'_z. If u_x, u_y u_z are displacement components of point P, then the displacement components of point Q are.

$$u_x + \Delta u_x, \qquad u_y + \Delta u_y, \qquad u_z + \Delta u_z$$

where Δu_x, Δu_y and Δu_z are given by Eq. (7a) – (7c).

From Eq. (2.12), remembering that in the linear range $E_{PQ} = \varepsilon_{PQ}$,

$$\Delta s' = \Delta s\ (1 + \varepsilon_{PQ}) \tag{28}$$

The coordinates of P, Q, P' and Q are as follows:

P: (x, y, z)

Q: $(x + \Delta x, y + \Delta y, z + \Delta z)$

P': $(x + u_x, y + u_y, z + u_z)$

Q': $(x + \Delta x + u_x + \Delta u_x, y + \Delta y + u_y + \Delta u_y, z + \Delta z + u_z + \Delta u_z)$

Hence,

$$n_x = \frac{\Delta x}{\Delta s}, \qquad n_y = \frac{\Delta y}{\Delta s}, \qquad n_z = \frac{\Delta z}{\Delta s}$$

$$n_x' = \frac{\Delta x + \Delta u_x}{\Delta s'}, \qquad n'_y = \frac{\Delta y + \Delta u_y}{\Delta s'}, \qquad n_z = \frac{\Delta z + \Delta u_z}{\Delta s'}$$

Substituting for $\Delta s'$ from Eq. (28) and for Δu_x, Δu_y, Δu_z from Eq. (7a) – (7c)

$$n_x' = \frac{1}{1+\varepsilon_{PQ}}\left[\left(1+\frac{\partial u_x}{\partial x}\right)n_x + \frac{\partial u_x}{\partial y}n_y + \frac{\partial u_z}{\partial z}n_z\right]$$

$$n_y' = \frac{1}{1+\varepsilon_{PQ}}\left[\frac{\partial u_y}{\partial x}n_x + \left(1+\frac{\partial u_y}{\partial y}\right)n_y + \frac{\partial u_y}{\partial z}n_z\right]$$

$$n_z' = \frac{1}{1+\varepsilon_{PQ}}\left[\frac{\partial u_z}{\partial x}n_x + \frac{\partial u_z}{\partial y}n_y + \left(1+\frac{\partial u_z}{\partial z}\right)n_z\right] \tag{29}$$

The value of ε_{PQ} is obtained suing Eq. (20).

CUBICAL DILATATION

Consider a point A with coordinates (x, y, z) and a neighbouring point B with coordinates $(x + \Delta x, y + \Delta y, z + \Delta z)$. After deformation, the points A and B move to A' and B' with coordinates

A' : $(x + u_x, y + u_y, z + u_z)$

B' : $(x + \Delta x + u_x + \Delta u_x, y + \Delta y + u_y + \Delta u_y, z + \Delta z + u_z + \Delta u_z)$

where u_x, u_y and u_z are the components of displacements of point A, and from Eqs (7a)-(7c)

$$\Delta u_x = \frac{\partial u_x}{\partial x}\Delta x + \frac{\partial u_x}{\partial y}\Delta y + \frac{\partial u_x}{\partial z}\Delta z$$

$$\Delta u_y = \frac{\partial u_y}{\partial x}\Delta x + \frac{\partial u_y}{\partial y}\Delta y + \frac{\partial u_y}{\partial z}\Delta z$$

$$\Delta u_z = \frac{\partial u_z}{\partial x}\Delta x + \frac{\partial u_z}{\partial y}\Delta y + \frac{\partial u_z}{\partial z}\Delta z$$

The displaced segment $A'B'$ will have the following components along the x, y and z axes:

$$x \text{ axis: } \Delta x + \Delta u_x = \left(1 + \frac{\partial u_x}{\partial x}\right)\Delta x + \frac{\partial u_x}{\partial y}\Delta y + \frac{\partial u_x}{\partial z}\Delta z$$

$$y \text{ axis: } \Delta y + \Delta u_y = \frac{\partial u_x}{\partial y}\Delta x + \left(1 + \frac{\partial u_y}{\partial y}\right)\Delta y + \frac{\partial u_y}{\partial z}\Delta z \qquad (30)$$

$$z \text{ axis: } \Delta z + \Delta u_z = \frac{\partial u_z}{\partial y}\Delta x + \frac{\partial u_z}{\partial y}\Delta y + \left(1 + \frac{\partial u_z}{\partial z}\right)\Delta z$$

Consider now an infinitesimal rectangular parallelepiped with sides Δx, Δy, and Δz (Fig. 5). When the body undergoes deformation, the right parallelepiped *PQRS* becomes an oblique parallelepiped $p'\ Q'\ R'\ S'$.

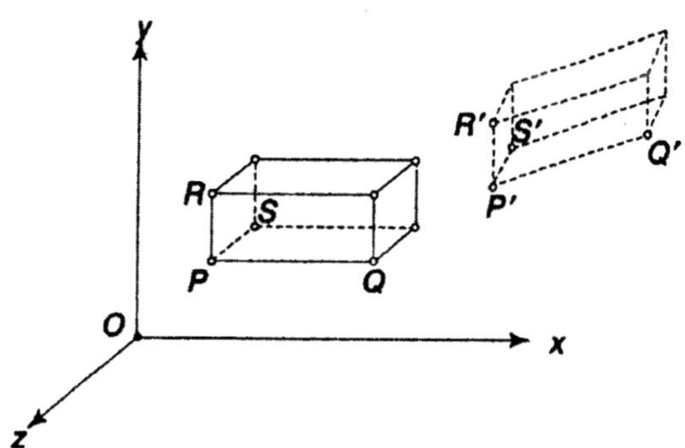

Fig. 5

Identifying *PQ* of Fig. 5 with *AB* of Eqs (30), one has $\Delta y = \Delta z = 0$. Then, from Eqs (30) the projections of $P'Q'$ will be

$$\text{along } x \text{ axis: } \left(1 + \frac{\partial u_x}{\partial x}\right)\Delta x$$

$$\text{along } y \text{ axis: } \frac{\partial u_y}{\partial x}\Delta x$$

$$\text{along } z \text{ axis: } \frac{\partial u_z}{\partial x}\Delta x$$

Hence, one can successively identify AB with PQ ($\Delta y = \Delta z = 0$), PR ($\Delta y = \Delta z = 0$), PR ($\Delta x = \Delta z = 0$) and get the components of $P'Q'$, $P'R'$ and $P'S'$ along the x, y and z axes as

	$P'Q'$	$P'R'$	$P'S'$
x axis:	$\left(1+\frac{\partial u_x}{\partial x}\right)\Delta x$	$\frac{\partial u_x}{\partial y}\Delta y$	$\frac{\partial u_x}{\partial z}\Delta z$
y axis:	$\frac{\partial u_y}{\partial x}\Delta x$	$\left(1+\frac{\partial u_y}{\partial y}\right)\Delta y$	$\frac{\partial u_y}{\partial z}\Delta z$
z axis:	$\frac{\partial u_z}{\partial x}\Delta x$	$\frac{\partial u_z}{\partial y}\Delta y$	$\left(1+\frac{\partial u_z}{\partial z}\right)\Delta z$

The volume of the right parallelepiped before deformation is equal to $V = \Delta x\ \Delta y\ \Delta z$. The volume of the deformed parallelepiped is obtained from the well–known formula from analytic geometry as

$$V' = V + \Delta V = D \cdot \Delta x\ \Delta y\ \Delta z$$

where D is the following determinant:

$$D = \begin{vmatrix} \left(1+\frac{\partial u_x}{\partial x}\right) & \frac{\partial u_x}{\partial y} & \frac{\partial u_x}{\partial z} \\ \frac{\partial u_y}{\partial x} & \left(1+\frac{\partial u_y}{\partial y}\right) & \frac{\partial u_y}{\partial z} \\ \frac{\partial u_z}{\partial x} & \frac{\partial u_z}{\partial y} & \left(1+\frac{\partial u_z}{\partial z}\right) \end{vmatrix} \tag{31}$$

If we assume that the strains are small quantities such that their squares and products can be neglected, the above determinant becomes

$$D = 1+\frac{\partial u_x}{\partial x}+\frac{\partial u_y}{\partial y}+\frac{\partial u_z}{\partial z}$$

$$= 1 + \varepsilon_{xx} + \varepsilon_{yy} + \varepsilon_{zz} \tag{32}$$

Hence, the new volume according to the linear strain theory will be

$$V' = V + \Delta V = (1 + \varepsilon_{xx} + \varepsilon_{yy} + \varepsilon_{zz})\ \Delta x\ \Delta y\ \Delta z \tag{33}$$

The volumetric strain is defined as

$$\Delta = \frac{\Delta V}{V} = \varepsilon_{xx} + \varepsilon_{yy} + \varepsilon_{zz} \tag{34}$$

Therefore, according to the linear theory, the volumetric strain, also known as cubical dilatation, is equal to the sum of three linear strains.

Example 5:

The following state of strain exists at a point P

$$[\varepsilon_{ij}] = \begin{bmatrix} 0.02 & -0.04 & 0 \\ -0.04 & 0.06 & 0.02 \\ 0 & -0.02 & 0 \end{bmatrix}$$

Solution:

In the direction PQ having direction cosines n_x 0.6, $n_y = 0$ and $n_z = 0.08$, determine ε_{PQ}.

From Eq. (20)

$\varepsilon_{PQ} = 0.02\ (0.36) + 0.06\ (0) + 0\ (0.64) - 0.04\ (0) - 0.02\ (0) + 0\ (0.48) = 0.007$

Example 6:

In Example 5, what is the cubical dilatation at point P?

Solution:

From Eq. (34)

$$\Delta = \varepsilon_{xx} + \varepsilon_{yy} + \varepsilon_{zz}$$
$$= 0.02 + 0.06 + 0 = 0.08$$

CHANGE IN THE ANGLE BETWEEN TWO LINE ELEMENTS

Let PQ be a line element with direction cosines n_{x1}, n_{y1}, n_{z1} and PR be another line element with direction cosines n_{x2}, n_{y2}, n_{z2} (Fig. 6). Let θ be the angle between the two line elements before deformation. After deformation, the line segments become $P'Q'$ and $P'R'$ with an included angle θ. We can determine θ easily from the results.

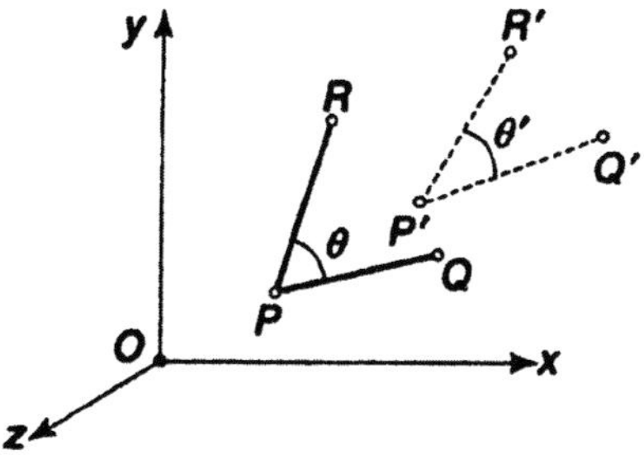

Fig. 6

From analytical geometry

$$\cos\theta' = n'_{x1}\, n'_{x2} + n'_{y1}\, n'_{y2} + n'_{z1}\, n'_{z2}$$

The values of $\cos\theta' = n'_{x1}\, n'_{y1}, n'_{z1}\, n'_{x2} + n'_{y2}$ and n'_{z2} can be substituted from Eq. (29). Neglecting squares and products of small strain components.

$$\cos\theta' = \frac{1}{(1+\varepsilon_{PQ})\,(1+\varepsilon_{PR})}[(1+2\varepsilon_{xx})n_{x1}\, n_{x2} + (1+2\varepsilon_{yy})n_{y1}\, n_{y2}$$

$$+(1+2\varepsilon_{zz})n_{z1}\, n_{z2} + \gamma_{xy}(n_{x1}\, n_{y2} + n_{x2}\, n_{y1})$$

$$+\gamma_{yz}(n_{y1}\, n_{z2} + n_{y2}\, n_{z1}) + \gamma_{zx}(n_{x1}\, n_{z2} + n_{x2}\, n_{y1})] \qquad (35)$$

In particular, if the two line segments *PQ* and *RS* are at right angles to each other before strain, then after strain,

$$\cos\theta' = \frac{1}{(1+\varepsilon_{PQ})\,(1+\varepsilon_{PR})}[2\varepsilon_{xx}n_{x1}\, n_{x2} 2\varepsilon_{yy}n_{y1}\, n_{y2} + 2\varepsilon_{zz}n_{z1}\, n_{z2}$$

$$+\gamma_{xy}(n_{x1}\, n_{y2} + n_{x2}\, n_{y1}) + \gamma_{yz}(n_{y1}\, n_{z2} + n_{y2}\, n_{z1})$$

$$+\gamma_{zx}(n_{x1}\, n_{z2} + n_{x2}\, n_{z1})] \qquad (36a)$$

Now (90°–θ) represents the change in the initial right angle. If this is denoted by α, then

$$\theta = 0° - \alpha \qquad (36b)$$

or $\qquad \cos\theta' = \cos(90° - \alpha) = \sin\alpha = \alpha$

since α is small. Therefore Eq. (2.36a) gives the shear strain α between *PQ* and *PR*. If we represent the directions of *PQ* and *PR* at *P* by *x'* and *y'* axes, then

$\gamma_{x'y'}$ at P = $\cos\theta$ = expression given in Eqs (36a), (36b) and (36c)

Example 7:

The displacement field for a body is given by

u = k (x² + y) i + k (y + z)j + k (x² – 2z²)k where k = 10⁻³

Solution:

At a point *P* (2, 2, 3), consider two line segments *PQ* and *PR* having the following direction cosines before deformation

$$PQ: n_{x1} = n_{y1} = n_{z1} = \frac{1}{\sqrt{3}}, \qquad PR: n_{x2} = n_{y2} = \frac{1}{\sqrt{2}}, \qquad n_{z2} = 0$$

Determine the angle between the two segments before and after deformation.

Before deformation, the angle θ between *PQ* and *PR* is

$$\cos\theta = n_{x1}\, n_{x2} + n_{y1}\, n_{y2} + n_{z1}\, n_{z2} = \frac{1}{\sqrt{6}} + \frac{1}{\sqrt{6}} = 0.8165$$

$\therefore \quad \theta = 35.3°$

The strain components at P after deformation are

$$\varepsilon_{xx} = \frac{\partial u_x}{\partial x} = 2kx = 4k, \quad \varepsilon_{yy} = \frac{\partial u_y}{\partial y} = k, \; \varepsilon_{zz} = \frac{\partial u_z}{\partial z} = 4kz = 12k$$

$$\gamma_{xy} = \frac{\partial u_x}{\partial y} + \frac{\partial u_y}{\partial x} = k, \quad \gamma_{yz} = \frac{\partial u_y}{\partial z} + \frac{\partial u_z}{\partial y} = k, \; \gamma_{zx} = \frac{\partial u_z}{\partial x} + \frac{\partial u_x}{\partial z} = 4k$$

The linear strains in directions PQ and PR are from Eq. (20)

$$\varepsilon_{PQ} = k\left[\left(4\times\frac{1}{3}\right) + \frac{1}{3} + \left(12\times\frac{1}{3}\right) + \left(1\times\frac{1}{3}\right) + \left(1\times\frac{1}{3}\right) + \left(4\times\frac{1}{3}\right)\right] = \frac{23}{3}k$$

$$\varepsilon_{PR} = k\left[\left(4\times\frac{1}{2}\right) + \left(1\times\frac{1}{2}\right) + (12\times 0) + \left(1\times\frac{1}{2}\right) + 0 + 0\right] = 3k$$

After deformation, the angle between $P'\,Q'$ and $P'\,R'$ is from Eq. (35)

$$\cos\theta' = \frac{1}{(1+23/3k)\,(1+3k)}\left[(1+8k)\frac{1}{\sqrt{6}} + (1+2k)\frac{1}{\sqrt{6}} + 0 + \left(\frac{1}{\sqrt{6}} + \frac{1}{\sqrt{6}}\right)k + \left(0 + \frac{1}{\sqrt{6}}\right)k + \left(0 + \frac{1}{\sqrt{6}}\right)k\right]$$

$= 0.8144 \qquad$ and $\qquad \theta = 35.5°$

PRINCIPAL AXES OF STRAIN AND PRINCIPAL STRAINS

When a displacement field is defined at a point P, the relative extension (i.e. strain) at P in the direction PQ is given by Eq. (20) as

$$\varepsilon_{PQ} = \varepsilon_{xx}n_x^2 + \varepsilon_{yy}n_y^2 + \varepsilon_{zz}n_z^2 + \gamma_{xy}n_x n_y + \gamma_{yz}n_y n_z + \gamma_{xz}n_x n_z$$

As the values of n_x, n_y and n_z change, we get different values of strain ε_{PQ}. Now we ask ourselves the following questions:

What is the direction (n_x, n_y n_z) along which the strain is an extremum (i.e. maximum or minimum) and what is the corresponding extremum value?

According to calculus, in order to find the maximum or the minimum, we would have to equate,

$$\partial\varepsilon_{PQ}/\partial n_x, \qquad \partial\varepsilon_{PQ}/\partial n_y, \qquad \partial\varepsilon_{PQ}/\partial n_z,$$

to zero, if n_x, n_y and n_z were all independent. However, n_x, n_y and n_z are not all independent since they are related by the condition

$$n_x^2 + n_y^2 + n_z^2 = 1 \tag{37}$$

Taking n_x, and n_y as independent and differentiating with respect to n_x and n_y we get

$$2n_x + 2n_z \frac{\partial n_z}{\partial n_x} = 0$$

$$2n_y + 2n_z \frac{\partial n_z}{\partial n_y} = 0 \qquad (38)$$

Differentiating ε_{PQ} with respect to n_x and n_y, and equating them to zero for extremum

$$0 = 2n_x\varepsilon_{xx} + n_y\gamma_{xy} + n_z\gamma_{zx} + \frac{\partial n_z}{\partial n_x}(n_x\gamma_{zx} + n_y\gamma_{zy} + 2n_z\varepsilon_{zz})$$

$$0 = 2n_y\varepsilon_{yy} + n_y\gamma_{xy} + n_z\gamma_{yz} + \frac{\partial n_z}{\partial n_y}(n_x\gamma_{zx} + n_y\gamma_{zy} + 2n_z\varepsilon_{zz})$$

Substituting for $\partial n_z/\partial n_x$ and $\partial n_z/\partial n_y$ from Eqs (38),

$$\frac{2n_x\varepsilon_{xx} + n_y\gamma_{xy} + n_z\gamma_{zx}}{n_x} = \frac{n_x\gamma_{zx} + n_y\gamma_{zy} + 2n_z\varepsilon_{zz}}{n_z}$$

$$\frac{2n_y\varepsilon_{yy} + n_y\gamma_{xy} + n_z\gamma_{yz}}{n_y} = \frac{n_x\gamma_{zx} + n_y\gamma_{zy} + 2n_z\varepsilon_{zz}}{n_z}$$

Denoting the right–hand side expression in the above two equations by 2ε and rearranging,

$$2\varepsilon_{xx}n_x + \gamma_{xy}n_y + \gamma_{xz}n_z - 2\varepsilon n_x = 0 \qquad (39a)$$

$$\gamma_{xy}n_x + 2\varepsilon_{yy}n_y + \gamma_{yz}n_z - 2\,\varepsilon n_y = 0 \qquad (39b)$$

and $$\gamma_{zx}n_x + \gamma_{zy}n_y + 2\varepsilon_{zz}n_z - 2\varepsilon n_z = 0 \qquad (39c)$$

One can solve Eqs (39a) – (39c) to get the values of n_x, n_y and n_z, which determine the direction along which the relative extension is an extremum. Let us assume that this direction has been determined. Multiplying the first equation by n_x, second by n_y and the third by n_z and adding them, we get

$$2(\varepsilon_{xx}\, n_x^2 + \varepsilon_{yy}\, n_y^2 + \varepsilon_{zz}n_z^2 + \gamma_{xy}n_xn_y + \gamma_{yz}n_yn_z + \gamma_{zx}n_zn_x) = 2\varepsilon(n_x^2 + n_y^2 + n_z^2)$$

If we impose the condition $n_x^2 + n_y^2 + n_z^2 = 1$, the right–hand side becomes equal to 2ε. From Eq. (20), the left–hand side is the expression for $2\varepsilon_{PQ}$. Therefore

$$\varepsilon_{PQ} = \varepsilon$$

This means that in Eqs (39a) – (39c) the values of n_x, n_y and n_z determine the direction along which the relative extension is an extremum and further, the value of ε is equal to this extremum. Equations (39a) – (39c) can be written as

$$\begin{aligned}
&(\varepsilon_{xx}-\varepsilon)n_x+\frac{1}{2}\gamma_{xy}n_y+\frac{1}{2}\gamma_{xz}n_z=0\\
&\frac{1}{2}\gamma_{yx}n_x+(\varepsilon_{yy}-\varepsilon)n_y+\frac{1}{2}\gamma_{yz}n_z=0\\
&\frac{1}{2}\gamma_{zx}n_x+\frac{1}{2}\gamma_{zy}n_y+(\varepsilon_{zz}-\varepsilon)n_z=0
\end{aligned} \tag{40a}$$

If we adopt the notation given in Eq. (22), i.e. put

$$\frac{1}{2}\gamma_{xy}=e_{xy},\quad \frac{1}{2}\gamma_{yz}=e_{yz},\quad \frac{1}{2}\gamma_{zx}=e_{zx}$$

then Eqs (40a) can be written as

$$\begin{aligned}
&(\varepsilon_{xx}-\varepsilon)n_x+e_{xy}n_y+e_{xz}n_z=0\\
&e_{zx}n_x+(\varepsilon_{yy}-\varepsilon)n_y+e_{yz}n_z=0\\
&e_{zx}n_x+e_{zx}n_y+(\varepsilon_{zz}-\varepsilon)n_z=0
\end{aligned} \tag{40b}$$

The above set of equations is homogeneous in n_x, n_y and n_z. For the existence of a non–trivial solution, the determinant of its coefficient must be equal to zero, i.e.

$$\begin{vmatrix}(\varepsilon_{xx}-\varepsilon) & e_{xy} & e_{xz}\\ e_{yx} & (\varepsilon_{yy}-\varepsilon) & e_{yz}\\ e_{zx} & e_{zy} & (\varepsilon_{zz}-\varepsilon)\end{vmatrix}=0 \tag{41}$$

Expanding the determinant, we get

$$\varepsilon^3-J_1\varepsilon^2+J_2\varepsilon-J_3=0 \tag{42}$$

where

$$J_1=\varepsilon_{xx}+\varepsilon_{yy}+\varepsilon_{zz} \tag{43}$$

$$J_2=\begin{vmatrix}\varepsilon_{xx} & e_{xy}\\ e_{yx} & \varepsilon_{yy}\end{vmatrix}+\begin{vmatrix}\varepsilon_{yy} & e_{yz}\\ e_{zy} & \varepsilon_{zz}\end{vmatrix}+\begin{vmatrix}\varepsilon_{xx} & e_{xz}\\ e_{zx} & \varepsilon_{zz}\end{vmatrix} \tag{44}$$

$$J_3=\begin{vmatrix}\varepsilon_{xx} & e_{xy} & e_{xz}\\ e_{yx} & \varepsilon_{yy} & e_{yz}\\ e_{zx} & e_{zy} & \varepsilon_{zz}\end{vmatrix} \tag{45}$$

If is important to observe that J_2 and J_3 involve e_{xy}, e_{yz}, and e_{zx} not γ_{xy}, γ_{yz}, and γ_{zx}. Equations (41) – (45) are all similar to Eqs (8), (9), (12), (13) and (14). The problem posed and its analysis are similar to the analysis of principal stress axes and principal stresses. The results of Sec. 1.10 – 1.15

can be applied to the case of strain.

For a given state of strain at point P, if the relative extension (i.e. strain) ε is an extremum in a direction n, then ε is the principal strain at P and n is the principal strain direction associated with ε.

In every state of strain there exist at least three mutually perpendicular principal axes and at most three distinct principal strains. The principal strains ε_1, ε_2, and ε_3 are the roots of the cubic equation.

$$\varepsilon^3 - J_1\varepsilon^2 + J_2\varepsilon - J_3 = 0 \tag{46}$$

where J_1, J_2, J_3 are the first, second and third invariants of strain. The principal directions associated with ε_1, ε_2 and ε_3 are obtained by substituting ε_i $(i = 2, 3)$ in the following equations and solving for n_x, n_y and n_z.

$$\begin{aligned} &(\varepsilon_{xx} - \varepsilon_i)n_x + e_{xy'}n_y + e_{xz}n_z = 0 \\ &e_{xy'}n_x + (\varepsilon_{yy} - \varepsilon_i)n_y + e_{yz}n_z = 0 \\ &n_x^2 + n_y^2 + n_z^2 = 1 \end{aligned} \tag{47}$$

If ε_1, ε_2, and ε_3 are distinct, then the axes of n_1, n_2 and n_3 are unique and mutually perpendicular. If, say $\varepsilon_1 = \varepsilon_2 \neq \varepsilon_3$, then the axis of n_3 is unique and every direction perpendicular to n_3 is a principal direction associated with $\varepsilon_1 = \varepsilon_2$.

If $\varepsilon_1 = \varepsilon_2 = \varepsilon_3$, then every direction is a principal direction.

Example 8:

The displacement field in micro units for a body is given by

$$u = (x^2 + y)i + (3 + z)j + (x^2 + 2y)k$$

Determine the principal strains at (3,1, – 2) and the direction of the minimum principal strain.

Solution:

The displacement components in micro units are,

$$u_x = x^2 + y, \qquad u_y = 3 + z, \qquad u_z = x^2 + 2y.$$

The rectangular strain components are

$$\varepsilon_{xx} = \frac{\partial u_x}{\partial x} = 2x, \qquad \varepsilon_{yy} = \frac{\partial u_y}{\partial y} = 0, \qquad \varepsilon_{zz} = \frac{\partial u_z}{\partial z} = 0$$

$$\gamma_{xy} = \frac{\partial u_x}{\partial y} + \frac{\partial u_y}{\partial x} = 1, \; \gamma_{yz} = \frac{\partial u_y}{\partial z} + \frac{\partial u_z}{\partial y} = 3, \; \gamma_{zx} = \frac{\partial u_z}{\partial y} + \frac{\partial u_z}{\partial z} = 2x$$

At point (3,1, –2) the strain components are therefore,

$$\varepsilon_{xx} = 6, \qquad \varepsilon_{yy} = 0, \qquad \varepsilon_{zz} = 0$$

$$\gamma_{xy} = 1, \qquad \gamma_{yz} = 3, \qquad \varepsilon_{zx} = 6$$

The strain invariants from Eqs (43) – (45) are

$$J_1 = \varepsilon_{xx} + \varepsilon_{yy} + \varepsilon_{zz} = 6$$

$$J_2 = \begin{vmatrix} 6 & \frac{1}{2} \\ \frac{1}{2} & 0 \end{vmatrix} + \begin{vmatrix} 0 & \frac{3}{2} \\ 3 & 0 \end{vmatrix} + \begin{vmatrix} 6 & 3 \\ 3 & 0 \end{vmatrix} = -\frac{23}{2}$$

Note that J_1 and J_3 involve $e_{xy} = \frac{1}{2}\gamma_{xy}$, $\quad e_{yz} = \frac{1}{2}\gamma_{yz}$, $\quad e_{zx} = \frac{1}{2}\gamma_{zx}$

$$J_3 = \begin{vmatrix} 6 & \frac{1}{2} & 3 \\ \frac{1}{2} & 0 & \frac{3}{2} \\ 3 & \frac{3}{2} & 0 \end{vmatrix} = -9$$

The cubic from Eq. (46) is

$$\varepsilon^3 - 6\varepsilon^2 - \frac{23}{2}\varepsilon + 9 = 0$$

Following the standard method suggested in Sec. 1.15

$$a = \frac{1}{3}\left(-\frac{69}{2} - 36\right) = -\frac{47}{2}$$

$$b = \frac{1}{27}(-432 - 621 + 243) = -30$$

$$\cos\phi = -\frac{-30}{2 \times \sqrt{-a^3/27}} = 0.684$$

$$\therefore \qquad \phi = 46°48'$$

$$g = 2\sqrt{-a/3} = 5.6$$

The principal strains in micro units are

$$\varepsilon_1 = g\ cos\ \phi/3 + 2 = +\ 7.39$$

$$\varepsilon_2 = g\ cos\ (\phi/3 + 120°) + 2 = -2$$

$$\varepsilon_3 = g\ cos\ (\phi/3 + 240°) + 2 = +0.61$$

As a check the first invariant J_1 is

$$\varepsilon_{xx} + \varepsilon_{yy} + \varepsilon_{zz} = \varepsilon_1 + \varepsilon_2 + \varepsilon_3 = 7.39 - 2 + 0.61 = 6$$

The second invariant J_2 is

$$\varepsilon_1\varepsilon_2 + \varepsilon_2\varepsilon_3 + \varepsilon_3\varepsilon_1 = -14.78 - 1.22 + 4.51 = -11.49$$

The third invariant J_3 is

$$\varepsilon_1\varepsilon_2\varepsilon_3 = 7.39 \times 2 \times 0.61 = -9$$

These agree with the earlier values.

The minimum principal strain is – 2. For this, from Eq. (47)

$$(6+2)n_x + \frac{1}{2}n_y + 3n_z = 0$$

$$\frac{1}{2}n_x + 2n_y + \frac{3}{2}n_z = 0$$

$$n_x^2 + n_y^2 + n_z^2 = 1$$

The solutions are $n_x = 0.267$, $n_y = 0.534$ and $n_z = -0.801$

Example 9:

For the state of strain given in Example 5, determine the principal strains and the directions of the maxi:num and minimum principal strains.

Solution:

From the strain matrix given, the invariants are

$$J_1 = \varepsilon_{xx} + \varepsilon_{yy} + \varepsilon_{zz} = 0.02 + 0.06 + 0 = 0.08$$

$$J_2 = \begin{vmatrix} 0.02 & -0.02 \\ -0.02 & 0.06 \end{vmatrix} + \begin{vmatrix} 0.06 & -0.01 \\ -0.01 & 0 \end{vmatrix} + \begin{vmatrix} 0.02 & 0 \\ 0 & 0 \end{vmatrix}$$

$$= (0.0012 - 0.0004) + (-0.00001) + 0 = 0.0007$$

$$J_3 = \begin{vmatrix} 0.02 & -0.02 & 0 \\ -0.02 & 0.06 & -0.01 \\ 01 & -0.01 & 0 \end{vmatrix} = 0.02(-0.0001) + 0 + 0 = -0.000002$$

The cubic equation is

$$\varepsilon^3 - 0.08\varepsilon^2 + 0.0007\varepsilon + 0.000002 = 0$$

Following the standard procedure described in Sec. 1.15, one can determine the principal strains. However, observing that the constant J_3 in the cubic is very small, one can ignore it and write the cubic as

$$\varepsilon^2 - 0.08\varepsilon^2 + 0.0007\varepsilon = 0$$

One of the solutions obviously is $\varepsilon = 0$. For the other two solutions (ε not equal to zero), dividing by ε

$$\varepsilon^2 - 0.08\ \varepsilon + 0.0007\varepsilon = 0$$

The solutions of this quadratic equation are

$$\varepsilon = 0.4 \pm 0.035, \text{ i.e. } 0.075 \quad \text{and} \quad 0.005$$

Rearranging such that $\varepsilon_1 \geq \varepsilon_2 \geq \varepsilon_3$, the principal strains are

$$\varepsilon_1 = 0.07, \qquad \varepsilon_2 = 0.01, \qquad \varepsilon_3 = 0$$

As a check:

$$J_1 = \varepsilon_1 + \varepsilon_2 + \varepsilon_3 = 0.07 + 0.01 = 0.08$$

$$J_2 = \varepsilon_1\varepsilon_2 + \varepsilon_2\varepsilon_2 + \varepsilon_3\varepsilon_1 = (0.07 \times 0.01) = 0.0007$$

$$J_3 = \varepsilon_1\varepsilon_2\varepsilon_3 = 0 \qquad \text{(This was assumed as zero)}$$

Hence, these values agree with their previous values. To determine the direction of $\varepsilon_1 = 0.07$, from Eqs (47)

$$(0.02 - 0.07)\ n_x - 0.02n_y = 0$$

$$-\ 0.02n_x + (0.06 - 0.07)\ n_y - 0.01n_z = 0$$

$$n_x^2 + n_y^2 + n_z^2 = 1$$

The solutions are $n_x = 0.44$, $n_y = -0.176$ and $n_z = 0.88$.

Similarly, for $\varepsilon_3 = 0$, from Eqs (47)

$$0.02n_x - 0.02n_y = 0$$

$$-\ 0.02n_x + 0.06n_y - 0.01n_z = 0$$

$$n_x^2 + n_y^2 + n_z^2 = 1$$

The solutions are $n_x = n_y = 0.236$ and $n_z = 0.944$.

PLANE STATE OF STRAIN

If, in a given state of strain, there exists a coordinate system *Oxyz,* such that for this system

$$\varepsilon_{zz} = 0, \qquad \gamma_{yz} = 0, \qquad \gamma_{zx} = 0 \tag{48}$$

then the state is said to have a plane state of strain parallel to the *xy* plane. The non–vanishing strain components are ε_{xx}, ε_{yy} and γ_{xy}.

If *PQ* is a line element in this *xy* plane, with direction cosines n_x, n_y, then

$$\varepsilon_{PQ} = \varepsilon_{xx}\ n_x^2 + \varepsilon_{yy}\ n_y^2 + \gamma_{xy}\ n_x n_y$$

or if *PQ* makes an angle θ with the *x* axis, then

$$\varepsilon_{PQ} = \varepsilon_{xx} \cos^2\theta + \varepsilon_{yy} \sin^2\theta + \frac{1}{2}\gamma_{xy} \sin 2\theta \tag{49}$$

If ε_1 and ε_2 are the principal strains, then

$$\varepsilon_1, \varepsilon_2 = \frac{\varepsilon_{xx} + \varepsilon_{yy}}{2} \pm \left[\left(\frac{\varepsilon_{xx} - \varepsilon_{yy}}{2} \right)^2 + \left(\frac{\gamma_{xy}}{2} \right)^2 \right]^{1/2} \tag{50}$$

Note the $\varepsilon_3 = \varepsilon_{zz}$ is also a principal strain. The principal strain axes make angles ϕ and $\phi + 90°$ with the x axis, such that

$$\tan 2\phi = \frac{\gamma_{xy}}{\varepsilon_{xx} - \varepsilon_{yy}} \tag{51}$$

The discussions and conclusions will be identical with the analysis of stress if we use ε_{xx}, ε_{yy}, and ε_{zz} in place of σ_x, σ_y, and σ_z respectively, and $e_{xy} = \frac{1}{2}\gamma_{xy}, e_{yz} = \frac{1}{2}\gamma_{yz}, e_{zx} = \frac{1}{2}\gamma_{zx}$ in place of τ_{xy}, τ_{yz} and τ_{zx} respectively.

THE PRINCIPAL AXES OF STRAIN REMAIN ORTHOGONAL AFTER STRAIN

Let PQ be one of the principal extensions or strain axes with direction cosines n_{x1}, n_{y1} and n_{z1}. Then according to Eqs (40b)

$$(\varepsilon_{xx} - \varepsilon_1)n_{x1} + e_{xy}n_{y1} + e_{xz}n_{z1} = 0$$

$$e_{xy}n_{x1} + (\varepsilon_{yy} - \varepsilon_1)n_{y1} + e_{yz}n_{z1} = 0$$

$$e_{xz}n_{x1} + e_{yz}n_{y1} + (\varepsilon_{zz} - \varepsilon_1)n_{z1} = 0$$

Let n_{x2}, n_{y2} and n_{z2} be the direction cosines of a line PR, perpendicular to PQ before strain. Therefore,

$$n_{x1}n_{x2} + n_{y1}n_{y1} + n_{z1}n_{z2} = 0$$

Multiplying Eq. (40b), given above, by n_{x2}, n_{y2} and n_{z2} respectively and adding, we get,

$$\varepsilon_{xx}n_{x1}n_{x2} + \varepsilon_{yy}n_{y1}n_{y2} + \varepsilon_{zz}n_{z1}n_{z2} + e_{xy}(n_{x1}n_{y2} + n_{y1}n_{x2}) + e_{yz}(n_{y1}n_{z2} + n_{y2}n_{z1}) + e_{zx}(n_{x1}n_{z2} + n_{x2}n_{z1}) = 0$$

Multiplying by 2 and putting

$$2e_{xy} = \gamma_{xy}, \quad 2e_{yz} = \gamma_{yz}, \qquad 2e_{zx} = \gamma_{zx}$$

we get

$$2\varepsilon_{xx}n_{x1}n_{x2} + 2\varepsilon_{yy}n_{y1}n_{y2} + 2\varepsilon_{zz}n_{z1}n_{z2} + \gamma_{xy}(n_{x1}n_{y2} + n_{y1}n_{x2}) + \gamma_{yz}(n_{y1}n_{z2} + n_{y2}n_{z1}) + \gamma_{zx}(n_{x1}n_{z2} + n_{z1}n_{x2}) = 0$$

Comparing the above with Eq. (36a), we get

$$\cos\theta\,(1 + \varepsilon_{PQ})(1 + \varepsilon_{PR}) = 0$$

where θ is the new angle between PQ and PR after strain.

Since ε_{PQ} and ε_{PR} are quite general, to satisfy the equation, $\theta = 90°$, i.e. the line segments remain perpendicular after strain also. Since *PR* is an arbitrary perpendicular line to the principal axis *PQ*, every line perpendicular to *PQ* before strain remains perpendicular after strain. In particular, *PR* can be the second principal axis of strain.

Repeating the above steps, if *PS* is the third principal axis of strain perpendicular to *PQ* and *PR*, it remains perpendicular after strain also. Therefore, at point *P*, we can identify a small rectangular element, with faces normal to the principal axes of strain, that will remain rectangular after strain also.

PLANE STRAINS IN POLAR COORDINATES

We now consider displacements and deformations of a two–dimensional radial element in polar coordinates. The polar coordinates of a point *a* are *r* and θ. The radial and circumferential displacements are denoted by u_r and u_θ. Consider an elementary radial element *abcd*, as shown in Fig. 7.

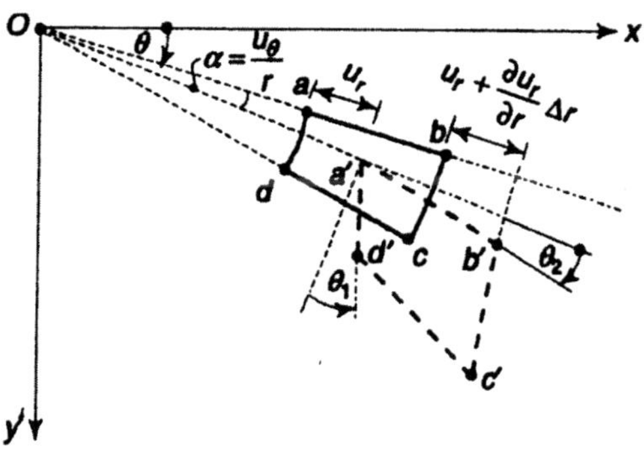

Fig. 7

Point *a* with coordinates (r, θ) gets displaced after deformation to position *a*' with coordinates $(r + u_r, \theta + \alpha)$. The neighbouring point *b* $(r + \Delta r, \theta)$ gets moved to *b*′ with coordinates

$$\left(r + \Delta r + u_r + \frac{\partial u_r}{\partial r}\Delta r, \theta + \alpha + \frac{\partial \alpha}{\partial r}\Delta r\right)$$

The length of *a*′ *b*' is therefore

$$\Delta r + \frac{\partial u_r}{\partial r}\Delta r$$

The radial strain ε_r is therefore

$$\varepsilon_r = \frac{\partial u_r}{\partial r} \tag{52}$$

The circumferential strain ε_θ is caused in two. If the element *abcd* undergoes a purely radial displacement, then the length $ad = r\,\Delta\theta$ changes to $(r + u_r)\Delta\theta$. The strain due to this radial movement alone is

$$\frac{u_r \Delta\theta}{r\Delta\theta} = \frac{u_r}{r}$$

In addition to this, the point *d* moves circumferentially to *d'* through the distance

$$u_\theta + \frac{\partial u_\theta}{\partial \theta}\Delta\theta$$

Since point *a* moves circumferentially through u_θ, the change in *ab* is $\frac{\partial u_\theta}{\partial \theta}\Delta\theta$. The strain due to this part is

$$\frac{\partial u_\theta}{\partial \theta}\frac{\Delta\theta}{r\Delta\theta} = \frac{1}{r}\frac{\partial u_\theta}{\partial \theta}$$

The total circumferential strain is therefore

$$\varepsilon_\theta = \frac{u_r}{r} + \frac{1}{r}\frac{\partial u_\theta}{\partial \theta} \tag{53}$$

To determine the shear strain we observe the following:

The circumferential displacement of *a* is u_θ, whereas that of *b* is $u_\theta + \frac{\partial u_\theta}{\partial r}\Delta r$. The magnitude of θ_2 is

$$\left[\left(u_\theta + \frac{\partial u_\theta}{\partial r}\Delta r\right) - \alpha(r + \Delta r)\right]\frac{1}{\Delta r}$$

But $\alpha = \frac{u_\theta}{r}$.

Hence, $$\theta_2 = \left(u_\theta + \frac{\partial u_\theta}{\partial r}\Delta r - u_\theta - \frac{u_\theta}{r}\Delta r\right)\frac{1}{\Delta r}$$

$$= \frac{\partial u_\theta}{\partial r} - \frac{u_\theta}{r}$$

Similarly, the radial displacement of *a* is u_r whereas that of *d* is

$u_r + \frac{\partial u_r}{\partial \theta}\Delta\theta$. Hence,

$$\theta_1 = \frac{1}{r\Delta\theta}\left[\left(u_r + \frac{\partial u_r}{\partial \theta}\Delta\theta\right) - u_r\right]$$

$$= \frac{1}{r}\frac{\partial u_r}{\partial \theta}$$

Hence, the shear strain $\gamma_{r\theta}$ is

$$\gamma_{r\theta} = \theta_1 + \theta_2 = \frac{1}{r}\frac{\partial u_r}{\partial \theta} + \frac{\partial u_r}{\partial \theta} - \frac{u_\theta}{r} \tag{54}$$

COMPATIBILITY CONDITIONS

It was observed that the displacement of a point in a solid body can be represented by a displacement vector u, which has components,

$$u_x, u_y, u_z.$$

along the three axes x, y and z respectively. The deformation at a point is specified by the six strain components,

$$\varepsilon_{xx}, \varepsilon_{yy}, \varepsilon_{zz}, \gamma_{xy}, \gamma_{yz} \text{ and } \gamma_{zx}.$$

The three displacement components and the six rectangular strain components are related by the six strain displacement relations of Cauchy, given by Eqs (18) and (19). The determination of the six strain components from the three displacement functions is straightforward since it involves only differentiation. However, the reverse operation, i.e. determination of the three displacement functions from the six strain components is more complicated since it involves integrating six equations to obtain three functions. One may expect, therefore, that all the six strain components cannot be prescribed number of these relations are six and they fall into two groups

First group: We have

$$\varepsilon_{xx} = \frac{\partial u_x}{\partial x}, \qquad \varepsilon_{yy} = \frac{\partial u_y}{\partial y}, \qquad \gamma_{xy} = \frac{\partial u_x}{\partial y} + \frac{\partial u_y}{\partial x}$$

Differentiate the first two of the above equations as follows:

$$\frac{\partial^2 \varepsilon_{xx}}{\partial y^2} = \frac{\partial^3 u_x}{\partial x \partial y^2} = \frac{\partial^2}{\partial x \partial y}\left(\frac{\partial u_x}{\partial y}\right)$$

$$\frac{\partial^2 \varepsilon_{yy}}{\partial x^2} = \frac{\partial^3 u_y}{\partial y \partial x^2} = \frac{\partial^2}{\partial x \partial y}\left(\frac{\partial u_y}{\partial x}\right)$$

Adding these two, we get

$$\frac{\partial^2}{\partial x \partial y} = \left(\frac{\partial u_x}{\partial y} + \frac{\partial u_y}{\partial x}\right) = \frac{\partial^2 \gamma_{xy}}{\partial x \partial y^2}$$

i.e.
$$\frac{\partial^2 \varepsilon_{xx}}{\partial y^2} + \frac{\partial^2 \varepsilon_{yy}}{\partial x^2} = \frac{\partial^2 \gamma_{xy}}{\partial x \partial y}$$

Similarly, by considering ε_{yy}, ε_{zz} and γ_{yz} and ε_{zz}, ε_{xx} and γ_{zx}, we get two more conditions. This leads us to the first group of conditions.

$$\frac{\partial^2 \varepsilon_{xx}}{\partial y^2} + \frac{\partial^2 \varepsilon_{yy}}{\partial x^2} = \frac{\partial^2 \gamma_{xy}}{\partial x \partial y}$$

$$\frac{\partial^2 \varepsilon_{yy}}{\partial z^2} + \frac{\partial^2 \varepsilon_{zz}}{\partial y^2} = \frac{\partial^2 \gamma_{yz}}{\partial y \partial z} \qquad (55)$$

$$\frac{\partial^2 \varepsilon_{zz}}{\partial x^2} + \frac{\partial^2 \varepsilon_{xx}}{\partial z^2} = \frac{\partial^2 \gamma_{zx}}{\partial z \partial x}$$

Second group: This group establishes the conditions among the shear strains. We have

$$\gamma_{xy} = \frac{\partial u_x}{\partial y} + \frac{\partial u_y}{\partial x}$$

$$\gamma_{yz} = \frac{\partial u_y}{\partial z} + \frac{\partial u_z}{\partial y}$$

$$\gamma_{xz} = \frac{\partial u_z}{\partial x} + \frac{\partial u_x}{\partial z}$$

Differentiating

$$\frac{\partial \gamma_{xy}}{\partial z} = \frac{\partial^2 u_x}{\partial z \partial y} + \frac{\partial^2 u_y}{\partial z \partial x}$$

$$\frac{\partial \gamma_{yz}}{\partial x} = \frac{\partial^2 u_y}{\partial x \partial z} + \frac{\partial^2 u_z}{\partial x \partial y}$$

$$\frac{\partial \gamma_{zx}}{\partial y} = \frac{\partial^2 u_z}{\partial x \partial y} + \frac{\partial^2 u_x}{\partial y \partial z}$$

Adding the last two equations and subtracting the first

$$\frac{\partial \gamma_{yz}}{\partial x} + \frac{\partial \gamma_{zx}}{\partial y} - \frac{\partial \gamma_{xy}}{\partial z} = 2\frac{\partial^2 u_z}{\partial x \partial y}$$

Differentiating the above equation once more with respect to z and observing that

$$\frac{\partial^3 u_z}{\partial x \partial y \partial z} = \frac{\partial^2 \varepsilon_{zz}}{\partial x \partial y}$$

we get,

$$\frac{\partial}{\partial z}\left(\frac{\partial \gamma_{yz}}{\partial x} + \frac{\partial \gamma_{zx}}{\partial y} - \frac{\partial \gamma_{xy}}{\partial z}\right) = 2\frac{\partial^3 u_z}{\partial x \partial y \partial z} = 2\frac{\partial^2 \varepsilon_{zz}}{\partial x \partial y}$$

This is one of the required relations of the second group. By a cyclic change of the letters we get the other two equations. Collecting all equations, the six strain compatibility relations are

$$\frac{\partial^2 \varepsilon_{xx}}{\partial y^2} + \frac{\partial^2 \varepsilon_{yy}}{\partial x^2} = \frac{\partial^2 \gamma_{xy}}{\partial x \partial y} \tag{56a}$$

$$\frac{\partial^2 \varepsilon_{yy}}{\partial z^2} + \frac{\partial^2 \varepsilon_{zz}}{\partial y^2} = \frac{\partial^2 \gamma_{yz}}{\partial y \partial z} \tag{56b}$$

$$\frac{\partial^2 \varepsilon_{zz}}{\partial x^2} + \frac{\partial^2 \varepsilon_{xx}}{\partial z^2} = \frac{\partial^2 \gamma_{zx}}{\partial z \partial x} \tag{56c}$$

$$\frac{\partial}{\partial z}\left(\frac{\partial \gamma_{yz}}{\partial x} + \frac{\partial \gamma_{zx}}{\partial y} - \frac{\partial \gamma_{xy}}{\partial z}\right) = 2\frac{\partial^2 \varepsilon_{zz}}{\partial x \partial y} \tag{56d}$$

$$\frac{\partial}{\partial x}\left(\frac{\partial \gamma_{zx}}{\partial y} + \frac{\partial \gamma_{xy}}{\partial z} - \frac{\partial \gamma_{yz}}{\partial x}\right) = 2\frac{\partial^2 \varepsilon_{xx}}{\partial y\, \partial z} \tag{56e}$$

$$\frac{\partial}{\partial y}\left(\frac{\partial \gamma_{xy}}{\partial z} + \frac{\partial \gamma_{yz}}{\partial x} - \frac{\partial \gamma_{zx}}{\partial y}\right) = 2\frac{\partial^2 \varepsilon_{yy}}{\partial x\, \partial y} \tag{56f}$$

The above six equations are called Saint–Venant's equations of compatibility. We can give a geometrical interpretation to the above equations. For this purpose, imagine an elastic body cut into small parallelepipeds and give each of them the deformation defined by the six strain components. It is easy to conceive that if the components of strain are not connected by certain relations, it is impossible to make a continuous deformed solid from individual deformed parallelepipeds. Saint–Venant's compatibility relations furnish these conditions. Hence, these equations are also known as continuity equations.

Example 10:

For a circular rod subjected to a torque, the displacement components at any point (x, y, z) are obtained as

$$u_x = -\tau yz + ay + bz + c$$

$$u_y = -\tau xz - ax + ez + f$$
$$u_z = -bx - ey + k$$

where a, b, c, e, f and k are constants, and τ is the shear stress.

Solution:

(i) Select the constants a, b, c, e, f, k such that the end section $z = 0$ is fixed in the following manner:

(a) Point o has no displacement.

(b) The element Δz of the axis does not rotate in the plane xoy nor in the plane yoz

(c) The element Δy of the axis does not rotate in the plane xoy.

(ii) Determine the strain components.

(iii) Verify whether these strain components satisfy the compatibility conditions.

(i) Since point 'o' does not have any displacement

$$u_x = c = 0, \quad u_y = f = 0, \quad u_z = k = 0$$

The displacements of a point Δz from 'o' are

$$\frac{\partial u_x}{\partial z}\Delta z, \quad \frac{\partial u_y}{\partial z}\Delta z \quad \text{and} \quad \frac{\partial u_z}{\partial z}\Delta z$$

Similarly, the displacements of a point Δy from 'o' are

$$\frac{\partial u_x}{\partial y}\Delta y, \quad \frac{\partial u_y}{\partial y}\Delta y \quad \text{and} \quad \frac{\partial u_z}{\partial y}\Delta y$$

Hence, according to condition (b)

$$\frac{\partial u_y}{\partial z}\Delta z = 0 \quad \text{and} \quad \frac{\partial u_x}{\partial z}\Delta z = 0$$

and according to condition (c)

$$\frac{\partial u_x}{\partial y}\Delta y = 0$$

Applying these requirements

$$\frac{\partial u_y}{\partial z} \text{ at '}o\text{' is } e \text{ and hence, } e = 0$$

$$\frac{\partial u_x}{\partial z} \text{ at '}o\text{' is } b \text{ and hence, } b = 0$$

$$\frac{\partial u_x}{\partial y} \text{ at 'o' is } a \text{ and hence, } a = 0$$

Consequently, the displacement components are

$$u_x = -\, tyz, \qquad u_y = txz \qquad \text{and} \qquad u_z = 0$$

(ii) The strain components are

$$\varepsilon_{xx} = \frac{\partial u_x}{\partial x} = 0, \quad \varepsilon_{yy} = \frac{\partial u_y}{\partial y} = 0, \quad \varepsilon_{zz} = 0;$$

$$\gamma_{xy} = \frac{\partial u_x}{\partial y} + \frac{\partial u_y}{\partial x} = -\tau z + \tau z = 0$$

$$\gamma_{yz} = \frac{\partial u_y}{\partial z} + \frac{\partial u_z}{\partial x} = \tau x$$

$$\gamma_{zx} = \frac{\partial u_z}{\partial x} + \frac{\partial u_x}{\partial z} = -\tau y$$

(iii) Since the strain components are linear in x, y and z, the Saint–Venant's compatibility requirements are automatically satisfied.

STRAIN DEVIATOR AND ITS INVARIANTS

Similar to the analysis of stress, we can resolve the e_{ij} matrix into a spherical (i.e. isotropic) and a deviatoric part. The e_{ij} matrix is

$$\left[e_{ij}\right] = \begin{bmatrix} \varepsilon_{xx} & e_{xy} & e_{xz} \\ e_{xy} & \varepsilon_{yy} & e_{yz} \\ e_{xz} & e_{yz} & \varepsilon_{zz} \end{bmatrix}$$

This can be resolved into two parts as

$$\left[e_{ij}\right] = \begin{bmatrix} \varepsilon_{xx} - e & e_{xy} & e_{xz} \\ e_{xy} & \varepsilon_{yy} - e & e_{yz} \\ e_{xz} & e_{yz} & \varepsilon_{zz} - e \end{bmatrix} + \begin{bmatrix} e & 0 & 0 \\ 0 & e & 0 \\ 0 & 0 & e \end{bmatrix} \tag{57}$$

where $$e = \frac{1}{3}(\varepsilon_{xx} + \varepsilon_{yy} + \varepsilon_{zz}) \tag{58}$$

represents the mean elongation at a given point. The second matrix on the right–hand side of Eq. (57) is the spherical part of the strain matrix. The first matrix represents the deviatoric part or the strain deviator. If an isolated element of the body is subjected to the strain deviator only, then according to Eq. (34), the volumetric strain is equal to

$$\frac{\Delta V}{V} = (\varepsilon_{xx} - e) + (\varepsilon_{yy} - e) + (\varepsilon_{zz} - e)$$

$$= \varepsilon_{xx} + \varepsilon_{yy} + \varepsilon_{zz} - 3e \tag{59}$$

$$= 0$$

This means that an element subjected to deviatoric strain undergoes pure deformation without a change in volume, Hence, this part is also known as the pure shear part of the strain matrix. The spherical part of the strain matrix, i.e. the second matrix on the right–hand side of Eq. (57) is an isotropic state of strain. It is called isotropic because when a body is subjected to this particular state of strain, then every direction is a principal strain direction, with a strain of magnitude e, according to Eq, (20). A sphere subjected to this state of strain will uniformly expand or contract and remain spherical.

Consider the invariants of the strain deviator. These are constructed in the same way as the invariants of the stress and strain matrices with an appropriate replacement of notations.

(i) Linear invariant is zero since

$$J'_1 = (\varepsilon_{xx} - e) + (\varepsilon_{yy} - e) + (\varepsilon_{zz} - e) = 0 \tag{60}$$

(ii) Quadratic invariant is

$$J'_2 = \begin{bmatrix} \varepsilon_{xx} - e & e_{xy} \\ e_{xy} & \varepsilon_{yy} - e \end{bmatrix} + \begin{bmatrix} \varepsilon_{yy} - e & e_{yz} \\ e_{yz} & \varepsilon_{zz} - e \end{bmatrix} + \begin{bmatrix} \varepsilon_{xx} - e & e_{xz} \\ e_{xz} & \varepsilon_{zz} - e \end{bmatrix}$$

$$= -\frac{1}{6}\Big[(\varepsilon_{xx} - \varepsilon_{yy})^2 + (\varepsilon_{yy} - \varepsilon_{zz})^2 + (\varepsilon_{zz} - \varepsilon_{xx})^2$$

$$+6(e_{xy} + e_{yx} + e_{zx})^2\Big]$$

(iii) Cubic invariant is

$$J'_3 = \begin{vmatrix} \varepsilon_{xx} - e & e_{xy} & e_{xz} \\ e_{xy} & \varepsilon_{yy} - e & e_{yz} \\ e_{xz} & e_{zy} & \varepsilon_{zz} - e \end{vmatrix}$$

The second and third invariants of the deviatoric strain matrix describe the two types of distortions that an isolated element undergoes when subjected to the given strain matrix e_{ij}

EXERCISES

1. The displacement field for a body is given by

$$u = (x^2 + y)i + (3 + z)j + (x^2 + 2y)k$$

Write down the displacement gradient matrix at point (2, 3, 1).

$$\left[Ans. \begin{bmatrix} 4 & 1 & 0 \\ 0 & 0 & 1 \\ 4 & 2 & 0 \end{bmatrix} \right]$$

2. The displacement field for a body is given by

$$u = [(x^2 + y^2 + 2)i + (3x + 4y^2)j + (2x^3 + 4z)k]10^{-4}$$

What is the displaced position of a point originally at (1, 2, 3)?

[*Ans.* (1.0007, 2.0019, 3.0014)]

3. For the displacement field given in problem 2.2, what are the strain components at (1, 2, 3). Use only linear terms.

$$\left[\begin{array}{l} Ans.\ \varepsilon_{xx} = 0.0002, \varepsilon_{yy} = 0.0016, \varepsilon_{zz} = 0.0004 \\ \gamma_{xy} = 0.0007, \gamma_{yz} = 0, \gamma_{zx} = 0.0006 \end{array} \right]$$

4. What are the strain a components for problem 2.3, if non–linear terms are also included?

$$\left[\begin{array}{l} Ans.\ E_{xx} = 2p + 24.5p^2,\ E_{yy} = 16p + 136p^2,\ E_{zz} = 4p + 8p^2 \\ E_{xy} = 7p + 56p^2,\ E_{yz} = 0,\ E_{zx} = 6p + 24p^2 \text{ where } p = 10^{-4} \end{array} \right]$$

5. If the displacement field is given by

$$u_x = kxy, \qquad u_y = kxy, \qquad u_z = 2k(x + y)z$$

where k is a constant small enough to ensure applicability of the small deformation theory,

(a) write down the strain matrix

(b) what is the strain in the direction $n^x = n_y = n_z = 1/\sqrt{3}$?

$$\left[\begin{array}{l} Ans.\ (a) \left[\varepsilon_{ij} \right] = k \begin{bmatrix} y & x+y & 2z \\ x+y & x & 2z \\ 2z & 2z & 2(x+y) \end{bmatrix} \\ (b)\ \varepsilon_{PQ} = \dfrac{4k}{3}(x+y+z) \end{array} \right]$$

6. The displacement field is given by

$$u_x = k(x^2 + 2z), \quad u_y = k(4x + 2y^2 + z), \quad u_z = 4kz^2$$

k is a very small constant. What are the strains at (2, 2, 3) in directions

(a) $n_x = 0, n_y = 1/\sqrt{2}, n_z = 1/\sqrt{2}$

(b) $n_x = 1, n_y = n_z = 0$

(c) $n_x = 0.6, n_y = 0, n_z = 08$

$$\left[Ans.\ (a)\ \frac{33}{2}k, (b)\ 4k, (c)\ 17.76k\right]$$

7. For the displacement field given in problem 2.6, with $k = 0.001$, determine the change in angle between two line segments PQ and PR at $P(2, 2, 3)$ having direction cosines before deformation a

(a) PQ: $n_{x1} = 0,\ n_{y1} = n_{z1} = \frac{1}{\sqrt{2}}$

PR: $n_{x2} = 1,\ n_{y2} = n_{z2} = 0$

(b) PQ: $n_{x1} = 0,\ n_{y1} = n_{z1} = \frac{1}{\sqrt{2}}$

PR: $n_{x2} = 0.6,\ n_{y2} = 0,\ n_{z2} = 0.8$

$$\begin{bmatrix} Ans.\ (a)\ 90° - 89.8° = 0.2° \\ (b)\ 55.5° - 50.7° = 4.8° \end{bmatrix}$$

8. The rectangular components of a small strain at a point is given by the following matrix. Determine the principal strains and the direction of the maximum unit strain (i.e. ε_{max}).

$$\left[\varepsilon_{ij}\right] = P\begin{bmatrix} 1 & 0 & 0 \\ 0 & 0 & -4 \\ 0 & -4 & 3 \end{bmatrix} \text{ where } p = 10^{-4}$$

$$\begin{bmatrix} Ans.\ \varepsilon_1 = 4p, \varepsilon_2 = p, \varepsilon_3 = -p \\ \text{for } \varepsilon_1 : n_x = 0, n_y = 0.447, n_z = 0.894 \\ \text{for } \varepsilon_2 : n_x = 1, n_y = n_z = 0 \\ \text{for } \varepsilon_3 : n_x = 0, n_y = 0.894, n_z = 0.447 \end{bmatrix}$$

9. For the following plane strain distribution, verify whether the compatibility condition is satisfied:

$$\varepsilon_{xx} = 3x^2y, \quad \varepsilon_{yy} = 4y^2x + 10^{-2}, \quad \gamma_{xy} = 2xy + 2x^3$$

[*Ans.* Not satisfied]

10. Verify whether the following strain field satisfies the equations of compatibility. P is a constant:

$$\varepsilon_{xx} = py, \quad \varepsilon_{yy} = px, \quad \varepsilon_{zz} = 2p(x + y)$$

$$\gamma_{xy} = p(x + y), \quad \varepsilon_{yz} = 2pz, \quad \varepsilon_{zx} = 2pz$$

[*Ans.* Yes]

11. State the conditions under which the following is a possible system of strains:

$$\varepsilon_{xx} = a + b(x^2 + y^2)\, x^4 + y^4, \qquad \gamma_{yz} = 0$$
$$\varepsilon_{yy} = \alpha + \beta(x^2 + y^2) + x^4 + y^4, \qquad \gamma_{zx} = 0$$
$$\gamma_{xy} = A + Bxy\,(x^2 + y^2 - c^2), \qquad \varepsilon_{zz} = 0$$

[*Ans.* $B = 4;\ b + b + 2c^2 = 0$]

12. Given the following system of strains

$$\varepsilon_{xx} = 5 + x^2 + y^2 + x^4 + y^4$$
$$\varepsilon_{yy} = 6 + 3x^2 + 3y^2 + x^4 + y^4$$
$$\gamma_{xy} = 10 + 4xy\,(x^2 + y^2 + 2)$$
$$\varepsilon_{zz} = \gamma_{yz} = \gamma_{zx} = 0$$

determine whether the above strain field is possible. If it is possible, determine the displacement components in terms of x and y, assuming that $u_x = u_y = 0$ and $\omega_{xy} = 0$ at the origin.

$$\left[\begin{array}{l} \textit{Ans.}\ \text{It is possible.}\ u_x = 5x + \frac{1}{3}x^3 + xy^2 + \frac{1}{5}x^5 + xy^4 + cy \\ \qquad\qquad u_y = 6y + 3x^2y + y^3 + x^4y + \frac{1}{5}x^5 + cx \end{array}\right]$$

13. For the state of strain given in problem 12, write down the spherical part and the deviatoric part and determine the volumetric strain.

$$\left[\begin{array}{l} \textit{Ans.}\ \text{Components of spherical part are} \\ e = \frac{1}{3}[11 + 4(x^2 + y^2) + 2(x^4 + y^4)] \\ \text{Volumetric strain } = 11 + 4(x^2 + y^2) + 2(x^4 + y^4) \end{array}\right]$$

2

Stress–Strain Relations for Linearly Elastic Solids

INTRODUCTION

In the preceding two chapters we dealt with the state of stress at a point and the state of strain at a point. The strain components were related to the displacement components through six of Cauchy's strain–displacement relationships. In this chapter, the relationships between the stress and strain components will be established. Such equations are termed constitutive equations. They depend on the manner in which the material resists deformation.

The constitutive equations are mathematical descriptions of the physical phenomena based on experimental observations and established principles. Consequently, they are approximations of the true behavioural pattern, since an accurate mathematical representation of the physical phenomena would be too complicated and unworkable.

The constitutive equations describe the behaviour of a material, not the behaviour of a body. Therefore, the equations relate the state of stress at a point to the state of strain at the point.

GENERALISED STATEMENT OF HOOKE'S LAW

Consider a uniform cylindrical rod of diameter d subjected to a tensile force P. As is well known from experimental observations, when P is gradually increased from zero to some positive value, the length of the rod also increases.

Based on experimental observations, it is postulated in elementary strength of materials that the axial stress σ is proportional to the axial strain ε up to a limit called the proportionality limit. The constant of proportionality is the Young's Modulus E, i.e.

$$\varepsilon = \frac{\sigma}{E} \quad \text{or} \quad \sigma = E\varepsilon \tag{1}$$

It is also well known that when the uniform rod elongates, its lateral dimensions, i.e. its diameter, decreases. In elementary strength of materials, the ratio of lateral strain to longitudinal strain was termed as Poisson's ratio V. We now extend this information or knowledge to relate the six rectangular components of stress to the six rectangular components of strain. We assume that each of the six independent components of stress may be expressed as a linear function of the six components of strain and vice versa.

The mathematical expressions of this statement are the six stress–strain equations:

$$\begin{aligned}
\sigma_x &= a_{11}\varepsilon_{xx} + a_{12}\varepsilon_{yy} + a_{13}\varepsilon_{zz} + a_{14}\gamma_{xy} + a_{15}\gamma_{yz} + a_{16}\gamma_{zx} \\
\sigma_y &= a_{21}\varepsilon_{xx} + a_{22}\varepsilon_{yy} + a_{23}\varepsilon_{zz} + a_{24}\gamma_{xy} + a_{25}\gamma_{yz} + a_{26}\gamma_{zx} \\
\sigma_z &= a_{31}\varepsilon_{xx} + a_{32}\varepsilon_{yy} + a_{33}\varepsilon_{zz} + a_{34}\gamma_{xy} + a_{35}\gamma_{yz} + a_{36}\gamma_{zx} \\
\tau_{xy} &= a_{41}\varepsilon_{xx} + a_{42}\varepsilon_{yy} + a_{43}\varepsilon_{zz} + a_{44}\gamma_{xy} + a_{45}\gamma_{yz} + a_{46}\gamma_{zx} \\
\tau_{yz} &= a_{51}\varepsilon_{xx} + a_{52}\varepsilon_{yy} + a_{53}\varepsilon_{zz} + a_{54}\gamma_{xy} + a_{55}\gamma_{xy} + a_{56}\gamma_{zx} \\
\tau_{zx} &= a_{61}\varepsilon_{xx} + a_{62}\varepsilon_{yy} + a_{63}\varepsilon_{zz} + a_{64}\gamma_{xy} + a_{65}\gamma_{yz} + a_{66}\gamma_{zx}
\end{aligned} \tag{2}$$

Or conversely, six strain–stress equations of the type

$$\begin{aligned}
\varepsilon_{xx} &= b_{11}\sigma_x + b_{12}\sigma_y + b_{13}\sigma_z + b_{14}\tau_{xy} + b_{15}\tau_{yz} + b_{16}\tau_{zx} \\
\varepsilon_{yy} &= \ldots \text{ etc}
\end{aligned} \tag{3}$$

where a_{11}, a_{12}, b_{11}, b_{12}, . . . , are constants for a given material. Solving Eq. (2) as six simultaneous equations, one can get Eq. (3), and vice versa. For homogeneous, linearly elastic material, the six Eqs (2) or (3) are known as Generalised Hooke's Law. Whether we use the set given by Eq. (2) or that given by Eq. (3), 36 elastic constants are apparently involved.

STRESS–STRAIN RELATIONS FOR ISOTROPIC MATERIALS

We now make a further assumption that the ideal material we are dealing with has the same properties in all directions so far as the stress–stain relations are concerned. This means that the material we are dealing with is isotropic, i.e. it has no directional property.

Care must be taken to distinguish between the assumption of isotropy, which is a particular statement regarding the stress–strain properties at a given point, and that of homogeneity, which is a statement that the stress–strain properties, whatever they may be, are the same at all points. For example, timber of regular grain is homogeneous but not isotropic.

Assuming that the material is isotropic, one can show that only two independent elastic constants are involved in the generalised statement of Hooke's law. It was shown that at any point there are three faces (or planes) on which the resultant stresses are wholly normal, i.e. there are no shear stresses on these planes.

These planes were termed the principal planes and the stresses on these planes the principal stresses. It was shown that at any point one can identify before strain, a small rectangular parallelepiped or a box which remains rectangular after strain. The normals to the faces of this box were called the principal axes of strain. Since in an isotropic material, a small rectangular box the faces of which are subjected to pure normal stresses, will remain rectangular after deformation (no asymmetrical deformation), the normal to these faces coincide with the principal strain axes.

Hence, for an isotropic material, one can relate the principal stresses σ_1, σ_2, σ_3 with the three principal strains ε_1, ε_2 and ε_3 through suitable elastic constants. Let the axes x, y and z coincide with the principal stress and principal strain directions. For the principal stress σ_1 the equation becomes

$$\sigma_1 = a\varepsilon_1 + b\varepsilon_2 + c\varepsilon.$$

where a, b and c are constants. But we observe that b and c should be equal since the effect of σ_1 in the directions of ε_2 and ε_3, which are both at right angles to σ_1, must be the same for an isotropic material. In other words, the effect of σ_1 in any direction transverse to it is the same in an isotropic material. Hence, for σ_1 the equation becomes

$$\begin{aligned}\sigma_1 &= a\varepsilon_1 + b\,(\varepsilon_2 + \varepsilon_3) \\ &= (a - b)\varepsilon_1 + b(\varepsilon_1 + \varepsilon_2 + \varepsilon_3)\end{aligned}$$

by adding and subtracting $b\varepsilon_1$. But $(\varepsilon_1 + \varepsilon_2 + \varepsilon_3)$ is the first invariant of strain J_1 or the cubical dilatation Δ. Denoting b by λ and $(a - b)$ by 2μ, the equation for σ_1 becomes

$$\sigma_1 = \lambda\Delta + 2\mu\varepsilon_2 \tag{4a}$$

Similarly, for σ_2 and σ_3 we get

$$\sigma_2 = \lambda\Delta + 2\mu\varepsilon_2 \tag{4b}$$

$$\sigma_3 = \lambda\Delta + 2\mu\varepsilon_3 \tag{4c}$$

The constants λ and μ are called Lame's coefficients. Thus, there are only two elastic constants involved in the relations between the principal stresses and principal strains for an isotropic material. As the next sections show, this can be extended to the relations between rectangular stress and strain components also.

MODULUS OF RIGIDITY

Let the co-ordinate axes *Ox, Oy, Oz* coincide with the principal stress axes. For an isotropic body, the principal strain axes will also be along *Ox, Oy, Oz*. Consider another frame of reference *Ox', Oy', Oz'*, such that the direction cosines of *Ox'* are n_{x1}, n_{y1}, n_{z1} and of *Oy'* are n_{x2}, n_{y2}, n_{z2}. Since *Ox'* and *Oy'* are at right angles to each other.

$$n_{x1}n_{x2} + n_{y1}, n_{y2} + n_{z1}n_{z2} = 0 \tag{5}$$

The normal stress $\sigma_{x'}$ and the shear stress $\tau_{x'y'}$ are obtained from Cauchu's formula, Eqs. (9). The resultant stress vector on the *x'* plane will have components as

$$\overset{x'}{T_x} = n_{x1}\sigma_1, \quad \overset{x'}{T_y} = n_{y1}\sigma_2, \quad \overset{x'}{T_z} = n_{z1}\sigma_3$$

These are the components in *x, y* and *z* directions. The normal stress on this *x'* plane is obtained as the sum of the projections of the components along the normal, i.e.

$$\sigma_n = \sigma_x = n_{x1}^2\sigma_1 + n_{y1}^2\sigma_2 + n_{z1}^2\sigma_3 \tag{6a}$$

Similarly, the shear stress component on this *x'* plane in *y'* direction is obtained as the sum of the projections of the components in *y'* direction, which has direction cosines n_{x2}, n_{y2}, n_{z2}. Thus

$$\tau_{x'y'} = n_{x1}n_{x2}\sigma_1 + n_{y1}n_{y2}\sigma_2 + n_{z1}n_{z2}\sigma_3 \tag{6b}$$

On the same lines, if ε_1, ε_2 and ε_3 are the principal strains, which are also along *x, y, z* directions, the normal strain in *x'* direction, from Eq. (20), is

$$\varepsilon_{x'x'} = n_{x1}^2\varepsilon_1 + n_{y1}^2\varepsilon_2 + n_{z1}^2\varepsilon_3 \tag{7a}$$

The shear strain $\gamma_{x'y'}$ is obtained from Eq. (36c) as

$$\gamma_{x'y'} = \frac{1}{(1+\varepsilon_{x'})(1+\varepsilon_{y'})}\Big[2(n_{x1}n_{x2}\,\varepsilon_1 + n_{y1}n_{y2}\varepsilon_2 + n_{z1}n_{z2}\varepsilon_3)$$

$$+n_{x1}n_{x2} + n_{y1}n_{y2}\varepsilon_2 + n_{z1}n_{z2}\Big]$$

Using Eq. (5), and observing that $\varepsilon_{x'}$ and $\varepsilon_{y'}$ are small compared to unity in the denominator,

$$\gamma_{x'y'} = 2\ (n_{x1}n_{x2}\varepsilon_1 + n_{y1}n_{y2}\varepsilon_2 + n_{z1}n_{z2}\varepsilon_3 \tag{7b}$$

substituting the values of σ_1, σ_2 and σ_3 from Eqs (4a) – (4c) into Eq. (6b)

$$\tau_{x'y'} = n_{x1}n_{x2}\,(\lambda\Delta + 2\ \mu\ \varepsilon_1) + n_{y1}n_{y2}\ (\lambda\Delta + 2\ \mu\varepsilon_2) + n_{z1}n_{z2}\ (\lambda\Delta + 2\ \mu\ \varepsilon_3)$$
$$= \lambda\Delta(n_{x1}n_{x2} + n_{y1}n_{y2} + n_{z1}n_{z2}) + 2\mu(\ n_{x1}n_{x2}\varepsilon_2 + n_{y1}n_{y2}\varepsilon_2 + n_{z1}n_{z2}\varepsilon_3)$$

Hence, from Eqs (5) and (7b)

$$\tau_{x'y'} = \mu\gamma_{x'y'} \tag{8}$$

Equation (8) relates the rectangular shear stress component $\tau_{x'y'}$ with the rectangular shear strain component $\gamma_{x'y'}$. Comparing this with the relation used in elementary strength of materials, one observes that μ is the modulus of rigidity, usually denoted by G.

By taking another axis Oz' with direction cosines n_{x3}, n_{y3} and n_{z3} and at right angles to Ox' and Oy' (so that $O'x'y'z'$ forms an orthogonal set of axes), one can get equations similar to (6a) and (6b) for the other rectangular stress components. Thus,

$$\sigma_{y'} = n_{x2}^2\sigma_1 + n_{y2}^2\sigma_2 + n_{z2}^2\sigma_3 \tag{9a}$$

$$\sigma_{z'} = n_{x3}^2\sigma_1 + n_{y3}^2\sigma_2 + n_{z3}^2\sigma_3 \tag{9b}$$

$$\tau_{y'z'} = n_{x2}n_{x3}\sigma_1 + n_{y2}n_{y3}\sigma_2 + n_{z2}n_{z3}\sigma_3 \tag{9c}$$

$$\tau_{z'x'} = n_{x3}n_{x1}\sigma_1 + n_{y3}n_{y1}\sigma_2 + n_{z3}n_{z3}\sigma_3 \tag{9d}$$

Similarly, following Eqs (7a) and (7b) for the other rectangular strain components, one gets

$$\varepsilon_{y'y'} = n_{x2}^2\varepsilon_1 + n_{y2}^2\varepsilon_2 + n_{z2}^2\varepsilon_3 \tag{10a}$$

$$\varepsilon_{z'z'} = n_{x3}^2\varepsilon_1 + n_{y3}^2\varepsilon_2 + n_{z3}^2\varepsilon_3 \tag{10b}$$

$$\gamma_{y'z'} = 2(n_{x2}n_{x3}\varepsilon_1 + n_{y2}n_{y3}\varepsilon_2 + n_{z2}n_{z3}\varepsilon_3) \tag{10c}$$

$$\gamma_{z'x'} = 2(n_{x3}n_{x1}\varepsilon_1 + n_{y3}n_{y1}\varepsilon_2 + n_{z3}n_{z1}\varepsilon_3) \tag{10d}$$

Form Eqs (6a), (4c) and (7a)

$$\begin{aligned}\sigma_{x'} &= n_{x1}^2\sigma_1 + n_{x2}^2\sigma_2 + n_{x3}^2\sigma_3 \\ &= \lambda\Delta\left(n_{x1}^2 + n_{x2}^2 + n_{x3}^2\right) + 2\mu\left(\varepsilon_1 n_{x1}^2 + \varepsilon_2 n_{x2}^2 + \varepsilon_3 n_{x3}^2\right) \\ &= \lambda\Delta + 2\mu\varepsilon_{x'x'}\end{aligned} \tag{11a}$$

Similarly, one gets

$$\sigma_{y'} = \lambda\Delta + 2\mu\varepsilon_{y'y'} \tag{11b}$$

$$\sigma_{z'} = \lambda\Delta + 2\mu\varepsilon_{z'z'} \tag{11c}$$

Similar to Eq. (8),

$$\tau_{y'z'} = \mu\gamma_{y'z'} \tag{12a}$$

$$\tau_{x'z'} = \mu\gamma_{z'x'} \tag{12b}$$

Equations (11a) – (11c), (8) and (3.12a) and (12b) relate the six rectangular stress components to six rectangular strain components and in these only two elastic constants are involved. Therefore, the Hooke's law for an isotropic material will involve two independent elastic constants λ and μ (or G).

BULK MODULUS

Adding Eqs (11a) – (11c)

$$\sigma_{x'} + \sigma_{y'} + \sigma_{z'} = 3\lambda\Delta + 2\mu\,(\varepsilon_{x'x'} + \varepsilon_{y'y'} + \varepsilon_{z'z'}) \tag{13a}$$

Observing that

$$\sigma_{x'} + \sigma_{y'} + \sigma_{z'} = I_1 = \sigma_1 + \sigma_2 + \sigma_3 \quad \text{(first invariant of stress)}$$

and

$$\varepsilon_{x'x'} + \varepsilon_{y'y'} + \varepsilon_{z'z'} = J_1 = \varepsilon_1 + \varepsilon_2 + \varepsilon_3 \quad \text{(first invariant of strain)}$$

Equation (13a) can be written in several alternative forms as

$$\sigma_1 + \sigma_2 + \sigma_3 = (3\lambda\Delta + 2\mu)\Delta \tag{13b}$$

$$\sigma_{x'} + \sigma_{y'} + \sigma_{z'} = (3\lambda\Delta + 2\mu)\Delta \tag{13c}$$

$$I_1 = (3\lambda\Delta + 2\mu)J_1 \tag{13d}$$

Noting from Eq. (34) that Δ is the volumetric strain, the definition of bulk modulus K is

$$K = \frac{\text{Pressure}}{\text{volumetric strain}} = \frac{p}{\Delta} \tag{14a}$$

If $\sigma_1 = \sigma_2 = \sigma_3 = p$, then from Eq. (13b)

$$3p = (3\lambda\Delta + 2\mu)\Delta$$

or

$$3\frac{p}{\Delta} = (3\lambda + 2\mu)$$

and from Eq. (14a)

$$K\frac{1}{3} = (3\lambda + 2\mu) \tag{14b}$$

Thus, the bulk modulus for an isotropic solid is related to Lame's constants through Eq. (14b).

YOUNG'S MODULUS AND POISSON'S RATIO

From Eq. (13b), se have

$$\Delta = \frac{\sigma_1 + \sigma_2 + \sigma_3}{(3\lambda + 2\mu)}$$

Substituting this in Eq. (4a)

$$\sigma_1 = \frac{\lambda}{(3\lambda + 2\mu)}(\sigma_1 + \sigma_2 + \sigma_3) + 2\mu\varepsilon_1$$

or $$\varepsilon_1 = \frac{\lambda+\mu}{\mu(3\lambda+2\mu)}\left[\sigma_1 - \frac{\lambda}{2(\lambda+\mu)}(\sigma_2+\sigma_3)\right] \tag{15}$$

Form elementary strength of materials

$$\varepsilon_1 = \frac{1}{E}[(\sigma_1 - v(\sigma_2+\sigma_3))]$$

where E is Young's modulus, and v is Poisson's ratio. Comparing this with Eq. (15),

$$E = \frac{\mu(3\lambda+2\mu)}{(\lambda+\mu)}; \qquad v = \frac{\lambda}{2(\lambda+\mu)} \tag{16}$$

RELATIONS BETWEEN THE ELASTIC CONSTANTS

In elementary strength of materials, we are familiar with Young's modulus E, Poisson's ratio v, shear modulus or modulus of rigidity G and bulk modulus K. Among these, only two are independent and E and v are generally taken as the independent constants. The other two, namely, G and K, are expressed as

$$G = \frac{E}{2(1+v)}, \qquad K = \frac{E}{3(1-2v)} \tag{17}$$

It has been shown in this chapter, that for an isotropic material, the 36 elastic constants involved in the Generalised Hooke's law, can be reduced to two independent elastic constants. These two elastic constants are Lame's coefficients λ and μ. The second coefficient μ is the same as the rigidity modulus G. In terms of these, the other elastic constants can be expressed as

$$E = \frac{\mu(3\lambda+2\mu)}{(\lambda+\mu)}, \qquad v = \frac{\lambda}{2(\lambda+\mu)}$$

$$K = \frac{(3\lambda+2\mu)}{3} \qquad G = \mu, \qquad \lambda = \frac{vE}{(1+v)(1-2v)} \tag{18}$$

It should be observed from Eq. (17) that for the bulk modulus to be positive, the value of Poisson's ratio v cannot exceed 1/2. This is the upper limit for v. For $v = 1/2$.

$$3G = E \qquad \text{and} \qquad K = \infty$$

A material having Poisson's ratio equal to 1/2 is known as an incompressible material, since the volumetric strain for such an isotropic material is zero.

For easy reference one can collect the equations relating stresses and strains that have been obtained so far.

(i) In terms of principal stresses and principal strains:

$$\begin{aligned} \sigma_1 &= \lambda\Delta + 2\mu\varepsilon_1 \\ \sigma_2 &= \lambda\Delta + 2\mu\varepsilon_2 \\ \sigma_3 &= \lambda\Delta + 3\mu\varepsilon_3 \end{aligned} \tag{19}$$

where $\Delta = \varepsilon_1 + \varepsilon_2 + \varepsilon_3 = J_1$.

$$\begin{aligned} \varepsilon_1 &= \frac{\lambda+\mu}{\mu(3\lambda+2\mu)}\left[\sigma_1 - \frac{\lambda}{2(\lambda+\mu)}(\sigma_2+\sigma_3)\right] \\ \varepsilon_2 &= \frac{\lambda+\mu}{\mu(3\lambda+2\mu)}\left[\sigma_2 - \frac{\lambda}{2(\lambda+\mu)}(\sigma_3+\sigma_1)\right] \\ \varepsilon_3 &= \frac{\lambda+\mu}{\mu(3\lambda+2\mu)}\left[\sigma_3 - \frac{\lambda}{2(\lambda+\mu)}(\sigma_1+\sigma_2)\right] \end{aligned} \tag{20}$$

(ii) In terms of rectangular stress and strain components referred to an orthogonal coordinate system *Oxyz*:

$$\begin{aligned} \sigma_x &= \lambda\Delta + 2\mu\varepsilon_{xx} \\ \sigma_y &= \lambda\Delta + 2\mu\varepsilon_{yy} \\ \sigma_z &= \lambda\Delta + 3\mu\varepsilon_{zz} \quad (3.21a) \end{aligned}$$

where $\Delta = \varepsilon_{xx} + \varepsilon_{yy} + \varepsilon_{zz} = J_1$.

$$\tau_{xy} = \mu\gamma_{xy}, \qquad \tau_{yz} = \mu\gamma_{yz}, \qquad \tau_{zx} = \mu\gamma_{zx} \tag{21b}$$

$$\begin{aligned} \varepsilon_{xx} &= \frac{\lambda+\mu}{\mu(3\lambda+2\mu)}\left[\sigma_x - \frac{\lambda}{2(\lambda+\mu)}(\sigma_y+\sigma_z)\right] \\ \varepsilon_{yy} &= \frac{\lambda+\mu}{\mu(3\lambda+2\mu)}\left[\sigma_y - \frac{\lambda}{2(\lambda+\mu)}(\sigma_z+\sigma_x)\right] \\ \varepsilon_{zz} &= \frac{\lambda+\mu}{\mu(3\lambda+2\mu)}\left[\sigma_z - \frac{\lambda}{2(\lambda+\mu)}(\sigma_x+\sigma_y)\right] \end{aligned} \tag{22a}$$

$$\gamma_{xy} = \frac{1}{\mu}\tau_{xy}, \qquad \gamma_{yz} = \frac{1}{\mu}\tau_{yz}, \qquad \gamma_{zx} = \frac{1}{\mu}\tau_{zx} \tag{22b}$$

In the proceeding sets of equations, *l* and *m* are Lame's constants. In terms of the more familiar elastic constants *E* and *v*, the stress–strain relations are

(iii) With $\varepsilon_{xx} + \varepsilon_{yy} + \varepsilon_{zz} = J_1 = \Delta$

$$\sigma_x = \frac{E}{(1+v)}\left[=\frac{v}{(1-2v)}\Delta + \varepsilon_{xx}\right]$$

$$= \lambda J_1 + 2G\varepsilon_{xx}$$

$$\sigma_y = \frac{E}{(1+v)}\left[= \frac{v}{(1-2v)}\Delta + \varepsilon_{yy}\right] \tag{23a}$$

$$= \lambda J_1 + 2G\varepsilon_{yy}$$

$$\sigma_z = \frac{E}{(1+v)}\left[= \frac{v}{(1-2v)}\Delta + \varepsilon_{zz}\right]$$

$$= \lambda J_1 + 2G\varepsilon_{zz}$$

$$\tau_{xy} = G\gamma_{xy}, \qquad \tau_{yz} = G\gamma_{yz}, \qquad \tau_{xx} = G\gamma_{zx} \tag{23b}$$

$$\varepsilon_{xx} = \frac{1}{E}\left[(\sigma_x - v(\sigma_y + \sigma_z)\right]$$

$$\varepsilon_{yy} = \frac{1}{E}\left[(\sigma_y - v(\sigma_z + \sigma_x)\right] \tag{24a}$$

$$\varepsilon_{zz} = \frac{1}{E}\left[(\sigma_z - v(\sigma_x + \sigma_y)\right]$$

$$\gamma_{xy} = \frac{1}{G}\tau_{xy}, \qquad \gamma_{yz} = \frac{1}{G}\tau_{yz}, \qquad \gamma_{zx} = \frac{1}{G}\tau_{zx} \tag{24b}$$

DISPLACEMENT EQUATIONS OF EQUILIBRIUM

If a solid body is in equilibrium, the six rectangular stress components have to satisfy the three equations of equilibrium. In this chapter, we have shown how to relate the stress components to the strain components using the stress–strain relations. Hence, stress equations of equilibrium can be converted to strain equations of equilibrium. The strain components were related to the displacement components. Therefore, the strain equations of equilibrium can be converted to displacement equations of equilibrium. In this section, this result will be derived.

The first equation from Eq. (65) is

$$\frac{\partial \sigma_x}{\partial x} + \frac{\partial \tau_{xy}}{\partial y} + \frac{\partial \tau_{zx}}{\partial z} = 0$$

For an isotropic material

$$\sigma_x = \lambda\Delta + 2\mu\varepsilon_{xx}; \qquad \tau_{xy} = \mu\gamma_{xy}; \qquad \tau_{xz} = \mu\gamma_{xz}$$

Hence, the above equation becomes

$$\lambda\frac{\partial \Delta}{\partial x} + \mu\left(2\frac{\partial \varepsilon_{xx}}{\partial x} + \frac{\partial \gamma_{xy}}{\partial y} + \frac{\partial \gamma_{xz}}{\partial z}\right) = 0$$

From Cauchy's strain–displacement relations

$$\varepsilon_{xx} = \frac{\partial u_x}{\partial x}, \quad \gamma_{xy} = \frac{\partial u_x}{\partial y} + \frac{\partial u_y}{\partial x}, \quad \gamma_{zx} = \frac{\partial u_x}{\partial z} + \frac{\partial u_z}{\partial x}$$

Substituting these

$$\lambda \frac{\partial \Delta}{\partial x} + \mu \left(2\frac{\partial^2 u_x}{\partial x^2} + \frac{\partial^2 u_x}{\partial y^2} + \frac{\partial^2 u_y}{\partial x \partial y} + \frac{\partial^2 u_x}{\partial z^2} + \frac{\partial^2 u_z}{\partial x \partial z} \right) = 0$$

or $$\lambda \frac{\partial \Delta}{\partial x} + \mu \left(\frac{\partial^2 u_x}{\partial x^2} + \frac{\partial^2 u_x}{\partial y^2} + \frac{\partial^2 u_x}{\partial z^2} \right) + \mu \left(\frac{\partial^2 u_x}{\partial x^2} + \frac{\partial^2 u_y}{\partial x \partial y} + \frac{\partial^2 u_z}{\partial x \partial z} \right) = 0$$

or $$\lambda \frac{\partial \Delta}{\partial x} + \mu \left(\frac{\partial^2 u_x}{\partial x^2} + \frac{\partial^2 u_x}{\partial y^2} + \frac{\partial^2 u_x}{\partial z^2} \right) + \mu \frac{\partial}{\partial x} \left(\frac{\partial u_x}{\partial x} + \frac{\partial u_y}{\partial y} + \frac{\partial u_z}{\partial z} \right) = 0$$

Observing that

$$\Delta = \varepsilon_{xx} + \varepsilon_{yy} + \varepsilon_{zz} = \frac{\partial u_x}{\partial x} + \frac{\partial u_y}{\partial y} + \frac{\partial u_z}{\partial z}$$

$$(\lambda + \mu) \frac{\partial}{\partial x} \left(\frac{\partial u_x}{\partial x} + \frac{\partial u_y}{\partial y} + \frac{\partial u_z}{\partial z} \right) + \mu \left(\frac{\partial^2 u_x}{\partial x^2} + \frac{\partial^2 u_x}{\partial y^2} + \frac{\partial^2 u_x}{\partial z^2} \right) = 0$$

This is one of the displacement equations of equilibrium. Using the notation

$$\nabla^2 = \frac{\partial^2}{\partial x^2} + \frac{\partial^2}{\partial y^2} + \frac{\partial^2}{\partial z^2}$$

the displacement equation of equilibrium becomes

$$(\lambda + \mu) \frac{\partial \Delta}{\partial x} + \mu \nabla^2 u_x = 0 \tag{25a}$$

Similarly, from the second and third equations of equilibrium, one gets

$$(\lambda + \mu) \frac{\partial \Delta}{\partial x} + \mu \nabla^2 u_y = 0 \tag{25b}$$

$$(\lambda + \mu) \frac{\partial \Delta}{\partial z} + \mu \nabla^2 u_z = 0$$

These are known as Lame's displacement equations of equilibrium. They involve a synthesis of the analysis of stress, analysis of strain and the relations between stresses and strains.

These equations represent the mechanical, geometrical and physical characteristics of an elastic solid. Consequently, Lame's equations play a very prominent role in the solutions of problems.

Example 1

A rubber cube is inserted in a cavity of the same form and size in a steel block and the top of the cube is pressed by a steel block with a pressure of P pascals. Considering the steel to be absolutely hard and assuming that there is no friction between steel and rubber, find (i) the pressure of rubber against the box walls, and (ii) the extremum shear stresses in rubber.

Solution:

(i) Let l be the dimension of the cube. Since the cube is constrained in x and y directions

$$\varepsilon_{xx} = 0 \text{ and } \varepsilon_{yy} = 0$$

and $\quad \sigma_z = -p$

Therefore

$$\varepsilon_{xx} = \frac{1}{E}[\sigma_x - v(\sigma_y + \sigma_z)] = 0$$

$$\varepsilon_{yy} = \frac{1}{E}[\sigma_y - v(\sigma_x + \sigma_z)] = 0$$

Solving

$$\sigma_x = \sigma_y = \frac{v}{1-v}\sigma_z = -\frac{v}{1-v}p$$

If poisson's ratio = 0.2, then

$$\sigma_x = \sigma_y = \sigma_z = -p$$

(ii) The extremum shear stresses are

$$\tau_2 = \frac{\sigma_1 - \sigma_3}{2}, \quad \tau_3 = \frac{\sigma_1 - \sigma_2}{2}, \quad \tau_1 = \frac{\sigma_2 - \sigma_3}{2}$$

If $v \le 0.5$, then σ_x and σ_y are numerically less than or equal to σ_z. Since σ_x, σ_y and σ_z are all compressive

$$\sigma_1 = \sigma_x = \frac{v}{1-v}p$$

$$\sigma_2 = \sigma_y = \frac{v}{1-v}p$$

$$\sigma_3 = \sigma_z = -p$$

$$\therefore \quad \tau_1 = p\left(1 - \frac{v}{1-v}\right) = \frac{1-2v}{1-v}p, \ \tau_2 = \frac{1-2v}{1-v}p, \ \tau_3 = 0$$

If $v = 0.5$, the shear stresses are zero.

Example 2:

A cubical element is subjected to the following state of stress.

$\sigma_x = 100$ *MPa*, $\sigma_y = -20$ *MPa*, $\sigma_z = -40$ *MPa*, $\tau_{xy} = \tau_{yz} = \tau_{zx} = 0$

Assuming the material to be homogeneous and isotropic, determine the principal shear strains and the octahedral shear strain, if $E = 2 \times 10^5$ *MPa and* $v = 0.25$.

Solution:

Since the shear stresses on x, y and z planes are zero, the given stresses are principal stresses. Arranging such that $\sigma_1 \geq \sigma_2 \geq \sigma_3$

$$\sigma_1 = 100 \text{ MPa}, \; \sigma_2 = -20 \text{ MPa}, \; \sigma_3 = -40 \text{ MPa}$$

The external shear stresses are

$$\tau_1 = \frac{1}{2}(\sigma_2 - \sigma_3) = \frac{1}{2}(-20 + 40) = 10 \text{ Mpa}$$

$$\tau_2 = \frac{1}{2}(\sigma_3 - \sigma_1) = \frac{1}{2}(-40 - 100) = -70 \text{ Mpa}$$

$$\tau_3 = \frac{1}{2}(\sigma_1 - \sigma_2) = \frac{1}{2}(100 + 20) = 60 \text{ Mpa}$$

The modulus of rigidity G is

$$G = \frac{E}{2(1+v)} = \frac{2 \times 10^5}{2 \times 1.25} = 8 \times 10^4 \text{ MPa}$$

The principal shear strains are therefore

$$\gamma_1 = \frac{\tau_1}{G} = \frac{10}{8 \times 10^4} = 1.25 \times 10^{-4}$$

$$\gamma_2 = \frac{\tau_2}{G} = -\frac{70}{8 \times 10^4} = 8.75 \times 10^{-4}$$

$$\gamma_3 = \frac{\tau_3}{G} = \frac{60}{8 \times 10^4} = 7.5 \times 10^{-4}$$

From Eq. (44a), the octahedral shear stress is

$$\tau_0 = \frac{1}{3}\left[(\sigma_1 - \sigma_2)^2 + (\sigma_2 - \sigma_3)^2 + (\sigma_3 - \sigma_1)^2\right]^{1/2}$$

$$= \frac{1}{3}\left[120^2 + 20^2 + 140^2\right]^{1/2} = 61.8 \text{ MPa}$$

The octahedral shear strain is therefore

$$\gamma_0 = \frac{\tau_0}{G} = \frac{61.8}{8 \times 10^4} = 7.73 \times 10^{-4}$$

THEORIES OF FAILURE OR YIELD CRITERIA AND INTRODUCTION TO IDEALLY PLASTIC SOLID

Introduction : It is known from the results of material testing that when bars of ductile materials are subjected to uniform tension, the stress-strain curves show a linear range within which the materials behave in an elastic manner and a definite yield zone where the materials undergo permanent deformation. In the case of the so-called brittle materials, there is no yield zone. However, a brittle material, under suitable conditions, can be brought to a plastic state before fracture occurs. In general, the results of material testing reveal that the behaviour of various materials under similar test conditions, e.g. under simple tension, compression or torsion, varies considerably.

In the process of designing a machine element or a structural member, the designer has to take precautions to see that the member under consideration does not fail under service conditions. The word 'failure' used in this context may mean either fracture or permanent deformation beyond the operational range due to the yielding of the member. In Chapter 1, it was stated that the state of stress at any point can be characterised by the six rectangular stress components - three normal stresses and three shear stresses. Similarly, in Chapter 2, it was shown that the state of strain at a point can be characterised by the six rectangular strain components. When failure occurs, the question that arises is: what causes the failure? Is it a particular state of stress, or a particular state of strain or some other quantity associated with stress and strain? Further, the cause of failure of a ductile material need not be the same as that for a brittle material.

Consider, for example, a uniform rod made of a ductile material subject to tension. When yielding occurs.

(i) The principal stress s at a point will have reached a definite value, usually denoted by σ_y;

(ii) The maximum shearing stress at the point will have reached a value equal to $\tau = \frac{1}{2}\sigma_y$;

(iii) The principal extension will have become $\varepsilon = \sigma_y / E$;

(iv) The octahedral shearing stress will have attained a value equal to $(\sqrt{2}/3)\,\sigma_y$; and so on.

Any one of the above or some other factors might have caused the yielding. Further, as pointed out earlier, the factor that causes a ductile material to yield might be quite different from the factor that causes fracture in a brittle material under the same loading conditions. Consequently, there will be many criteria or yielding. Whatever may be the theory adopted, the information regarding it will have to be obtained from a simple test, like that of a uniaxial tension or a pure torsion test. This is so because the state of stress or strain which causes the failure of the material concerned can easily be calculated. The critical value obtained from this test will have to be applied for the stress or strain at a point in a general machine or a structural member so as not to initiate failure at that point.

There are six main theories of failure and these are discussed in the next section. Another theory, called Mohr's theory, is slightly different in its approach and will be discussed separately.

THEORIES OF FAILURE

Maximum Principal Stress Theory

This theory is generally associated with the name of Rankine. According to this theory, the maximum principal stress in the material determines failure regardless of what the other two principal stresses are, so long as they are algebraically smaller. This theory is not much supported by experimental results. Most solid materials can withstand very high hydrostatic pressures without fracture or without much permanent deformation if the pressure acts uniformly from all sides as is the case when a solid material is subjected to high fluid pressure.

Materials with a loose or porous structure such as wood, however, undergo considerable permanent deformation when subjected to high hydrostatic pressures. On the other hand, metals and other crystalline solids (including consolidated natural rocks) which are impervious, are elastically compressed and can withstand very high hydrostatic pressures. In less compact solid materials, a marked evidence of failure has been observed when these solids are subjected to hydrostatic pressures. Further, it has been observed that even brittle materials, like glass bulbs, which are subject to high hydrostatic pressure do not fail when the pressure is acting, but fail either during the period the pressure is being reduced or later when the pressure is rapidly released. It is stated that the liquid could have penetrated through the fine invisible surface cracks and when the pressure was released, the entrapped liquid may not have been able to escape fast enough. Consequently, high pressure gradients are caused on the surface of the material which tend to burst or explode the glass.

As Karman pointed out, this penentration and the consequent failure of the material can be prevented if the latter is covered by a thin flexible metal foil and then subjected to high hydrostatic pressures. Further noteworthy observations on the bursting action of a liquid which is used to transmit pressure were made by Bridgman who found that cylinders of hardened chrome-nickel steel were not able to withstand an internal pressure well if the liquid transmitting the pressure was mercury instead of viscous oil. It appears that small atoms of mercury are able to penentrate the cracks, whereas the large molecules of oil are not able to penentrate so easily.

From these observations, we the conclusion that a pure state of hydrostatic pressure [$\sigma_1 = \sigma_2 = \sigma_3 = -P$ ($P > 0$)] cannot produce permanent deformation in compact crystalline or amorphous solid materials but produces only a small elastic contraction, provided the liquid is prevented from entering the fine surface cracks or crevices of the solid. This contradicts the maximum principal stress theory. Further evidence to show that the maximum principal stress theory cannot be a good criterion for failure can be demonstrated in the following manner:

Consider the block shown in Fig. 1, subjected to stress σ_1 and σ_2, where σ_1 is tensile and σ_2 is compressive.

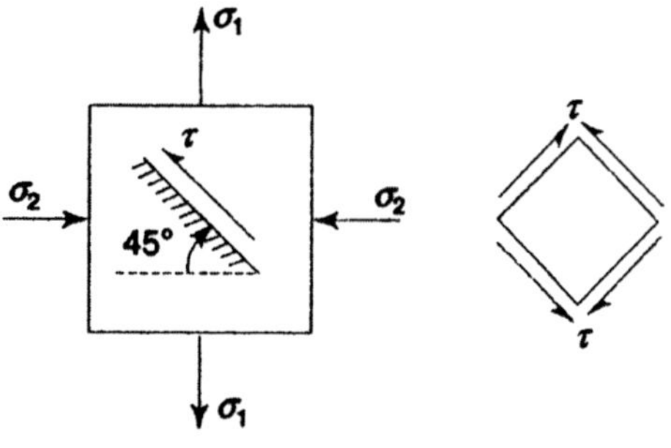

Fig. 1

If σ_1 is equal to σ_2 in magnitude, then on a 45° plane, from Eq. (63b), the shearing stress will have a magnitude equal to σ_1. Such a state of stress occurs in a cylindrical bar subjected to pure torsion. If the maximum principal stress theory was valid, σ_1 in magnitude.

Notwithstanding all these, the maximum principal stress theory, because of its simplicity, is considered to be reasonably satisfactory for brittle materials which do not fail by yielding. Using information from a uniaxial tension (or compression) test, we say that failure occurs when the maximum principal stress at any point reaches a value equal to the tensile (or compressive) elastic limit or yield strength of the material obtained from the uniaxial test. Thus, if $\sigma_1 > \sigma_2 > \sigma_3$ are the principal stresses at a point and σ_y the yield stress or tensile elastic limit for the material under a uniaxial test, then failure occurs when

$$\sigma_1 \geq \sigma_y \tag{1}$$

Maximum Shearing Stress Theory

Observations made in the course of extrusion tests on the flow of soft metals through orifices lend support to the assumption that the plastic state in such metals is created when the maximum shearing stress just reaches the value of the resistance of the metal against shear. Assuming $\sigma_1 = \sigma_2 = \sigma_3$, yielding, according to this theory, occurs when the maximum shearing stress

$$\tau_{max} = \frac{\sigma_1 - \sigma_3}{2}$$

reaches a critical value. The maximum shearing stress theory is accepted to be fairly well justified for ductile materials. In a bar subject to uniaxial tension or compression, the maximum shear stress occurs on a plane at 45° to the load axis. Tension tests conducted on mild steel bars show that at the time of yielding, the so-called slip lines occur approximately at 45°, thus supporting the theory. On the other hand, for brittle crystalline materials which cannot be brought into the plastic state under tension but which may yield a little before fracture under compression, the angle of the slip planes or of the shear fracture surfaces, which usually develop along these planes, differs considerably from the planes of maximum shear. Further, in these brittle materials, the values of the maximum shear in tension and compression are not equal. Failure of material under triaxial tension (of equal magnitude) also does not support this theory, since equal triaxial tensions cannot produce any shear.

However, as remarked earlier, for ductile load carrying members where large shears occur and which are subject to unequal triaxial tensions, the maximum shearing stress theory is used because of its simplicity.

If $\sigma_1 > \sigma_2 > \sigma_3$ are the three principal stresses at a point, failure occurs when

$$\tau_{max} = \frac{\sigma_1 - \sigma_3}{2} \geq \frac{\sigma_y}{2} \tag{2}$$

where $\sigma_y/2$ is the shear stress at yield point in a uniaxial test.

Maximum Elastic Strain Theory

According to theory, failure occurs at a point in a body when the maximum strain at that point exceeds the value of the maximum strain in a uniaxial test of the material at yield point. Thus, if σ_1, σ_2 and σ_3 are the principal stresses at a point, failure occurs when

$$\varepsilon_1 = \frac{1}{E}[\sigma_1 - \nu(\sigma_2 + \sigma_3)] \geq \frac{\sigma_y}{E} \tag{3}$$

We have observed that a material subjected to triaxial compression does not suffer failure, thus contradicting this theory. Also, in a block subjected to a biaxial tension, as shown in Fig. 5, the principal strain ε_1 is

$$\varepsilon_2 = \frac{1}{E}(\sigma_1 - v\sigma_2)$$

and is smaller than σ_1/E because of σ_2. Therefore, according to this theory, σ_1 can be increased more than σ_y without causing failure, whereas, if σ_2 were compressive, the magnitude of σ_1 to cause failure would be less than σ_y. However, this is not supported by experiments.

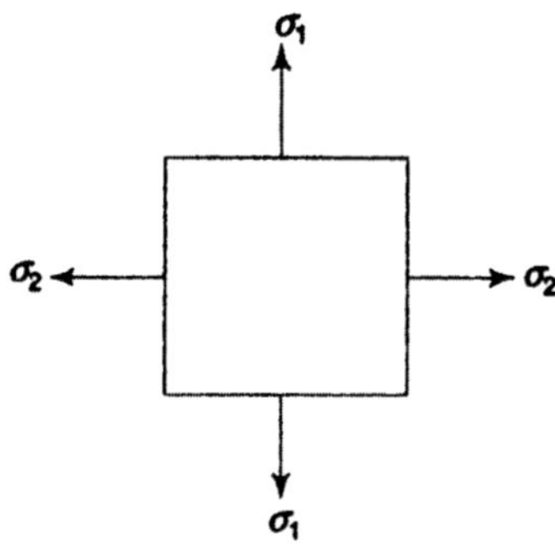

Fig. 2

While the maximum strain theory is an improvement over the maximum stress theory, it is not a good theory for ductile materials. Fir materials which fail by brittle fracture, one may prefer the maximum strain theory to the maximum stress theory.

Octahedral Shearing Stress Theory

According to this theory, the critical quantity is the shearing stress on the octahedral plane. The plane which is equally inclined to all the three principal axes *Ox*, *Oy* and *Oz* is called the octahedral plane. The normal to this plane has direction cosines n_x, n_y and $n_z = 1/\sqrt{3}$. The tangential stress on this plane is the octahedral shearing stress. If σ_1, σ_2 and σ_3 are the principal stresses at a point, then from Eqs (44a) and (44c)

$$\tau_{oct} = \frac{1}{3}[(\sigma_1 - \sigma_2)^2 + (\sigma_2 - \sigma_3)^2 + (\sigma_3 - \sigma_1)^2]^{1/2}$$

$$= \frac{\sqrt{2}}{3}\left(l_1^2 - 3l_2\right)^{1/2}$$

In a uniaxial test, at yield point, the octahedral stress $(\sqrt{2}/3)\,\sigma_y = 0.47\sigma_y$. Hence, according to the present theory, failure occurs at point where the values of principal stresses are such that

$$\tau_{oct} = \frac{1}{3}\left[(\sigma_1 - \sigma_2)^2 + (\sigma_2 - \sigma_3)^2 + (\sigma_3 - \sigma_1)^2\right]^{1/2} \geq \frac{\sqrt{2}}{3}\sigma_y \quad \text{(4a)}$$

or $$\left(I_1^2 - 3I_2\right) \geq {}_y^2 \quad \text{(4b)}$$

This theory is supported quite well by experimental evidences. Further, when a material is subjected to hydrostatic pressure, $\sigma_1 = \sigma_2 = \sigma_3 = -P$, and τ_{oct} is equal to zero. Consequently, according to this theory, failure cannot occur and this, as stated earlier, is supported by experimental results. This theory is equivalent to the maximum distortion energy theory.

Maximum Elastic Energy Theory

This theory is associated with the names of Beltrami and Haigh. According to this theory, failure at any point in a body subject to a state of stress begins only when the energy per unit volume absorbed at the point is equal to the energy absorbed per unit volume by the material when subjected to the elastic limit under a uniaxial state of stress. To calculate the energy absorbed per unit volume we proceed as follows:

Let σ_1, σ_2 and σ_3 be the principal stresses and let their magnitudes increase uniformly from zero to their final magnitudes. If ε_1, ε_2 and ε_3 are the corresponding principal strains, then the work done by the forces, from Fig. 3(b), is

$$\Delta W = \frac{1}{2}\sigma_1\ \Delta y\ \Delta z\ (\delta\Delta x) + \frac{1}{2}\sigma_2\ \Delta x\ \Delta z\ (\delta\Delta y) + \frac{1}{2}\sigma_3\ \Delta x\ \Delta y\ (\delta\Delta z)$$

where $\delta\Delta x$, $\delta\Delta y$ and $\delta\Delta z$ are extensions in x, y and z directions respectively.

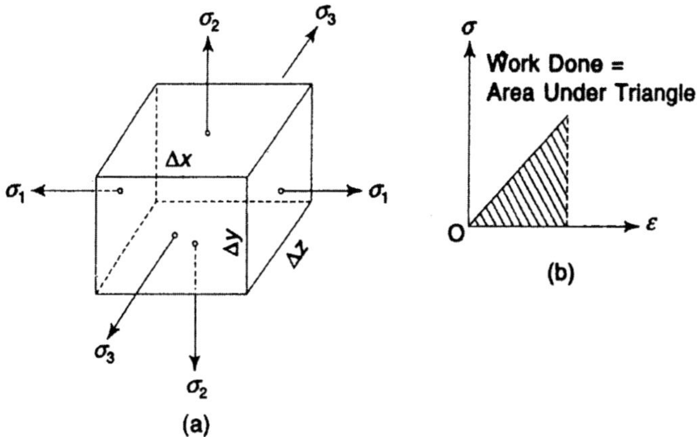

Fig. 3

From Hooke's law

$$\delta\Delta x = \varepsilon_1 \, \Delta x = \frac{1}{E}[\sigma_1 - v(\sigma_2 + \sigma_3)] \, \Delta x$$

$$\delta\Delta y = \varepsilon_2 \, \Delta y = \frac{1}{E}[\sigma_2 - v(\sigma_1 + \sigma_3)] \, \Delta y$$

$$\delta\Delta z = \varepsilon_3 \, \Delta z = \frac{1}{E}[\sigma_3 - v(\sigma_1 + \sigma_2)] \, \Delta z$$

Substituting these

$$\Delta W = \frac{1}{2E}[\sigma_1^2 + \sigma_2^2 + \sigma_3^2 - 2v(\sigma_1\sigma_2 + \sigma_2\sigma_3 + \sigma_3\sigma_1)] \, \Delta x \, \Delta y \, \Delta z$$

The above work is stored as internal energy if the rate of deformation is small. Consequently, the energy U per unit volume is

$$\frac{1}{2E}[\sigma_1^2 + \sigma_2^2 + \sigma_3^2 - 2v(\sigma_1\sigma_2 + \sigma_2\sigma_3 + \sigma_3\sigma_1)] \tag{5}$$

In a uniaxial test, the energy stored per unit volume at yield point or elastic limit is $1/2E \ \sigma_y^2$. Hence, failure occurs when

$$\sigma_1^2 + \sigma_2^2 + \sigma_3^2 - 2v(\sigma_1\sigma_2 + \sigma_2\sigma_3 + \sigma_3\sigma_1) \geq \sigma_y^2 \tag{6}$$

This theory is based on the work of Huber, von Mises and Hencky. According to this theory, it is not the total energy which is the criterion for failure; in fact the energy absorbed during the distortion of an element is responsible for failure.

The energy of distortion can be obtained by subtracting the energy of volumetric expansion from the total energy. Any given state of stress can be uniquely resolved into an isotropic state and a pure shear (or deviatoric) state. If σ_1, σ_2 and σ_3 are the principal stresses at a point then

$$\begin{bmatrix} \sigma_1 & 0 & 0 \\ 0 & \sigma_2 & 0 \\ 0 & 0 & \sigma_3 \end{bmatrix} = \begin{bmatrix} P & 0 & 0 \\ 0 & p & 0 \\ 0 & 0 & P \end{bmatrix} = \begin{bmatrix} \sigma_1 - p & 0 & 0 \\ 0 & \sigma_2 - P & 0 \\ 0 & 0 & \sigma_3 - p \end{bmatrix} \tag{7}$$

where $P = \frac{1}{3}(\sigma_1 + \sigma_2 + \sigma_3)$.

The first matrix on the right-hand side represents the isotropic state and the second matrix the pure shear state. Also, recall that the necessary and sufficient condition for a state to be a pure shear state is that its first invariant must be equal to zero. It was shown that any given state of strain can be resolved uniquely into an isotropic and a deviatoric state of strain. If ε_1, ε_2 and ε_3 are the principal strains at the point, we have

$$\begin{bmatrix} \varepsilon_1 & 0 & 0 \\ 0 & \varepsilon_2 & 0 \\ 0 & 0 & \varepsilon_3 \end{bmatrix} = \begin{bmatrix} e & 0 & 0 \\ 0 & e & 0 \\ 0 & 0 & e \end{bmatrix} = \begin{bmatrix} \varepsilon_1 - e & 0 & 0 \\ 0 & \varepsilon_2 - e & 0 \\ 0 & 0 & \varepsilon_3 - e \end{bmatrix} \tag{8}$$

where $e = \frac{1}{3}(\varepsilon_1 + \varepsilon_2 + \varepsilon_3)$.

If was also shown that the volumetric strain corresponding to the deviatoric state of strain is zero since its first invariant is zero.

It is easy to see from Eqs (7) and (8) that, by Hooke's law, the isotropic state of strain is related to the isotropic state of stress because

$$\varepsilon_1 = \frac{1}{E}[\sigma_1 - v(\sigma_2 + \sigma_3)]$$

$$\varepsilon_2 = \frac{1}{E}[\sigma_2 - v(\sigma_3 + \sigma_1)]$$

$$\varepsilon_3 = \frac{1}{E}[\sigma_3 - v(\sigma_2 + \sigma_1)]$$

Adding and taking the mean

$$\frac{1}{3}(\varepsilon_1 + \varepsilon_2 + \varepsilon_3) = e$$

$$= \frac{1}{3}[(\sigma_1 + \sigma_2 + \sigma_3) - 2v(\sigma_1 + \sigma_2 + \sigma_3)$$

or
$$e = \frac{1}{E}[(1 - 2v)P] \tag{9}$$

i. e. e is connected to P by Hooke's law. This states that the volumetric strain $3e$ is proportional to the pressure P, the proportionality constant being equal to

$$\frac{3}{E}(1 - 2v) = K, \text{ the bulk modulus, Eq. (14).}$$

Consequently, the work done or the energy stored during volumetric change is

$$U' = \frac{1}{2}pe + \frac{1}{2}pe + \frac{1}{2}pe = \frac{3}{2}pe$$

Substituting for e from Eq. (90)

$$U' = \frac{3}{2E}(1 - 2v)P^2 \tag{10}$$

$$= \frac{1 - 2v}{6E}(\sigma_1 + \sigma_2 + \sigma_3)^2$$

The total elastic strain energy density is given by Eq. (5). Hence, subtracting U' from U

$$U^* = \frac{1}{2E}(\sigma_1^2 + \sigma_2^2 + \sigma_3^2) - \frac{v}{E}(\sigma_1\sigma_2 + \sigma_2\sigma_3 + \sigma_3\sigma_1) -$$

$$\frac{1-2v}{6E}(\sigma_1 + \sigma_2 + \sigma_3)^2 \quad (11a)$$

$$= \frac{2(1+v)}{6E}(\sigma_1^2 + \sigma_2^2 + \sigma_3^2 - \sigma_1\,\sigma_2 - \sigma_2\,\sigma_3 - \sigma_3\,\sigma_1) \quad (11b)$$

$$= \frac{(1+v)}{6E}\left[(\sigma_1 - \sigma_2)^2 + (\sigma_2 - \sigma_3)^2 + (\sigma_3 - \sigma_1)^2\right] \quad (11c)$$

Substituting $G = \dfrac{E}{2(1+v)}$ for the shear modulus,

$$U^* = \frac{1}{6G}(\sigma_1^2 + \sigma_2^2 + \sigma_3^2 - \sigma_1\,\sigma_2 - \sigma_2\,\sigma_3 - \sigma_3\,\sigma_1) \quad (12a)$$

or $$U^* = \frac{1}{12G}\left[(\sigma_1 - \sigma_2)^2 + (\sigma_2 - \sigma_3)^2 + (\sigma_3 - \sigma_1)^2\right] \quad (12b)$$

This is the expression for the energy of distortion. In a uniaxial test, the energy of distortion is equal to $\dfrac{1}{6G}\sigma_y^2$. This is obtained by simply putting $\sigma_1 = \sigma_y$ and $\sigma_2 = \sigma_3 = 0$ in Eq. (12). This is also equal to $\dfrac{(1+v)}{3E}\sigma_y^2$ from Eq. (1c).

Hence, according to the distortion energy theory, failure occurs at that point where σ_1, σ_2 and σ_3 are such that

$$(\sigma_1 - \sigma_2)^2 + (\sigma_2 - \sigma_3)^2 + (\sigma_3 - \sigma_1)^2 \geq 2\sigma_y^2 \quad (13)$$

But we notice that the expression for the octahedral shearing stress from Eq. (22) is

$$\tau_{oct} = \frac{1}{3}\left[(\sigma_1 - \sigma_2)^2 + (\sigma_2 - \sigma_3)^2 + (\sigma_3 - \sigma_1)^2\right]^{1/2}$$

Hence, the distortion energy theory states that failure occurs when

$$9^2_{oct} = \geq 2\sigma_y^2$$

or $$\tau_{oct} = \geq \frac{\sqrt{2}}{3}\sigma_y \quad (14)$$

This is identical to Eq. (4). Therefore, the octahedral shearing stress theory and the distortion energy theory are identical. Experiments made on the flow of ductile metals under biaxial states of stress have shown that

Eq. (14) or equivalently, Eq. (13) express well the condition under which the ductile metals at normal temperatures start to yield. Further, as remarked earlier, the purely elastic deformation of a body under hydrostatic pressure ($\tau_{oct} = 0$) is also supported by this theory.

SIGNIFICANCE OF THE THEORIES OF FAILURE

The mode of failure of a member and the factor that is responsible for failure depend on a large number of factors such as the nature and properties of the material, type of loading, shape and temperature of the member, etc. We have observed, for example, that the mode of failure of a ductile material differs from that of a brittle material. While yielding or permanent deformation is the characteristic feature of ductile materials, fracture without permanent deformation is the characteristic feature of brittle materials. Further, if the loading conditions are suitably altered, a brittle material may be made to yield before failure.

Even ductile materials fail in a different manner when subjected to repeated loadings (such as fatigue) than when subjected to static loadings. All these factors indicate that any rational procedure of design of a member requires the determination of the mode of failure (either yielding or fracture), and the factor (such as stress, strain and energy) associated with it.

If tests could be performed on the actual member, subjecting it to all the possible conditions of loading that the member would be subjected to during operation, then one could determine the maximum loading condition that does not cause failure. But this may not be possible except in very simple cases. Consequently, in complex loading conditions, one has to identify the factor associated with the failure of a member and take precautions to see that this factor does not exceed the maximum allowable value. This information is obtained by performing a suitable test (uniform tension or torsion) on the material in the laboratory.

In discussing the various theories of failure, we have expressed the critical value associated with each theory in terms of the yield point stress σ_y obtained form a uniaxial tensile stress. This was done since it is easy to perform a uniaxial tensile stress and obtain the yield point stress value. It is equally easy to perform a pure torsion test on a round specimen and obtain the value of the maximum shear stress τ_y at the point of yielding.

Consequently, one can also express the critical value associated with each theory of failure in terms of the yield point shear stress τ_y. In a sense, using σ_y or τ_y is equivalent because during a uniaxial tension, the maximum shear stress τ at a point is equal to 1/2 σ, and in the case of pure shear, the normal stresses on a 45° element are σ and $-\sigma$, where σ is numerically equivalent to τ_y. These are shown in Fig. 4.

If one uses the yield point shear stress τ_y obtained from a pure torsion test, then the critical value associated with each theory of failure is as follows:

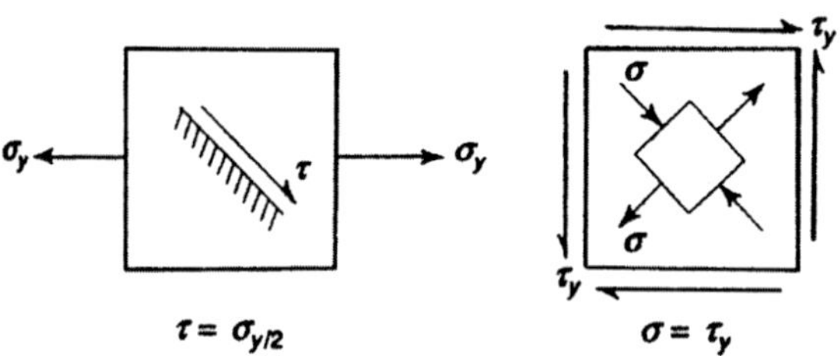

Fig. 4

(i) *Maximum normal stress theory:*

According to this theory, failure occurs when the normal stress σ at any point in the stressed member reaches a value

$$\sigma \geq \tau_y$$

This is because, in a pure torsion test when yielding occurs, the maximum normal stress s is numerically equivalent to τ_y.

(ii) *Maximum shear stress theory:*

According to this theory, failure occurs when the shear stress τ at a point in the member reaches a value

$$\tau \geq \tau_y$$

(iii) *Maximum strain theory:*

According to this theory, failure occurs when the maximum strain at any point in the member reaches a value

$$\varepsilon = \frac{1}{E}[\sigma_1 - v(\sigma_2 + \sigma_3)]$$

From Fig. 4, in the case of pure shear

$\sigma_1 = \sigma = \tau,\ \ \sigma_2 = 0,\ \sigma_3 = -\ \sigma = -\ \tau$

Hence, failure occurs when the strain e at any point in the member reaches a value

$$\varepsilon = \frac{1}{E}(\tau_y + v\tau_y) = \frac{1}{E}(1+v)\tau_y$$

(iv) *Octahedral shear stress theory:*

When an element is subjected to pure shear, the maximum and minimum normal stresses at a point are σ and $-\sigma$ (each numerically equal to the shear stress τ), as shown in Fig. 4 Corresponding to this, from Eq. (44a), the octahedral shear stress is

$$\tau_{oct} = \frac{1}{3}\left[(\sigma_1 - \sigma_2)^2 + (\sigma_2 - \sigma_3)^2 + (\sigma_3 - \sigma_1)^2\right]^{1/2}$$

Observing that $\sigma_1 = \sigma = \tau,\ \ \sigma_2 = 0,\ \sigma_3 = -\sigma = -\tau$

$$\tau_{oct} = \frac{1}{3}(\sigma^2 + \sigma^2 + 4\sigma^2)^{1/2}$$

$$= \frac{\sqrt{6}}{3}\sigma = \sqrt{\frac{2}{3}}\tau$$

So, failure occurs when the octahedral shear stress at any point is

$$\tau_{oct} = \sqrt{\frac{2}{3}}\,\tau_y$$

(v) *Maximum elastic energy theory:*

The elastic energy per unit volume stored at a point in a stressed body is, from Eq. (5),

$$U = \frac{1}{E}\left[\sigma_1^2 + \sigma_2^2 + \sigma_3^2 - 2v(\sigma_1\sigma_2 + \sigma_2\sigma_3 + \sigma_3\sigma_1\right]$$

In the case of pure shear, from Fig. 4,

$$\sigma_1 = \tau, \qquad \sigma_2 = 0 \qquad \sigma_3 = -\tau$$

Hence, $U = \dfrac{1}{2E}\left[\tau^2 + \tau^2 - 2v(-\tau^2)\right]$

$$= \frac{1}{E}(1+v)\tau^2$$

So, failure occurs when the elastic energy density at any point in a stressed body is such that

$$U = \frac{1}{E}(1+v)\tau_y^2$$

(vi) *Distortion energy theory:*

The distortion energy density at a point in a stressed body is, from Eq. (12).

$$U^* = \frac{1}{12G}\left[(\sigma_1 - \sigma_2)^2 + (\sigma_2 - \sigma_3)^2(\sigma_3 - \sigma_1)^2\right]$$

Once again, by observing that in the case of pure shear

$$\sigma_1 = \tau, \qquad \sigma_2 = 0 \qquad \sigma_3 = -\tau$$

$$U^* = \frac{1}{12G}\left[\tau^2 + \tau^2 + 4\tau^2\right]$$

$$= \frac{1}{12G}\tau^2$$

So, failure occurs when the distortion energy density at any point is equal to

$$U^* = \frac{1}{2G}\tau_y^2 = \frac{1}{2}.\frac{2(1+v)}{E}\tau_y^2$$

$$= \frac{(1+v)}{E}\tau_y^2$$

The foregoing results shown that one can express the critical value associated with each theory of failure either in terms of σ_y or in terms or τ_y. Assuming that a particular theory of failure is correct for a given material, then the values of σ_y and τ_y obtained from tests conducted on the material should be related by the corresponding expressions. For example, if the distortion energy is a valid theory for a material, then the value of the energy in terms of σ_y and that in terms of τ_y should be equal. Thus,

$$U^* = \frac{(1+v)}{E}\tau_y^2 = \frac{(1+v)}{3E}\sigma_y^2$$

or $$\tau_y = \frac{1}{\sqrt{3}}\sigma_y = 0.577\,\sigma_y$$

This means that the value of τ_y obtained from pure torsion test should be equal to 0.577 times the value of σ_y obtained from a uniaxial tension test conducted on the same material.

Table 1 summarizes these theories and the corresponding expressions. The first column lists the six theories of failure. The second column lists the critical value associated with each theory in terms of σ_y, the yield point stress in uniaxial tension test. For example, according to the octahedral shear stress theory, failure occurs when the octahedral shear stress at a point assumes a value equal to $\sqrt{2/3}\sigma_y$. The third column lists the critical value associated with each theory in terms of τ_y, the yield point shear stress value in pure torsion. For example, according to octahedral shear stress theory, failure occurs at a point when the octahedral shear stress equals a value $\sqrt{2/3}\tau_y$. The fourth column gives the relationship that should exist between τ_y and σ_y in each case if each theory is valid, Assuming octahedral shear stress theory is correct, then the value of τ_y obtained from pure torsion test should be equal to 0.577 times the yield point stress σ_y obtained from a uniaxial tension test.

Tests conducted on many ductile materials reveal that the values of τ_y lie between 0.50 and 0.60 of the tensile yield strength σ_y, the average value being about 0.57. This result agrees well with the octahedral shear stress theory and the distortion energy theory. The maximum shear stress theory predicts that shear yield value τ_y is 0.5 times the tensile yield value. This is about 15% less than the value predicted by the distortion energy (or the octahedral shear) theory. The maximum shear stress theory gives values for design on the safe side. Also, because of its simplicity, this theory is widely used in machine design dealing with ductile materials.

Table 1

Failure theory	*Tension*	*Shear*	*Relationship*
Max. normal stress	σ_y	$\sigma_y = \tau_y$	$\tau_y = \sigma_y$
Max. shear stress	$\tau = \frac{1}{2}\sigma_y$	τ_y	$\tau_y = 0.5\ \sigma_y$
Max. strain $\left(\nu = \frac{1}{4}\right)$	$\varepsilon = \frac{1}{E}\sigma_y$	$\varepsilon = \frac{5}{4}\frac{\tau_y}{E}$	$\tau_y = 0.8\ \sigma_y$
Octahedral shear	$\tau_{oct} = \frac{\sqrt{2}}{3}\sigma_y$	$\tau_{oct} = \sqrt{\frac{2}{3}}\tau_y$	$\tau_y = 0.577\ \sigma_y$
Max. energy $\left(\nu = \frac{1}{4}\right)$	$U = \frac{1}{2E}\sigma_y^2$	$U = \frac{5}{4}\frac{1}{E}\tau_y^2$	$\tau_y = 0.632\ \sigma_y$
Distortion energy	$U^* = \frac{1+\nu}{3}\frac{\sigma_y^2}{E}$	$U^* = (1+\nu)\frac{\tau_y^2}{E}$	$\tau_y = 0.577\ \sigma_y$

USE OF FACTOR OF SAFETY IN DESIGN

In designing a member to carry a given load without failure, usually a factor of safety N is used. The purpose is to design the member in such a way that it can carry N times the actual working load without failure. It has been observed that one can associate different factors for failure according to the particular theory of failure adopted. Consequently, one can use a factor appropriately reduced during the design process. Let X be a factor associated with failure and let F be the load. If X is directly proportional to F, then designing the member to safely carry a load equal to NF is equivalent to designing the member for a critical factor equal to X/N. However, if X is not directly proportional to F, but is, say, proportional to F^2, then designing the member to safely carry a load to equal to NF is equivalent to limiting the critical factor to $\sqrt{X/N}$. Hence, in using the factor of safety, care must be taken to see that the critical factor associated with failure is not reduced by N, but rather the load-carrying capacity is increased by N. This point will be made clear in the following example.

Example 1:

Determine the diameter d of a circular shaft subjected to a bending moment M and a torque T, according to the several theories of failure. Use a factor of safety N.

Solution:

Consider a point P on the periphery of the shaft. If d is the diameter, then owing to the bending moment M, the normal stress σ at P on a plane normal to the axis of the shaft is, from elementary strength of materials,

$$\sigma = \frac{My}{I} = M\frac{d}{2}\frac{64}{\pi d^4} \tag{15}$$

$$= \frac{32M}{\pi d^3}$$

The shearing stress on a transverse plane at P due to torsion T is

$$\tau = \frac{Td}{2I_p} = \frac{Td.32}{2\pi d^4} \tag{16}$$

$$= \frac{16T}{\pi d^3}$$

Therefore, the principal stresses at P are

$$\sigma_{1,3} = \frac{1}{2}\sigma \pm \frac{1}{2}\sqrt{(\sigma^2 + 4\tau^2)}, \sigma_2 = 0 \tag{17}$$

(i) *Maximum normal stress theory:* At point P, the maximum normal stress should not exceed σ_y, the yield point stress in tension. With a factor of safety N, when the load is increased N times, the normal and shearing stresses are $N\sigma$ and $N\tau$. Equating the maximum normal stress to σ_y,

$$\sigma_{max} = \sigma_1 = N\left[\frac{\sigma}{2} + \frac{1}{2}(\sigma^2 + 4\tau^2)^{1/2}\right] = \sigma_y$$

or $$\sigma + (\sigma^2 + 4\tau^2)^{1/2} = \frac{2\sigma_y}{N}$$

i.e., $$\frac{32M}{\pi d^3} + \frac{1}{\pi d^3} \times 32\left(M^2 + T^2\right)^{1/2} = \frac{2\sigma_y}{N}$$

i.e., $$16M + 16\left(M^2 + T^2\right)^{1/2} = \frac{\pi d^3 \sigma_y}{N}$$

From this, the value of d can be determined with the known values of M, T and σ_y.

(ii) *Maximum shear theory:* At point P, the maximum shearing stress from Eq. (17) is

$$\tau_{max} = \frac{1}{2}(\sigma_1 - \sigma_3) = \frac{1}{2}(\sigma^2 + 4\tau^2)^{1/2}$$

When the load is increased N times, the shear stress becomes N t.

Hence,

$$N\tau_{max} = \frac{1}{2}N(\sigma^2 + 4\tau^2)^{1/2} = \frac{\sigma_y}{2}$$

or, $$(\sigma^2 + 4\tau^2)^{1/2} = \frac{\sigma_y}{N}$$

Substituting for σ and τ

$$\frac{32}{\pi d^3}\left(M^2 + T^2\right)^{1/2} = \frac{\sigma_y}{N}$$

or, $$32\left(M^2 + T^2\right)^{1/2} = \frac{\pi d^3 \sigma_y}{N}$$

(iii) *Maximum strain theory:* The maximum elastic strain at point P with a factor of safety N is

$$\varepsilon_{max} = \frac{N}{E}[\sigma_1 - \nu(\sigma_2 + \sigma_3)]$$

From Eq. (3)

$$\sigma_1 - \nu(\sigma_2 + \sigma_3) = \frac{\sigma_y}{N}$$

Since $\sigma_2 = 0$, we have $\sigma_1 - \nu\,\sigma_3 = \dfrac{\sigma_y}{N}$

or $$\frac{\sigma}{2} + \frac{1}{2}(\sigma^2 + 4\tau^2)^{1/2} - \nu\frac{\sigma}{2} + \frac{\nu}{2}(\sigma^2 + 4\tau^2)^{1/2} = \frac{\sigma_y}{N}$$

Substituting for σ and τ

$$(1-\nu)\frac{32M}{\pi d^3} + (1+\nu)\frac{32}{\pi d^3} + \left(M^2 + T^2\right)^{1/2} = \frac{\sigma_y}{N}$$

or $$(1-v)16M+(1+v)16\left(M^2+T^2\right)^{1/2}=\frac{\pi d^3\sigma_y}{N}$$

(iv) *Octahedral shear stress theory:* The octahedral shearing stress at point P from Eq. (4a), and using a factor of safety N, is

$$N\tau_{oct}=\frac{N}{3}\left[(\sigma_1-\sigma_2)^2+(\sigma_2-\sigma_3)^2+(\sigma_3-\sigma_1)^2\right]^{1/2}=\frac{\sqrt{2}}{3}\sigma_y$$

or $$\left[(\sigma_1-\sigma_2)^2+(\sigma_2-\sigma_3)^2+(\sigma_3-\sigma_1)^2\right]^{1/2}=\frac{\sqrt{2}}{N}\sigma_y$$

with $\sigma_2=0$

$$\left[2\sigma_1^2+2\sigma_3^2-2\sigma_1\sigma_3\right]^{1/2}=\frac{\sqrt{2}}{N}\sigma_y$$

or $$\left[2\sigma_1^2+2\sigma_3^2-\sigma_1\sigma_3\right]^{1/2}=\frac{\sigma_y}{N}$$

Substituting for σ_1 and σ_3

$$\left[\frac{1}{4}\sigma^2+\frac{1}{4}(\sigma^2+4\tau^2)+\frac{1}{2}\sigma(\sigma^2+4\tau^2)^{1/2}+\frac{1}{4}\sigma^2+\frac{1}{4}(\sigma^2+4\tau^2)\right.$$
$$\left.-\frac{1}{2}\sigma(\sigma^2+4\tau^2)^{1/2}-\frac{1}{4}\sigma^2+\frac{1}{4}(\sigma^2+4\tau^2)\right]^{1/2}=\frac{\sigma_y}{N}$$

or $$(\sigma^2+3\tau^2)^{1/2}=\frac{\sigma_y}{N}$$

Substituting for σ and τ

$$\frac{16}{\pi d^3}\left(4M^2+3T^2\right)^{1/2}=\frac{\sigma_y}{N}$$

or $$16\left(4M^2+3T^2\right)^{1/2}=\frac{\pi d^3\sigma_y}{N}$$

(v) *Maximum energy theory*: The maximum elastic energy at P from Eq. (6) and with a factor of safety N is

$$U=\frac{N^2}{2E}\left[\sigma_1^2+\sigma_2^2+\sigma_3^2-2v(\sigma_1\sigma_2+\sigma_2\sigma_3+\sigma_3\sigma_1)\right]=\frac{\sigma_y^2}{2E}$$

Note: Since the stresses for design are $N\sigma_1$, $N\sigma_2$ and $N\sigma_3$, the factor N^2 appears in the expression for U. In the previous four case, only N appeared because of the particular form of the expression.

with $\sigma_2=0$,

$$(\sigma_1^2 + \sigma_3^2 - 2\nu\,\sigma_1\sigma_3) = \frac{\sigma_y^2}{N^2}$$

Substituting for σ_1 and σ_3

$$\left[\frac{1}{4}\sigma^2 + \frac{1}{4}(\sigma^2 + 4\tau^2) + \frac{1}{2}\sigma(\sigma^2 + 4\tau^2)^{1/2} + \frac{1}{4}\sigma^2 + \frac{1}{4}(\sigma^2 + 4\tau^2) - \frac{1}{2}\sigma(\sigma^2 + 4\tau^2)^{1/2} + 2\nu\tau^2\right] = \frac{\sigma_y^2}{N^2}$$

or $$\sigma^2 + (2 + 2\nu)\tau^2 = \frac{\sigma_y^2}{N^2}$$

i.e. $$\left[\sigma^2 + (2 + 2\nu)\tau^2\right]^{1/2} = \frac{\sigma_y}{N}$$

i.e. $$\frac{16}{\pi d^3}\left[4M^2 + (2 + 2\nu)T^2\right]^{1/2} = \frac{\sigma_y}{N}$$

or $$\left[4M^2 + 2(1 + \nu)T^2\right]^{1/2} = \frac{\pi d^3 \sigma_y}{16N}$$

(vi) *Maximum distortion energy theory:* The distortion energy associated with $N\sigma_1$, $N\sigma_2$ and $N\sigma_3$ at P is given by Eq. (11b). Equating this to distortion energy in terms of σ_y

$$U_d = \frac{N^2(1+\nu)}{6E}\left[(\sigma_1 - \sigma_2)^2 + (\sigma_2 - \sigma_3)^2 + (\sigma_3 - \sigma_1)^2\right]$$

$$= \frac{1+\nu}{3E}\sigma_y^2$$

with $\sigma_2 = 0$,

$$(2\sigma_1^2 + 2\sigma_3^2 - 2\sigma_1\sigma_3) = \frac{\sigma_y^2}{N^2}$$

or $$(\sigma_1^2 + \sigma_3^2 - \sigma_1\sigma_3)^{1/2} = \frac{\sigma_y}{N}$$

This yields the same result as the octahedral shear stress theory.

A NOTE ON THE USE OF FACTOR OF SAFETY

As remarked earlier, then a factor of safety N is prescribed, we may consider two ways of introducing it in design:

(i) Design the member so that it safely carries a load NF.

(ii) If the factor associated with failure is X, then see that this factor at any point in the member does not exceed X/N.

But the second method of using N is not correct, since by the definition of the factor of safety, the member is to be designed for N times the load. So long as X is directly proportional to F, whether one uses NF or X/N for design analysis, the result will be identical. If X is not directly proportional to F, method (ii) may give wrong results. For example, if we adopt method (ii) with the maximum energy theory, the result will be

$$U = \frac{1}{2E}\left[\sigma_1^2 + \sigma_2^2 + \sigma_3^2 - 2v(\sigma_1\sigma_2 + \sigma_2\sigma_3 + \sigma_3\sigma_1)\right] = \frac{1}{N}\frac{\sigma_y^2}{2E}$$

where X, the factor associated with failure, is $\frac{1}{2}\frac{\sigma_y^2}{E}$. But method (i) gives

$$U = \frac{N^2}{2E}\left[\sigma_1^2 + \sigma_2^2 + \sigma_3^2 - 2v(\sigma_1\sigma_2 + \sigma_2\sigma_3 + \sigma_3\sigma_1)\right] = \frac{\sigma_y^2}{2E}$$

The result obtained from method (i) is correct, since $N\sigma_1$, $N\sigma_2$ and $N\sigma_3$ are the principal stresses corresponding to the load NF. As one an see, the results are not the same. The result given by method (ii) is not the right one.

Example 2:

A force F = 45,000 N is necessary to rotate the shaft at uniform speed. The crank shaft is made of ductile steel whose elastic limit

is 207,000 KPa, both in tension and compression. With $E = 207 \times 10^6$ KPa, V = 0.25, determine the diameter of the shaft, using the octahedral shear stress theory and the maximum shear stress theory. Use a factor of safety N = 2. Consider a point on the periphery at section A for analysis.

Solution:

The moment at section A is

$$M = 45{,}000 \times 0.2 = 9000 \text{ Nm}$$

and the torque on the shaft is

$$T = 45{,}000 \times 0.15 = 6750 \text{ Nm}$$

The normal stress due to M at A is

$$\sigma = -\frac{64Md}{2\pi d^4} = -\frac{32M}{\pi d^3}$$

and the maximum shear stress due to T at A is

$$\tau = -\frac{32Td}{2\pi d^4} = -\frac{16T}{\pi d^3}$$

The shear stress due to the shear force F is zero at A. The principal stresses from Eq. (61) are

$$\sigma_{1,3} = \frac{1}{2}\sigma \pm \frac{1}{2}\left(\sigma^2 + 4\tau^2\right)^{1/2}, \qquad \sigma_2 = 0$$

(i) *Maximum shear stress theory:*

$$\tau_{max} = \frac{1}{2}(\sigma_1 - \sigma_3)$$

$$= \frac{1}{2}\left(\sigma^2 + 4\tau^2\right)^{1/2}$$

$$= \frac{1}{2}\frac{32}{\pi d^3}(M^2 + T^2)^{1/2}$$

$$= \frac{16}{\pi d^3}(9000^2 + 6750^2)^{1/2} = \frac{57295.8}{d^3}\text{ Pa}$$

With a factor of safety $N = 2$, the value of τ_{max} becomes

$$N\tau_{max} = \frac{114591.6}{d^3}\text{ Pa}$$

This should not exceed the maximum shear stress value at yielding in uniaxial tension test. Thus,

$$\frac{1}{d^3}(114591.6) = \frac{\sigma_y}{2} = \frac{207}{2}\times 10^6$$

∴ $d3 = 1107 \times 10^{-6}\text{ m}^3$

or $d = 10.35 \times 10^{-2}\text{ m} = 10.4\text{ cm}$

(ii) *Octahedral shear stress theory:*

$$\tau_{oct} = \frac{1}{3}\left[(\sigma_1 - \sigma_2)^2 + (\sigma_2 - \sigma_3)^2 + (\sigma_3 - \sigma_1)^2\right]^{1/2}$$

With $\sigma_2 = 0$,

$$\tau_{oct} = \frac{1}{3}\left[2\sigma_1^2 + 2\sigma_3^2 - 2\sigma_1\,\sigma_3\right]^{1/2}$$

Substituting for σ_1 and σ_3 and simplifying

$$\tau_{oct} = \frac{\sqrt{2}}{3}\left(\sigma^2 + 3\tau^2\right)^{1/2}$$

$$= \frac{\sqrt{2}}{3\pi d^3}\left[(32M)^2 + 3(16T)^2\right]^{1/2}$$

$$= \frac{16\sqrt{2}}{3\pi d^3}\left(4M^2 + 3T^2\right)^{1/2}$$

$$= \frac{16\sqrt{2}}{3\pi d^3}\left[4(9000)^2 + 3(6750)^2\right]^{1/2}$$

$$= \frac{\sqrt{2}}{3\pi d^3} \times 343418$$

Equating this to octahedral shear stress at yielding of a uniaxial tension bar, and using a factor of safety $N = 2$,

$$\frac{\sqrt{2}}{3\pi d^3} \times 2 \times 343418 = \frac{\sqrt{2}}{3}\sigma_y$$

or $\quad 2 \times 343418 = \pi d^3\ \sigma_y = \pi d^3 \times 207 \times 10^6$

$\therefore \quad d^3 = 1.056\ '\ 10^{-3}$

or $\quad d = 0.1018 \text{ m} = 10.18 \text{ cm}$

Example 3:

A cylindrical bar of 7 cm diameter is subjected to a torque equal to 3400 Nm, and a bending moment M. If the bar is at the point of failing in accordance with the maximum principal stress theory, determine the maximum bending moment it can support in addition to the torque. The tensile elastic limit for the material is 207 MPa, and the factor of safety to be used is 3.

Solution:

From Example 1(i)

$$16M + 16(M^2 + T^2)^{1/2} = \frac{\pi d^3}{N}\sigma_y$$

i.e. $$16M + 16(M^2 + 3400^2)^{1/2} = \frac{\pi \times 7^3 \times 10^{-6} \times 207 \times 10^6}{3}$$

or $\quad (M^2 + 3400^2)^{1/2} = 4647 - M$

or $\quad M^2 + 3400^2 = 4647^2 + M^2 - 9294\ M$

$\therefore \quad M = 1080$ Nm

Example 4:

In example 3, if failure is governed by the maximum strain theory, determine the diameter of the bar if it is subjected to a torque $T = 3040$ Nm and a bending moment $M = 1080$ Nm. The elastic modulus for the material is $E = 103 \times 10^6$ KPa, $V = 0.25$, factor of safety $N = 3$ and $\sigma_y = 207$ MPa.

Solution:

According to the maximum strain theory and Example 1 (iii)

$$16(1-v)M+16(1+v)(M^2+T^2)^{1/2}=\frac{\pi d^3}{N}\sigma_y$$

$$(16\times 0.75\times 1080)+(16\times 1.25)(1080^2+3400^2)^{1/2}=\frac{\pi d^3}{3}\times 207\times 10^6$$

i.e., $12960 + 71348 = 216.77 \times 10^6\ d^3$

or $d^3 = 389 \times 10^{-6}$

or $d = 7.3 \times 10^{-2}$ m $= 7.3$ cm.

Example 5:

An equipment used in deep sea investigation is immersed at a depth H. The weight of the equipment in water is W. The rope attached to the instrument has a specific weight γ_r and the water has a specific weight γ. Analyse the strength of the rope. The rope has a cross-sectional area A.

Solution:

The lower end of the rope is subjected to a triaxial state of stress. There is a tensile stress σ_1 due to the weight of the equipment and two hydrostatic compressions each equal to P, where

$$\sigma_1=\frac{W}{A},\quad \sigma_2=\sigma_3=-\gamma H \quad \text{(compression)}$$

At the upper section there is only a uniaxial tension σ_1' due to the weight of the equipment and rope immersed in water.

$$\sigma_1'=\frac{W}{A}+(\gamma_r-\gamma)\,H;\quad \sigma_2'=\sigma_3'=0$$

Therefore, according to the maximum shear stress theory, at lower section

$$\tau_{max}=\frac{\sigma_1-\sigma_3}{2}=\frac{1}{2}\left(\frac{W}{A}-\gamma H\right)$$

and at the upper section

$$\tau_{max}=\frac{\sigma_1'-\sigma_3'}{2}=\frac{1}{2}\left(\frac{W}{A}-\gamma H+\gamma_r H\right)$$

If the specific weight of the rope is more than twice that of water, then the upper section is the critical section. When the equipment is above the surface of the water, near the hoist, the stress is

$$\sigma_1=\frac{W'}{A}\qquad\text{and}\qquad \sigma_2=\sigma_3=0$$

$$\tau_{max} = \frac{1}{2}\frac{W'}{A}$$

W' is the weight of equipment in air and is more than W. It is also necessary to check the strength of the rope for this stress.

MOHR'S THEORY OF FAILURE

In the previous discussions failure, all the theories had one common feature. This was that the criterion of failure is unaltered by a reversal of sign of the stress. While the yield point stress σ_y for a ductile material is more or less the same in tension and compression, this is not true for a brittle material. In such a case, according to the maximum shear stress theory, we would get two different values for the critical shear stress. Mohr's theory is an attempt to extend the maximum shear stress theory (also known as the stress-difference theory) so as to avoid this objection.

To explain the basis of Mohr's theory, consider Mohr's circles, shown in Fig. 5, for a general state of stress.

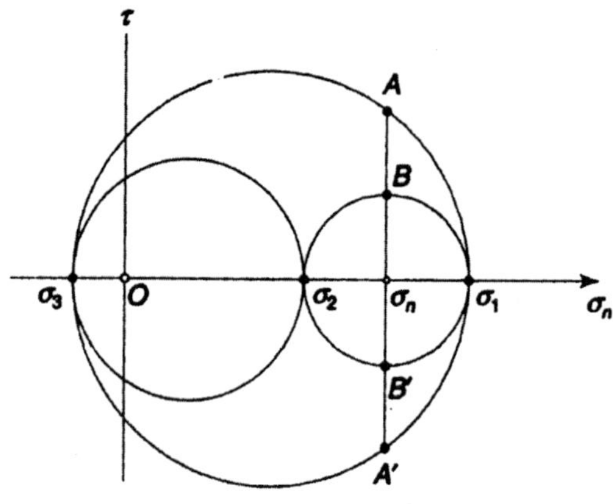

Fig. 5

σ_1, σ_2 and σ_3 are the principal stresses at the point. Consider the line $ABB'A'$. The points lying on BA and $B'A'$ represent a series of planes on which the normal stresses have the same magnitude σ_n but different shear stresses. The maximum shear stress associated with this normal stress value is τ, represented by point A or A'. The fundamental assumption is that if failure is associated with a given normal stress value, then the plane having this normal stress and a maximum shear stress accompanying it, will be the critical plane. Hence, the critical point for the normal stress σ_n will be the point A. From mohr's circle diagram, the planes having maximum shear stresses for given normal stresses, have their representative points on the outer circle. Consequently, as far as failure is concerned, the critical circle is the outermost circle in Mohr's circle diagram, with diameter $(\sigma_1 - \sigma_3)$.

Now, on a given material, we conduct three experiments in the laboratory, relating to simple tension, pure shear and simple compression. In each case, the test is conducted until failure occurs. In simple tension, $\sigma_1 = \sigma_{yt}$, $\sigma_2 = \sigma_3 = 0$. The outermost circle in the circle diagram (there is only one circle) corresponding to this state is shown as *T* in Fig. 6. The plane on which failure occurs will have its representative point on this outer circle. For pure shear, $\tau_{ys} = \sigma_1 = -\sigma_3$ and $\sigma_2 = 0$. The outermost circle for this state is indicated by *S*. In simple compression, $\sigma_1 = \sigma_2 = 0$ and $\sigma_3 = -\sigma_{yc}$. In general, for a brittle material, σ_{yc} will be greater than σ_{yt} numerically. The outermost circle in the circle diagram for this case is represented by *C*.

In addition to the three simple tests, we can perform many more tests (like combined tension and torsion) until failure occurs in each case, and

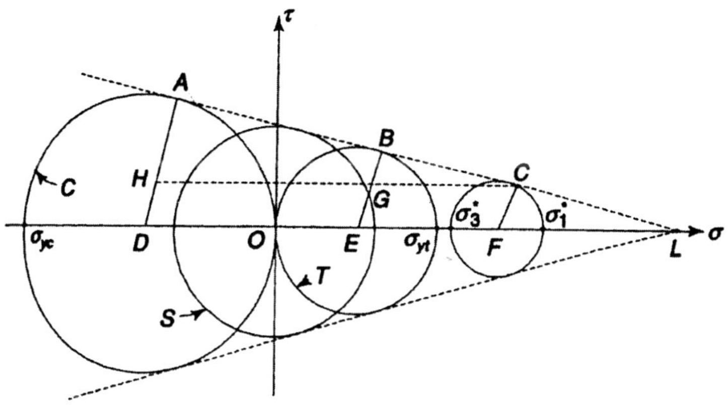

Fig. 6

correspondingly for each state of stress, we can construct the outermost circle. For all these circles, we can draw an envelope. The point of contact of the outermost circle for a given state with this envelope determines the combination of σ and τ, causing failure. Obviously, a large number of tests will have to performed on a single material to determine the envelope for it.

If the yield point stress in simple tension is small, compared to the yield point stress in simple compression, as shown in Fig. 8, then the envelope will cut the horizontal axis at point *L*, representing a finite limit for 'hydrostatic tension'. Similarly, on the left-hand side, the envelope rises indefinitely, indicating no elastic limit under hydrostatic compression.

For practical application of this theory, one assumes the envelopes to be straight lines, i.e. tangents to the circles as shown in Fig. 6. When a member is subjected to a general state of stress, for no failure to take place, the Mohr's circle with (σ_1 _ σ_3) as diameter should lie within the envelope. In the limit, the circle can touch the envelope. If one uses a factor of safety N, then the circle with N (σ_1 _ σ_3) as diameter can touch the envelopes. Figure 6 shows this limiting state of stress, where $\sigma_1^* = N\sigma_1$ and $\sigma_3^* = N\sigma_3$.

The envelopes being common tangents to the circles, triangles LCF, LBE and LAD are similar. Draw CH parallel to LO (the σ axis), making CBG and CAH similar. Then,

$$\frac{BG}{CG} = \frac{AH}{CH} \tag{a}$$

Now, $$BG = BE - GE = BE - CF = \frac{1}{2}\sigma_{yt} - \frac{1}{2}(\sigma_1^* - \sigma_3^*)$$

$$CG = FE = FO - EO = \frac{1}{2}(\sigma_1^* + \sigma_3^*) - \frac{1}{2}\sigma_{yt}$$

$$AH = AD - HD = AD - CF = \frac{1}{2}\sigma_{yc} - \frac{1}{2}(\sigma_1^* + \sigma_3^*)$$

$$CH = FD = FO - OD = \frac{1}{2}(\sigma_1^* + \sigma_3^*) + \frac{1}{2}\sigma_{yc}$$

Substituting these in Eq. (a), and after simplification,

$$\sigma_{yt} = \sigma_1^* - \frac{\sigma_{yt}}{\sigma_{yc}}\sigma_3^*$$

$$= N(\sigma_1 - K\sigma_3) \tag{18a}$$

where $$K - \frac{\sigma_{yt}}{\sigma_{yc}} \tag{18b}$$

Equation (18a) states that for a general state of stress where σ_1 and σ_3 are the maximum and minimum principal stresses, to avoid failure according to Mohr's theory, the condition is

$$\sigma_1 - K\sigma_3 \leq \frac{\sigma_{yt}}{N} = \sigma_{eq}$$

where N is the factor of safety used for design, and k is the ratio of σ_{yt} to σ_{yc} for the material. For a brittle material with no yield stress value, k is the ratio of σ ultimate in tension σ ultimate in compression, i.e.

$$K = \frac{\sigma_{ut}}{\sigma_{uc}} \tag{18c}$$

σ_{yt}/N is sometimes called the equivalent stress σ_{eq} in uniaxial tension corresponding to Mohr's theory of failure. When $\sigma_{yt} = \sigma_{yc}$, k will become equal to 1 and Eq. (18a) becomes identical to the maximum shear stress theory, Eq. (2).

Example 6:

Consider the problem discussed in Example 4.2. Let the crank-shaft material have σ_{yt} = 150 MPa and σ_{yc} = 330 MPa. If the diameter of the shaft is 10 cm, determine the allowable force F according to Mohr's theory of failure. Let the factor of safety be 2. Consider a point on the surface of the shaft where the stress due to bending is maximum.

Solution:

Bending moment at section A = $(20 \times 10^{-2} F)$ Nm

$$\text{Torque} = (15 \times 10^{-2} F) \text{ Nm}$$

$$\therefore \quad \sigma \text{ (bending)} = \frac{64Md}{2\pi d^4} = \frac{32M}{\pi d^3} \text{Nm}$$

$$\tau \text{ (torsion)} = \frac{32Td}{2\pi d^4} = \frac{16T}{\pi d^3} \text{Pa}$$

$$\sigma_{1,3} = \frac{1}{2}\sigma \pm \frac{1}{2}(\sigma^2 + 4\tau^2)^{1/2}, \qquad \sigma_2 = 0$$

$$\sigma_{1,3} = \frac{16M}{\pi d^3} \pm \frac{8}{\pi d^4}(4M^2 + T^2)^{1/2}$$

$$= \frac{8F}{\pi \times 10^{-3}}\left[2(20\times10^{-2}) \pm 10^{-2}(1600 + 22F)^{1/2}\right]$$

$$\frac{80F}{\pi}(40 \pm 42.7) = 2106F; \qquad -68.75F$$

$$k = \frac{\sigma_{yt}}{\sigma_{yc}} = \frac{150}{330} = 0.4545$$

$$\therefore \quad N(\sigma_1 - k\sigma_3) = 2F\,(2106 + 31.25) = 4274.5F$$

From Eq. (18a),

$$4274.5F = \sigma_{yt} = 150 \times 10^2 \text{ Pa}$$

$$\text{or} \quad F = 34092\text{N}$$

IDEALLY PLASTIC SOLID

If a rod of a ductile metal, such a mild steel, is tested under a simple uniaxial tension, the stress-strain diagram would be like the one shown in

Fig. 7(a). As can be observed, the curve has several distinct regions. Part *OA* is linear, signifying that in this region, the strain is proportional to the stress. If a specimen is loaded within this limit and gradually unloaded, it returns to its original length without any permanent deformation. This is the linear elastic region and point *A* denotes the limit of proportionality. Beyond

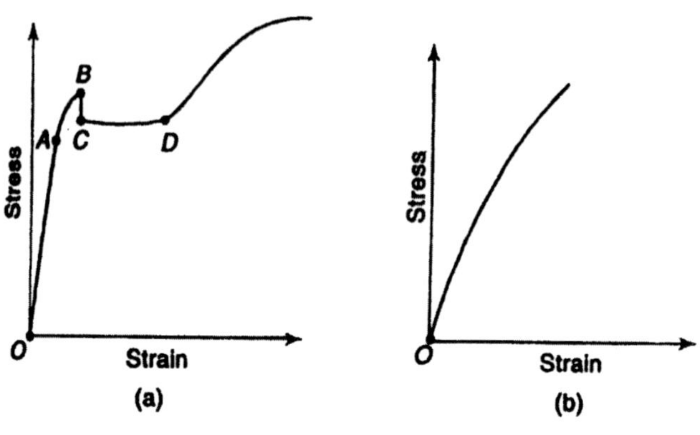

Fig. 7

A, the curve becomes slightly non-linear. However, the strain upto point *B* is still elastic. Point *B*, therefore, represents the elastic limit.

If the specimen is strained further, the stress drops suddenly (represented by point C) and thereafter the material yields at constant stress. After *D*, further straining is accompanied by increased stress, indicating work hardening. In the figure, the elastic region is shown exaggerated for clarity.

Most metals and alloys do not have a distinct yield point. The change from the purely elastic to the elastic-plastic state is gradual. Brittle materials, such as cast iron, titanium carbide or rock material, allow very little plastic deformation before reaching the breaking point. The stress-strain diagram for such a material would look like the one shown in Fig. 7(b).

In order to develop stress-strain relations during plastic deformation, the actual stress-strain diagrams are replaced by less complicated ones. These are shown in Fig. 8. In these, Fig. 8(a) represents a linearly elastic material, while Fig. 8(b) represents a material which is rigid (i.e. has no deformation) for stresses below σ_y and yields without limit when the stress level reaches the value σ_y. Such a material is called a rigid perfectly plastic material. Fig. 8(c) shows the behaviour of a material which is rigid for stresses below σ_y and for stress levels above σ_y a linear work hardening characteristics is exhibited.

A material exhibiting this characteristic behaviour is designated as rigid linear work hardening. Figure 8(d) and (e) represent respectively linearly elastic, perfectly plastic and linearly elastic-linear work hardening.

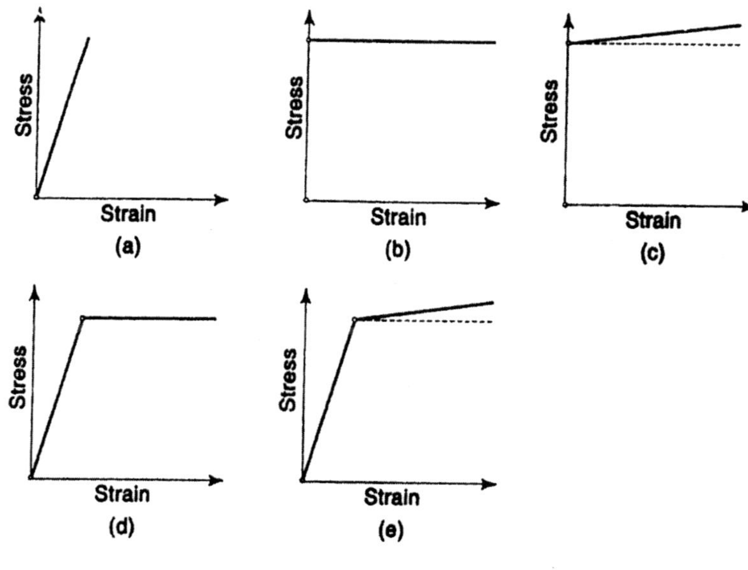

Fig. 8

In the following section, we shall very briefly discuss certain elementary aspects of the stress-strain relations for an ideally plastic solid. It is assumed that the material behaviour in tension or compression is identical.

STRESS SPACE AND STRAIN SPACE

The state of stress at a point can be represented by the six rectangular stress components τ_{ij} $(i, j = 1, 2, 3)$. One can imagine a six-dimensional space called the stress space, in which the state of stress can be represented by a point. Similarly, the state of strain at a point can be represented by a point in a six-dimensional strain space. In particular, a state of plastic strain $\varepsilon_{ij}^{(p)}$ can be so represented. A history of loading can be represented by a path in the stress space and the corresponding deformation or strain history as a path in the strain space.

A basic assumption that is now made is that there exists a scalar function called a stress function or loading function, represented by $f(\tau_{ij}, \varepsilon_{ij}, K)$, which depends on the states of stress and strain, and the history of loading. The function $f = 0$ represents a closed surface in the stress space. The function f characterises the yielding of the material as follows:

As long as $f < 0$ no plastic deformation or yielding takes place; $f > 0$ has no meaning. Yielding occurs when $f = 0$. For materials with no work hardening characteristics, the parameter $K = 0$.

In the previous sections of this chapter, several yield criteria have been considered. These criteria were expressed in terms of the principal stresses (σ_1, σ_2, σ_3) and the principal strains (ε_1, ε_2, ε_3). We have also observed that a material is said to be isotropic if the material properties do not depend on the particular coordinate axes chosen. Similarly, the plastic characteristics of the material are said to be isotropic if the yield function f depends only on the invariants of stress, strain and strain history. The isotropic stress theory of plasticity gives function f as an isotropic function of stresses alone. For such theories, the yield function can be expressed as $f(I_1, I_2, I_3)$ where I_1, I_2 and I_3 are the stress invariants. Equivalently, one may express the function as $f(\sigma_1, \sigma_2, \sigma_3)$. It is, therefore, possible to represent the yield surface in a three-dimensional space with coordinate axes σ_1, σ_2 and σ_3.

The Deviatoric Plane or the π Plane

In Section 2(a), it was stated that most metals can withstand considerable hydrostatic pressure without any permanent deformation. It has also been observed that a given state of stress can be uniquely resolved into a hydrostatic (or isotropic) state and a deviatoric (i.e. pure shear) state, i.e.

$$\begin{bmatrix} \sigma_1 & 0 & 0 \\ 0 & \sigma_2 & 0 \\ 0 & 0 & \sigma_3 \end{bmatrix} = \begin{bmatrix} p & 0 & 0 \\ 0 & p & 0 \\ 0 & 0 & P \end{bmatrix} = \begin{bmatrix} \sigma_1 - p & 0 & 0 \\ 0 & \sigma_2 - p & 0 \\ 0 & 0 & \sigma_3 - p \end{bmatrix}$$

or $$[\sigma_i] = [P] + [\sigma_i^{*}], \qquad (i = 1, 2, 3) \tag{19}$$

where $$P = \frac{1}{3}(\sigma_1 + \sigma_2 + \sigma_3)$$

is the mean normal stress, and

$$\sigma_i^{*} = \sigma_i - P, \qquad (i = 1, 2, 3)$$

Consequently, the yield function will be independent of the hydrostatic state. For the deviatoric state, $I_1^{*} = 0$. According to the isotropic stress theory, therefore, the yield function will be a function of the second and third invariants of the devatoric state, i.e. $f(I_2^{*}, I_2^{*})$. The equation

$$\sigma_1^{*} + \sigma_2^{*} + \sigma_3^{*} = 0 \tag{20}$$

represents a plane passing through the origin, whose normal OD is equally inclined (with direction cosines $1/\sqrt{3}$ to the axes σ_1, σ_2 and σ_3). This plane is called the deviatoric plane or the π plane. This is shown by point

P in Fig. 9. Since the addition or subtraction of an isotropic state does not affect the yielding process, point *P* can be moved parallel to *OD*. Hence, the yield function will represent a cylinder perpendicular to the π plane. The trace of this surface on the π plane is the yield locus.

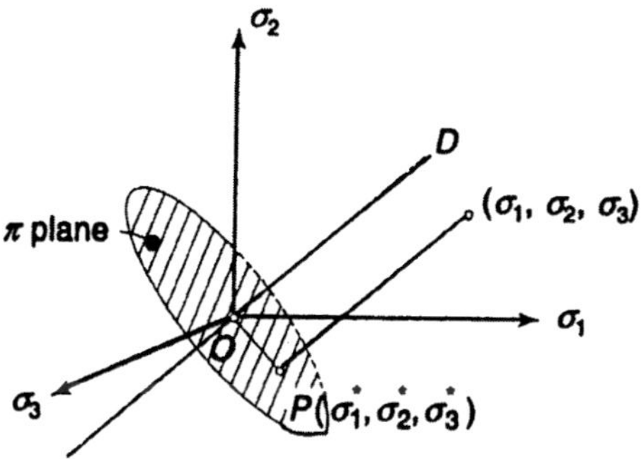

Fig. 9

GENERAL NATURE OF THE YIELD LOCUS

Since the yield surface is a cylinder perpendicular to the π plane, we can discuss its characteristics with reference to its trace on the π plane, i.e. with reference to the yield locus. Fig. 10 shows the π plane and the projections of the σ_1, σ_2 and σ_3 axes on this plane as σ'_1, σ'_2 and σ'_3. These projections make an angle of 120° with each other.

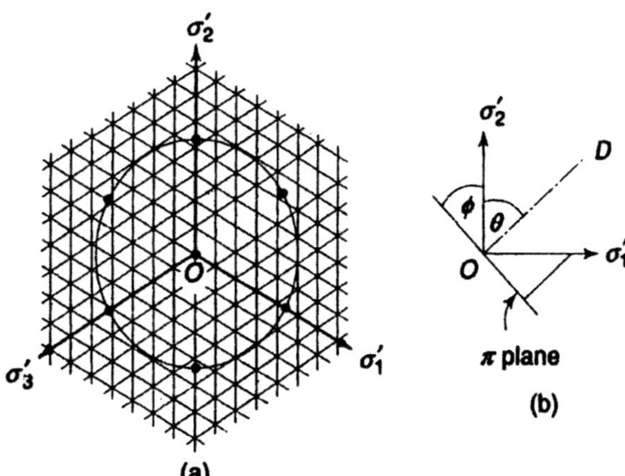

Fig. 10

Let us assume that the state (6, 0, 0) lies on the yield surface, i.e. the state $\sigma_1 = 6$, $\sigma_2 = 0$, $\sigma_3 = 0$, causes yielding. Since we have assumed isotropy, the states (0, 60, 0) and (0, 0, 6) also should cause yielding. Further, as we have assumed that the material behaviour in tension is identical to that in compression, the states (-6, 0, 0), (0, –6, 0) and (0, 0, –6) also cause yielding.

Thus, appealing to isotropy and the property of the material in tension and compression, one point on the yield surface locates five other points. If we choose a general point (*a, b, c*) on the yield surface, this will generate 11 other (or a total of 12) points on the surface. These are (*a, b, c*) (*c, a, b*) (*b, c, a*) (*a, c, b*) (*c, b, a*) (*b, a, c*) and the remaining six are obtained by multiplying these by –1.

Therefore, the yield locus is a symmetrical curve.

YIELD SURFACES OF TRESCA AND VON MISES

One of the yield conditions studied in Section 4.2 was stated by the maximum shear stress theory. According to this theory, if $\sigma_1 > \sigma_2 > \sigma_3$, the yielding starts when the maximum shear $\frac{1}{2}(\sigma_1 - \sigma_3)$ becomes equal to the maximum shear $\sigma_y/2$ in uniaxial tension yielding. In other words, yielding begins when $\sigma_1 - \sigma_3 = \sigma_y$. This condition is generally named after Tresca.

Let us assume that only σ_1 is acting. Then, yielding occurs when $\sigma_1 = \sigma_y$. The σ_1 axis is inclined at an angle of ϕ to its projection $\sigma_1^{'}$ axis on the π plane, and $\sin\ \phi = \cos\theta = 1/\sqrt{3}$, [Fig. 12 (b)]. Hence, the point $\sigma_1 = \sigma_y$ will have its projection on the π plane as $\sigma_y = \cos\phi = \sqrt{2/3}\,\sigma_y$ along the $\sigma_1^{'}$ axis. Similarly, other points on the π plane will be at distances of $\pm\sqrt{2/3}\,\sigma_y$ along the projections of σ_1, σ_2 and σ_3 axes on the π plane, i.e., along σ'_1, σ'_2 σ'_3 axes in Fig. 13. If σ_1, σ_2 and σ_3 are all acting (with $\sigma_1 > \sigma_2 > \sigma_3$), yielding occurs when $\sigma_1 - \sigma_3 = \sigma_y$. This defines a straight line joining points at a distance of σ_y along σ_1 and $-\sigma_3$ axes. The projection of this line on the π plane will be a straight line joining points at a distance of $\sqrt{2/3}\,\sigma_y$ along the σ'_1 and σ''_3 axes, as shown by *AB* in Fig. 13 Consequently, the yield locus is a hexagon.

Another yield criterion discussed in Section 2 was the octahedral shearing stress or the distortion energy theory. According to this criterion, Eq. (4b), yielding occurs when

$$f(l_1, l_2, l_3) = f(l_1^2 - 3l_2) = \sigma_y^2 \tag{21}$$

Since a hydrostatic state of stress does not have any effect on yielding, one can deal with the deviatoric state (for which $I_1^* = 0$) and write the above condition as

$$f(I_2^*, I_3^*) = f(I_2^*) - 3I_2^* = \sigma_y^2 \tag{22}$$

The yield function can, therefore, be written as

$$f = I_2^* + \frac{1}{3}\sigma_y^2 = I_2^* + s^2 \tag{23}$$

where s is a constant. This yield criterion is known as the von Mises condition for yielding. The yield surface is defined by

$$I_2^* + s^2 = 0$$

or $$\sigma_1\sigma_2 + \sigma_2\sigma_3 + \sigma_3\sigma_1 - 3p^2 = -s^2 \tag{24}$$

The other alternative forms of the above expression are

$$(\sigma_1 - p)^2 + (\sigma_2 - p)^2 + (\sigma_3 - p)^2 = 2s^2 \tag{25}$$

$$(\sigma_1 - \sigma_2)^2 + (\sigma_2 - \sigma_3)^2 + (\sigma_3 - \sigma_1)^2 = 6s^2 \tag{56}$$

Equation (25) can also be written as

$$\sigma_1^* + \sigma_2^* + \sigma_3^* = 2s^2 \tag{27}$$

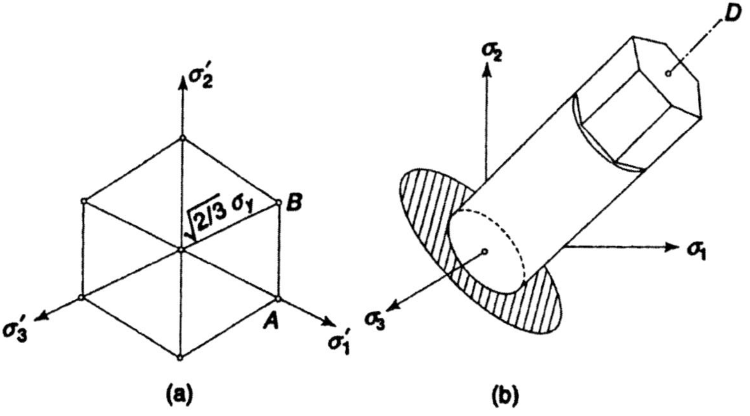

Fig. 11

This is the curve of intersection between the sphere $\sigma_1^2 + \sigma_2^2 + \sigma_3^2 = 2s^2$ and the π plane defined by $\sigma_1^* + \sigma_2^* + \sigma_3^* = 0$. This curve is, therefore, a circle with radius $\sqrt{2}s$ in the π plane. The yield surface according to the von Mises criterion is, therefore, a right circular cylinder. Form Eq. (23)

$$s^2 = \frac{1}{3}\sigma_y^2, \qquad \text{or} \qquad s = \frac{1}{\sqrt{3}}\sigma_y$$

Hence, the radius of the cylinder is $\sqrt{2/3}\sigma_y$ i.e. the cylinder of von Mises circumscribes Tresca's hexagonal cylinder. This shown in Fig. 11.

STRESS-STRAIN RELATIONS (PLASTIC FLOW)

The yield locus that has been discussed so far defines the boundary of the elastic zone in the stress space. When a stress point reaches this boundary, plastic deformation takes place. In this context, one can speak of only the change in the plastic strain rather than the total plastic strain because the latter is the sum total of all plastic strains that have taken place during the previous strain history of the specimen. Consequently, the stress-strain relations for plastic flow relate the strain increments. Another way of explaining this is to realise that the process of plastic flow is irreversible; that most of the deformation work is transformed into heat and that the stresses in the final state depend on the strain path. Consequently, the equations governing plastic deformation cannot, in principle, be finite relations concerning stress and strain components as in the case of Hooke's law, but must be differential relations.

The following assumptions are made:

(i) The body is isotropic

(ii) The volumetric strain is an elastic strain and is proportional to the mean pressure ($\sigma_m = p = \sigma$)

$$\varepsilon = 3k\sigma$$

or
$$d\varepsilon = 3k\sigma \tag{29}$$

(iii) The total strain increments de_{ij} are made up of the elastic strain increments $d\varepsilon_{ij}^e$ and plastic strain increments $d\varepsilon_{ij}^P$

$$d\varepsilon_{ij} = d\varepsilon_{ij}^e = d\varepsilon_{ij}^p \tag{30}$$

(iv) The elastic strain increments are related to stress components σ_{ij} through Hooke's law

$$d\varepsilon_{xx}^e = \frac{1}{E}[\sigma_x - v(\sigma_y + \sigma_z)]$$

$$d\varepsilon_{yy}^e = \frac{1}{E}[\sigma_y - v(\sigma_x + \sigma_z)]$$

$$d\varepsilon_{zz}^e = \frac{1}{E}[\sigma_z - v(\sigma_x + \sigma_y)] \tag{31}$$

$$d\varepsilon^e_{xy} = d\gamma^e_{xy} = \frac{1}{G}\tau_{xy}$$

$$d\varepsilon^e_{yz} = d\gamma^e_{yz} = \frac{1}{G}\tau_{yz}$$

$$d\varepsilon^e_{zx} = d\gamma^e_{zx} = \frac{1}{G}\tau_{zx}$$

(v) The deviatoric components of the plastic strain increments are proportional to the components of the deviatoric state of stress

$$d\left[\varepsilon^p_{xx} - \frac{1}{3}(\varepsilon^p_{xx} + \varepsilon^p_{yy} + \varepsilon^p_{zz})\right] = \left[\varepsilon_x - \frac{1}{3}(\sigma_x + \sigma_y + \sigma_z)\right] d\lambda \tag{32}$$

where $d\lambda$ is the instantaneous constant of proportionality.

Form (ii), the volumetric strain is purely elastic and hence

$$\varepsilon = \varepsilon^e_{xx} + \varepsilon^e_{yy} + \varepsilon^e_{zz}$$

But $$\varepsilon = \varepsilon^e_{xx} + \varepsilon^e_{yy} + \varepsilon^e_{zz} + (\varepsilon^p_{xx} + \varepsilon^p_{yy} + \varepsilon^p_{zz})$$

Hence,

$$\varepsilon^p_{xx} + \varepsilon^p_{yy} + \varepsilon^p_{zz} = 0 \tag{33}$$

Using this in Eq. (32)

$$d\varepsilon^p_{xx} = d\lambda\left[\sigma_x - \frac{1}{3}(\sigma_x + \sigma_y + \sigma_z)\right]$$

Denoting the components of stress deviator by σ_{ij}, the above equations and the remaining ones are

$$\begin{aligned} d\varepsilon^p_{xx} &= d\lambda\, s_{xx} \\ d\varepsilon^p_{yy} &= d\lambda\, s_{yy} \\ d\varepsilon^p_{zz} &= d\lambda\, s_{zz} \\ d\varepsilon^p_{xy} &= d\lambda\, s_{xy} \\ d\varepsilon^p_{yz} &= d\lambda\, s_{yz} \\ d\varepsilon^p_{zx} &= d\lambda\, s_{zx} \end{aligned} \tag{34}$$

Equivalently

$$d\varepsilon^p_{ij} = d\lambda\, s_{ij} \tag{35}$$

PRANDTL-REUSS EQUATIONS

Combining Eqs (30), (31) and (32)

$$d\varepsilon_{ij} = d\varepsilon_{ij}^{(e)} = d\lambda\, s_{ij} \tag{36}$$

where $d\varepsilon_{ij}^{(e)}$ is related to stress components through Hooke's law, as given in Eq, (31). Equations (30), (31) and (35) constitute the prandtl-Reuss equations. It is also observed that the principal axes of stress and plastic strain increments coincide. It is easy to show that $d\lambda$ is non-negative. For this, consider the work done during the plastic strain increment

$$dW_p = \sigma_x d\varepsilon_{xx}^p + \sigma_y d\varepsilon_{yy}^p + \sigma_z d\varepsilon_{zz}^p + \tau_{xy} d\gamma_{xy}^p + \tau_{yz} d\gamma_{yz}^p + \tau_{zx} d\gamma_{zx}^p$$

$$= d\lambda\left(\sigma_x s_{xx} + \sigma_y s_{yy} + \sigma_z s_{zz} + \tau_{xy} s_{xy} + \tau_{yz} s_{yz} + \tau_{zx} s_{zx}\right)$$

$$= d\lambda\left[\sigma_x(\sigma_x - P) + \sigma_y(\sigma_y - P) + \sigma_z(\sigma_z - P) + \tau_{xy}^2 + \tau_{yz}^2 + \tau_{zx}^2\right]$$

or $$dW_p = d\lambda\left[(\sigma_x - P)^2 + (\sigma_y - P)^2 + (\sigma_z - P)^2 + \tau_{xy}^2 + \tau_{yz}^2 + \tau_{zx}^2\right]$$

i.e. $$dW_p = d\lambda T^2 \tag{37}$$

Since $dW_p \geq 0$

we have $d\lambda \geq 0$

If the von Mises condition is applied, from Eqs (23) and (35)

$$dW_p = d\lambda 2s^2$$

or $$d\lambda \frac{dW_p}{2s^2} \tag{38}$$

i.e. $d\lambda$ is proportional to the increment of plastic work.

SAINT VENANT-VON MISES EQUATIONS

In a fully developed plastic deformation, the elastic components of strain are very small compared to plastic components. In such a case

$$d\varepsilon_{ij} \approx d\varepsilon_{ij}^P$$

and this gives the equations of the Saint Venant-von Mises theory of plasticity in the form

$$d\varepsilon_{ij} = d\lambda\, s_{ij} \tag{39}$$

Expanding this

$$d\varepsilon_{xx} = \frac{2}{3} d\lambda\left[\sigma_x - \frac{1}{2}(\sigma_y + \sigma_z)\right]$$

$$d\varepsilon_{yy} = \frac{2}{3} d\lambda\left[\sigma_y - \frac{1}{2}(\sigma_z + \sigma_x)\right]$$

$$d\varepsilon_{zz} = \frac{2}{3} d\lambda\left[\sigma_z - \frac{1}{2}(\sigma_x + \sigma_y)\right] \tag{40}$$

$$d\gamma_{xy} = d\lambda\tau_{xy}\,,\; d\gamma_{yz} = d\lambda\tau_{yz}$$
$$d\gamma_{zx} = d\lambda\tau_{zx}$$

The above equations are also called Levy-Mises equations. In this case, it should be observed that the principal axes of strain increments coincide with the axes of the principal stresses.

EXERCISES

1. Three elements *a, b, c* subjected to different states of stress. Which one of these three, do you think, will yield first according to
   (i) the maximum stress theory?
   (ii) the maximum strain theory?
   (iii) the maximum shear stress theory?

   Poisson's ratio $v = 0.25$ [*Ans.* (i) *b*, (ii) *a*, (iii) *c*]

2. Determine the diameter of a cold-rolled steel shaft, 0.6 m long, used to transmit 50 hp at 600 rpm. The shaft is simply supported at its ends in bearings. The shaft experiences bending owing to its own weight also. Use a factor of safety 2. The tensile yield limit is 280 $\times 10^3$ KPa (2.86 $\times 10^3$kgf/cm^2) and the shear yield limit is 140 $\times 10^3$ KPa (1.43 $\times$ 10^3 kgf/cm). Use the maximum shear stress theory.

   [*Ans.* d = 3.6 cm]

3. Determine the diameter of a ductile steel bar if the tensile load F is 35,000 N and the torsional moment T is 1800 Nm. Use a factor of safety N = 1.5.

   $E = 207 \times 10^6$ KPa (2.1 $\times$ 10^6 kgf/cm^2) and s_{yp} is 207,000 KPa (2100 kgf/cm^2).

   Use the maximum shear stress theory. [*Ans.* d = 4.1 cm]

4. For the problem discussed in problem 4.3, determine the diameter according to Mohr's theory if σ_{yt} = 207 MPa, σ_{yc} = 310 MPa. The factor of safety N = 1.5; F = 35,000 N and T = 1800 Nm.

   [*Ans.* d = 4.2 cm]

5. At a point in a steel member, the state of stress is as shown in Fig. 16. The tensile elastic limit is 413.7 kPa. If the shearing stress at the point is 206.85 kPa, when yielding starts, what is the tensile stress s at the point (a) according to the maximum shearing stress theory, and (b) according to the octahedral shearing stress theory?

   [*Ans.* (a) zero; (b) 206.85 kPa (2.1kgf/cm^2)]

6. A torque T is transmitted by means of a system of gears to the shaft. If $T = 2500$Nm (25,510 kgf cm), $R = 0.08$ m, $a = 0.8$ m and $b = 0.1$ m, determine the diameter of the shaft, using the maximum shear stress theory, $\sigma_y = 290000$ kPa. The factor of safety is 2. Note the when a torque is being transmitted, in addition to the tangential force, there occurs a radial force equal to $0.4F$, where F is the tangential force.

   Hint: The forces F and $0.4F$ acting on the gear A are shown in Fig. 17(b). The reactions at the bearings are also shown. There are two bending moments-one in the vertical plane and the other in the horizontal plane. In the vertical plane, the maximum moment is

   $\frac{(0.4Fab)}{(a+b)}$; in the horizontal plane the maximum moment is $\frac{(Fab)}{(a+b)}$;

   both these maximums occur at the gear section A. The resultant bending is

$$(M)_{\max} = \left[\left(\frac{0.4Fab}{a+b}\right)^2 + \left(\frac{Fab}{a+b}\right)^2\right]^{1/2}$$

$$= 1.08F\frac{ab}{a+b}$$

   The critical point to be considered is the circumferential point on the shaft subjected to this maximum moment.

   [*Ans.* $d = 65$ mm]

7. If the material of the bar has $\sigma_{yt} = 207 \times 10^6$ Pa and $\sigma_{yc} = 517 \times 10^6$ Pa determine the diameter of the bar according to Mohr's theory of failure. The other conditions are as given in Problem 4

8. Compute Lame's coefficients λ and μ for

   (a) steel having $E = 207 \times 10^6$ kPa (2.1×10^6 kgf/cm^2) and $v = 0.3$.

   (b) concrete having $E = 28 \times 10^6$ kPa(2.85×10^5 kgf/cm^2) and $v = 0.2$.

$$\left[\begin{array}{r} \textit{Ans.}\ \text{(a)}\ 120\times10^6\,\text{kPa}\ (1.22\times10^6\ \text{kgf/cm}^2),\ 80\times10^6\,\text{kPa} \\ (8.1680\times10^5\ \text{kgf/cm}^2) \\ \text{(b)}\ 7.8\times10^6\,\text{kPa}\ (7.96\times10^4\ \text{kgf/cm}^2),\ 11.7\times10^6\,\text{kPa} \\ (1.2\times10^5\ \text{kgf/cm}^2) \end{array}\right]$$

9. For steel, the following data is applicable:

$E = 207 \times 10^6$ kPa (2.1×10^6 kgf/cm²),

and $G = 80 \times 10^6$ kPa (0.82×10^6 kgf/cm²)

For the given strain matrix at a point, determine the stress matrix,

$$\left[\varepsilon_{ij}\right] = \begin{bmatrix} 0.001 & 0 & -0.002 \\ 0 & -0.003 & 0.0003 \\ -0.002 & 0.003 & 0 \end{bmatrix}$$

$$\left[Ans.\left[\tau_{ij}\right] = \begin{bmatrix} -68.4 & 0 & -160 \\ 0 & -708.4 & 24 \\ -160 & 24 & -228.4 \end{bmatrix} \times 10^3 \text{ kPa} \right]$$

10. A thin rubber sheet is enclosed between two fixed hard steel plates. Friction between the rubber and steel faces is negligible. If the rubber plate is subjected to stresses σ_x and σ_y as shown, determine the strains ε_{xx} and ε_{yy}, and also the stress ε_{zz}

$$\left[\begin{aligned} Ans.\ \sigma_z &= +v(\sigma_x + \sigma_y) \\ \varepsilon_{xx} &= +\frac{1+v}{E}\left[(1-v)\sigma_x - v\sigma_y\right] \\ \varepsilon_{yy} &= +\frac{1+v}{E}\left[(1-v)\sigma_y\ v\sigma_x\right] \end{aligned}\right]$$

3

Thermal Stresses and Energy Methods For Buckling Problems

INTRODUCTION

It is well known that changes in temperature cause bodies to expand or contract. The increase in the length of a uniform bar of length L, when its temperature is raised from T_0 to T, is

$$\Delta L = \alpha L(T - T_0)$$

where α is the coefficient of thermal expansion. If the bar is prevented from completely expanding in the axial direction, then average compressive stress induced is

$$\sigma = E\frac{\Delta L}{L}$$

where E is the modulus of elasticity. Thus, for complete restraint, the thermal stress is

$$\sigma = -\alpha E\,(T - T_0)$$

where the negative sign indicates the compressive nature of the stress. If the expansion is prevented only partially, then the stress induced is

$$\sigma = -k\alpha E\,(T - T_0)$$

where k represents a restraint coefficient. It is assumed in the above analysis that E and α are independent of temperature. In general, in an elastic continuum, the temperature change is not uniform throughout. It is a function of time and the space coordinates (x, y, x) i.e.

$$T = T\,(t, x, y, x)$$

The body under consideration may be restrained from expansion or movement in some regions and external tractions may be applied to other regions. The determination of stresses under such situations may be quite complex. In this chapter, we shall restrict ourselves to the analysis of the following problem;

(i) Thin circular disks with symmetrical temperature variation;

(ii) Long circular cylinders - hollow and solid;

(iii) Spheres with purely radial temperature variation - hollow and solid;

(iv) Straight beams of arbitrary cross-section;

(v) Curved beams.

Before these specific problems are analysed, we shall develop the general thermoelastic stress-strain relations and discuss two important general result.

THERMOELASTIC STRESS-STRAIN RELATIONS

Consider a body to be made up of a large number of small cubical elements. If the temperatures of all these elements are uniformly raised and if the boundary of the body is unconstrained, then all the cubical elements will expand uniformly and all will fit together to form a continuous body. If, however, the temperature rise is not uniform, each element will tend to expand by a different amount and if these elements have to fit together to form a continuous body, then distortions of the elements and consequently stresses should occur in the body.

The total strains at each point of a body are thus made up of two parts. The first part is a uniform expansion proportional to the temperature rise T. For any elementary cubical element of an isotropic body, this expansion is the same in all directions and in this manner only normal strains and no shearing strains occur. If the coefficient of linear thermal expansion is α, this normal strain in any direction is equal to αT. The second part of the strains at each point is due to the stress components. The total strains at each point can, therefore, be written as

$$\begin{aligned} \varepsilon_x &= \frac{1}{E}\left[\sigma_x - v\left(\sigma_y + \sigma_z\right)\right] + \alpha T \\ \varepsilon_y &= \frac{1}{E}\left[\sigma_y - v\left(\sigma_x + \sigma_z\right)\right] + \alpha T \\ \varepsilon_z &= \frac{1}{E}\left[\sigma_z - v\left(\sigma_x + \sigma_y\right)\right] + \alpha T \end{aligned} \tag{1a}$$

$$\gamma_{xy} = \frac{1}{G}\tau_{xy}, \quad \gamma_{yz} = \frac{1}{G}\tau_{yz}, \quad \gamma_{zx} = \frac{1}{G}\tau_{zx} \tag{1b}$$

The stresses can be expressed explicitly in terms of strains by solving Eq. (1a). These are

$$\sigma_x = \lambda e + 2\mu\varepsilon_x - (3\lambda + 2\mu)\,\alpha T$$
$$\sigma_y = \lambda e + 2\mu\varepsilon_y - (3\lambda + 2\mu)\,\alpha T \qquad (2a)$$
$$\sigma_z = \lambda e + 2\mu\varepsilon_z - (3\lambda + 2\mu)\,\alpha T$$
$$\tau_{xy} = \mu\gamma_{xy}, \qquad \tau_{yz} = \mu\gamma_{yz}, \qquad \tau_{zx} = \mu\gamma_{zx} \qquad (2b)$$

The Lame constants λ and μ (= G) are given by

$$\lambda = \frac{vE}{(1+v)(1-2v)}, \qquad \mu = G = \frac{E}{2(1+v)} \qquad (3)$$

EQUATIONS OF EQUILIBRIUM

The equations of equilibrium are the same as those of isothermal elasticity since they are based on purely mechanical considerations. In rectangular coordinates these are given by Eq. (65). These are repeated for convenience.

$$\frac{\partial\sigma_x}{\partial x} + \frac{\partial\tau_{xy}}{\partial y} + \frac{\partial\tau_{zx}}{\partial z} + \gamma_x = 0$$
$$\frac{\partial\tau_{xy}}{\partial x} + \frac{\partial\sigma_y}{\partial y} + \frac{\partial\tau_{yz}}{\partial z} + \gamma_y = 0 \qquad (9.4)$$
$$\frac{\partial\tau_{xz}}{\partial x} + \frac{\partial\tau_{yz}}{\partial y} + \frac{\partial\sigma_z}{\partial z} + \gamma_z = 0$$

where γ_x, γ_y and γ_z are body force components.

STRAIN-DISPLACEMENT RELATIONS

Only geometrical considerations are involved in deriving strain-displacement relations. Hence, the equations are the same as in isothermal elasticity. In rectangular coordinates, these are given by Eqs (18) and (19). To repeat, these are

$$\varepsilon_x = \frac{\partial u_x}{\partial x}, \quad \varepsilon_y = \frac{\partial u_y}{\partial y}, \quad \varepsilon_z = \frac{\partial u_z}{\partial z} \qquad (5a)$$

$$\gamma_{xy} = \frac{\partial u_x}{\partial y} + \frac{\partial u_y}{\partial x}, \quad \gamma_{yz} = \frac{\partial u_y}{\partial z} + \frac{\partial u_z}{\partial y}, \quad \gamma_{zx} = \frac{\partial u_z}{\partial x} + \frac{\partial u_x}{\partial z} \qquad (5b)$$

SOME GENERAL RESULTS

When the temperature distribution is known, the problem of thermoelasticity consists in determining the following 15 functions:

6 stress components σ_x, σ_y, σ_z, τ_{xy}, τ_{yz}, τ_{zx}

6 stress components ε_x, ε_y, ε_z, γ_{xy}, γ_{yz}, γ_{zx}

3 displacement components u_x, u_y, u_z

so as to satisfy the following 15 equations throughout the body

3 equilibrium equations, Eq. (4)

3 stress-strain relations, Eq. (1)

6 strain-displacement relations, Eq. (5)

and the prescribed boundary conditions. In most problems, the boundary conditions belong to one of the following two cases:

(i) **Traction Boundary Conditions:** In this case, the stress components determined must agree with the prescribed surface traction at the boundary.

(ii) **Displacement Boundary Conditions:** Here, the displacement components determined should agree with the prescribed displacements at the boundary.

In some cases, the prescribed boundary conditions may be a combination of the above two, i.e. on a part of the boundary, the surface tractions are prescribed and on the remaining part, displacements are prescribed.

(i) The method of arriving at a solution depends in general on the specific nature of the problem. It is shown in books on thermoelasticity that if the temperature distribution in a body is a linear function of the rectangular Cartesian space coordinates, i.e. if

$$T(x, y, z, t) = a(t) + b(t)x + c(t)y + d(t)z \tag{6}$$

where t represents time, then all the stress components are identically zero throughout the body, provided that all external restraints, body forces and displacement discontinuities are absent. Conversely, under those provisions, this is the only temperature distribution for which all stress components are identically zero. These results are obtained immediately by considering the stress compatibility relations.

It, therefore, follows from the above statement and from the linearity of the boundary-value problem as formulated through Eqs (1), (4) and (5) that a linear function may be added to or subtracted from a given temperature distribution without affecting the resulting stress distribution. However, the strains and displacements are elated, as is obvious.

(ii) We shall now show that if a body is subjected to a uniform temperature rise $T = T_0(t)$ and if the boundary of the body is prevented from having any displacements, then the solution of the corresponding thermoelastic problem is

$$u_x = 0, \quad u_y = 0, \quad u_z = 0$$

$$\varepsilon_x = \varepsilon_y = \varepsilon_z = 0, \quad \gamma_{xy} = \gamma_{yz} = \gamma_{zx} = 0$$

$$\tau_{xy} = \tau_{yz} = \tau_{zx} = 0, \quad \sigma_x = \sigma_y = \sigma_z = -\frac{E\alpha}{1-3v}T_0$$

To shown this, we shall apply the principle of superposition. We shall first allow from expansion of the body due to temperature rise T_0 with no restraint whatsoever. Since all cubical elements of the body expand freely, no stresses develop and all elements expand in an identical manner. Consequently,

$$\sigma_x = \sigma_y = \sigma_z = 0, \qquad \tau_{xy} = \tau_{yz} = \tau_{zx} = 0$$

$$\gamma_{xy} = \gamma_{yz} = \gamma_{zx} = 0 \qquad \varepsilon_x = \varepsilon_y = \varepsilon_z = aT_0$$

Therefore, $u_x = \alpha T_0 x, \quad u_y = \alpha T_0 y, \quad u_z = \alpha T_0 z$ (7a)

Now we apply boundary tractions to prevent this displacement. If the body is subject to a hydrostatic state of stress, then all elements of the body will experience the same state of stress $(-P)$.

With this state of stress, i.e. $\sigma_x = \sigma_y = \sigma_z = -P$ and $\tau_{xy} = \tau_{yz} = \tau_{zx} = 0$, the equations of equilibrium are identically satisfied.

Corresponding to this state of stress, the strain components are

$$\varepsilon_x = \varepsilon_y = \varepsilon_z = \frac{1}{E}[-p - v(-p-p)]$$

$$= \frac{1}{E}(1-2v)p$$

$$\gamma_{xy} = \gamma_{yz} = \gamma_{zx} = 0$$

Therefore, $u_x = \frac{1}{E}(1-2v)px : \quad u_y = -\frac{1}{E}(1-2v)py,$

$$u_z = -\frac{1}{E}(1-2v)pz$$

To get the original problem, the above values of u_x, u_y and u_z [Eq. (7a)] corresponding to free thermal expansion, should give zero displacements. Hence,

$$\alpha T_0 = x - \frac{1}{E}(1-2v)\ px = 0$$

or, $$p = \frac{E\alpha}{1-2v}T_0$$

Hence, $\sigma_x = \sigma_y = \sigma_z = -p = -\dfrac{E\alpha}{1-2v}T$ (8)

as stated earlier.

THIN CIRCULAR DISK: TEMPERATURE SYMMETRICAL ABOUT CENTRE

Consider a thin disk subjected to a temperature distribution which varies only with r and is independent of θ. It is assumed that it does not vary over the thickness and consequently, it is taken that the stresses and displacements also do not vary over the thickness. The stresses σ_r and σ_θ, therefore, satisfy the equilibrium equation

$$\frac{d\sigma_r}{dr} + \frac{\sigma_r - \sigma_\theta}{r} = 0 \tag{9}$$

Body forces are ignored. Also, because of symmetry, $\tau_{r\theta} = 0$. With $\sigma_z = 0$, the stress-strain relations given by Eq. (1a) take form, in polar coordinates,

$$\varepsilon_r = \frac{1}{E}(\sigma_r - v\sigma_\theta) + \alpha T \tag{10}$$

$$\varepsilon_\theta = \frac{1}{E}(\sigma_\theta - v\sigma_r) + \alpha T$$

Solving the above equation for σ_r and σ_θ, we find

$$\sigma_r = \frac{E}{1-v^2}[\varepsilon_r + v\varepsilon_\theta - (1+v)\alpha T]$$

$$\sigma_\theta = \frac{E}{1-v^2}[\varepsilon_\theta + v\varepsilon_r - (1+v)\alpha T]$$

Substituting these in the equation of equilibrium

$$r\frac{d}{dr}(\varepsilon_r + v\varepsilon_\theta) + (1+v)(\varepsilon_r - \varepsilon_\theta) = (1+v)\alpha r\frac{dT}{dr} \tag{12}$$

The strain-displacement relation for a symmetrically strained body, from Eqs (3) and (4), are

$$\varepsilon_r = \frac{du_r}{dr}, \quad \varepsilon_\theta = \frac{u_r}{r} \tag{13}$$

Substituting in Eq. (12)

$$\frac{d^2u_r}{dr^2} + \frac{1}{r}\frac{du_r}{dr} - \frac{u_r}{r^2} = (1+v)\alpha\frac{dT}{dr}$$

This may be written as

$$\frac{d}{dr}\left[\frac{1}{r}\frac{d}{dr}(ru_r)\right]=(1+v)\alpha\frac{dT}{dr}$$

Integration of the above equation yields,

$$u_r=(1+v)\alpha\frac{1}{r}\int_a^r Tr\,dr+C_1r+\frac{C_2}{r} \quad (14)$$

It can be observed that the above expression becomes identical to Eq. (8) if T is put equal to zero.

The lower limit a in the integral above depends on the disk. For a disk with a hole, a is the inner radius and for a solid disk, a is zero.

The stress components are determined by substituting the value of u_r in Eq. (13) and using the results in Eq. (11). The result are

$$\sigma_r=-\alpha E\frac{1}{r^2}\int_a^r Tr\,dr+\frac{E}{1-v^2}\left[C_1(1+v)-C_2(1-v)\frac{1}{r^2}\right] \quad (15)$$

$$\sigma_\theta=-\alpha E\frac{1}{r^2}\int_a^r Tr\,dr-\alpha ET+\frac{E}{1-v^2}\left[C_1(1+v)+C_2(1-v)\frac{1}{r^2}\right] \quad (16)$$

The constants C_1 and C_2 are determined by the boundary conditions. We shall now consider two specific cases. It should be observed that a linear variation of temperature with r will also induce stresses. This does not contradict the statement made in Sec. 5 that the stresses in a body are zero if the temperature distribution is linear with respect to a Cartesian frame of reference and if the body is free from external restraints and body forces. A linear radial variation will not give a linear variation with respect to the x, y and z-axes. In fact

$$T=kr=k\sqrt{x^2+y^2+z^2}$$

(i) Solid Disk of Radius *b* : In the case of a solid circular disk, a = 0 and in Eq. (14) it is observed from L' Hospital's rule that

$$\lim_{r\to 0}\frac{1}{r}\int_0^r Tr\,dr=0$$

Hence, the constant C_2 should be equal to zero, as otherwise u_r would become infinite at r = 0. The remaining constant C_1 is determined from the condition that s_r = 0 at r = b, the outer radius of the disk. From Eq. (15), therefore, we get

$$C_1(1+v)\frac{\alpha}{b^2}\int_0^b Tr\,dr$$

Substituting this, the stresses are

$$\sigma_r = \alpha E\left(\frac{1}{b^2}\int_0^b Tr\,dr - \frac{1}{r^2}\int_0^r Tr\,dr\right) \tag{17}$$

$$\sigma_\theta = \alpha E\left(-T + \frac{1}{b^2}\int_0^b Tr\,dr + \frac{1}{r^2}\int_0^r Tr\,dr\right) \tag{18}$$

From L' Hospital's rule

$$\lim_{r\to 0}\frac{1}{r^2}\int_0^r Tr\,dr = \frac{1}{2}T_0$$

where T_0 is the temperature at the centre of disk. Hence, at $r = 0$

$$\sigma_r(0) = \sigma_\theta(0) = \alpha E\left(\frac{1}{b^2}\int_0^r Tr\,dr - \frac{1}{2}T_0\right) \tag{19}$$

(ii) Disk with a Hole of Radius *a*: For a disk with a hole and traction free surfaces, $\sigma_r = 0$ at $r = a$ and $r = b$. Substituting these in Eq. (15)

$$C_1(1+\nu) + C_2(1-\nu)\frac{1}{a^2} = 0$$

$$-\alpha E\frac{1}{b^2}\int_a^b Tr\,dr + \frac{E}{1-\nu^2}\left[C_1(1+\nu) - C_2(1-\nu)\frac{1}{b^2}\right] = 0$$

Solving we get

$$C_1 = \alpha(1-\nu)\frac{1}{b^2-a^2}\int_a^b Tr\,dr;\quad C_2 = \alpha(1-\nu)\frac{a^2}{b^2-a^2}\int_a^b Tr\,dr$$

Substituting in Eqs (15) and (16)

$$\sigma_r = \frac{\alpha E}{r^2}\left[\frac{r^2-a^2}{b^2-a^2}\int_a^b Tr\,dr - \int_a^r Tr\,dr\right] \tag{20}$$

$$\sigma_\theta = \frac{\alpha E}{r^2}\left[\frac{r^2+a^2}{b^2-a^2}\int_a^b Tr\,dr + \int_a^r Tr\,dr - Tr^2\right] \tag{21}$$

and from Eq. (14)

$$u_2 = \frac{\alpha}{r}\left[(1+\nu)\int_a^r Tr\,dr + \frac{(1+\nu)r^2 + (1+\nu)a^2}{b^2-a^2}\int_a^b Tr\,dr\right] \tag{22}$$

If the temperature T is constant, then all the stress components are zero and the radial displacement $u_r = \alpha rT$.

LONG CIRCULAR CYLINDER

We shall now consider the nature of the thermal stresses induced in a long circular cylinder when the temperature is symmetrical about the axis and does not vary along the axis. If the z-axis is the axis of the cylinder and r the radius, then T is a function of r alone and is independent of z. Since the cylinder is long, sections far from the ends can be considered to be in a state of plane strain and we can analyse this problem with u_z, the axial displacement, assumed to be zero.

Once again, owing to symmetry, all the shear stress components are zero and there are now three normal stress components σ_r, σ_θ and σ_z. The stress-strain relations are

$$\varepsilon_r = \frac{1}{E}\left[\sigma_r - v(\sigma_\theta + \sigma_z)\right] + \alpha T$$

$$\varepsilon_\theta = \frac{1}{E}\left[\sigma_\theta - v(\sigma_r + \sigma_z)\right] + \alpha T$$

$$\varepsilon_z = \frac{1}{E}\left[\sigma_z - v(\sigma_r + \sigma_\theta)\right] + \alpha T$$

Since $u_z = 0$, we have $\varepsilon_z = 0$. Hence, from the last equation we get

$$\sigma_z = v\,(\sigma_r + \sigma_\theta) - ET\alpha \tag{23}$$

Substituting this in the expressions for ε_z and ε_θ

$$\varepsilon_r = \frac{1-v^2}{E}\left(\sigma_r - \frac{v}{1-v}\sigma_\theta\right) + (1+v)\alpha T$$

(24)

$$\varepsilon_\theta = \frac{1-v^2}{E}\left(\sigma_\theta - \frac{v}{1-v}\sigma_r\right) + (1+v)\alpha T$$

Let $$\frac{E}{1-v^2} = E_1, \quad \frac{V}{1-v} = V_1, \quad (1+v)\alpha = \alpha_1 \tag{25}$$

Then, Eq. (24) can be written as

$$\varepsilon_r = \frac{1}{E_1}(\sigma_r - v_1\sigma_\theta) + \alpha_1 T$$

$$\varepsilon_q = \frac{1}{E_1}(\sigma_\theta - v_1\sigma_r) + \alpha_1 T \tag{26}$$

Comparing the above expressions with Eq. (10), it is immediately observed that the expressions for ε_r and ε_θ in the plane strain case is similar to those in the plane stress case if we use E_1, V_1 and α_1, given by Eq. (25), in place of E, V and α respectively. Since the equation of equilibrium is the same as in the plane stress case, further analysis is identical to that in the plane stress case. The expressions for u_r, σ_r and σ_z can, therefore, be written from Eqs (14) and (16) as

$$u_r = \frac{1+v}{1-v}\alpha\frac{1}{r}\int_a^r Tr\,dr + C_1 r + \frac{C_2}{r} \tag{27}$$

$$\sigma_r = \frac{\alpha E}{1-v}\frac{1}{r^2}\int_a^r Tr\,dr + \frac{E}{1+v}\left(\frac{C_1}{1-2v}+\frac{C_2}{r^2}\right) \tag{28}$$

$$\sigma_\theta = \frac{E}{1-v}\frac{1}{r^2}\int_a^r Tr\,dr - \frac{\alpha ET}{1-v} + \frac{E}{1+v}\left(\frac{C_1}{1-2v}+\frac{C_2}{r^2}\right) \tag{29}$$

and from Eq. (22)

$$\sigma_z = \frac{\alpha ET}{1-v} + \frac{2vEC_1}{(1+v)(1-2v)} \tag{30}$$

When $T = 0$, the equations become identical to Eqs (8.22a-c). Normal force given by Eq. (30) is necessary to keep $u_z = 0$ throughout. The constants C_1 and C_2 are determined from the boundary conditions. We shall now consider two particular cases.

(i) Solid Cylinder of Radius *b*: As before, from L'Hospital's rule

$$\lim_{r\to 0}\frac{1}{r}\int_0^r Tr\,dr = 0$$

Hence, from Eq. (27) we observe that C_2 should be equal to zero, as otherwise u_r would be infinite at $r = 0$. Since $\sigma_r = 0$ at $r = b$, we get from Eq. (28) with $C_2 = 0$ and $a = 0$ in the lower limit of the integration

$$\frac{C_1}{(1+v)(1-2v)} = \frac{\alpha}{1-v}\frac{1}{b^2}\int_0^b Tr\,dr$$

or $$C_1 = \frac{(1+v)(1-2v)}{(1-v)}\frac{\alpha}{b^2}\int_0^b Tr\,dr$$

Comparing this with the value of C_1 obtained for the plane stress case, we observe that C_1 for the plane strain case can be obtained from C_1 for the plane stress case, merely by changing E, V and α to E_1, V_1 and α_1, and then converting these according to Eq. (25). The values of σ_r and σ_θ are, accordingly,

$$\sigma_r = \frac{\alpha E}{(1-v)}\left(\frac{1}{b^2}\int_0^b Tr\,dr - \frac{1}{r^2}\int_0^r Tr\,dr\right) \tag{31}$$

$$\sigma_\theta = \frac{\alpha E}{(1-v)}\left(-T + \frac{1}{b^2}\int_0^b Tr\,dr + \frac{1}{r^2}\int_0^r Tr\,dr\right) \tag{32}$$

and at $r = 0$

$$\sigma_r(0) = \sigma_\theta = \frac{\alpha E}{(1-v)}\left(\frac{1}{b^2}\int_0^b Tr\,dr - + \frac{1}{2}T_0\right) \tag{33}$$

where T_0 is the temperature at $r = 0$. Further, from Eq. (23)

$$\sigma_z = \frac{\alpha E}{(1-v)}\left(\frac{2v}{b^2}\int_0^b Tr\,dr - T\right) \tag{34}$$

The radial displacement is given by

$$u_1 = \frac{1+v}{1-v}\alpha\left[(1-2v)\frac{r}{b^2}\int_0^b Tr\,dr + \frac{1}{r}\int_0^b Tr\,dr\right] \tag{35}$$

Note: In obtaining Eq. (34), we have assumed a plane strain condition with $\varepsilon_z = 0$. Consequently, a stress distribution σ_z as given by Eq. (30) was necessary to maintain $u_z = 0$. If, however, the ends of the cylinder are free, then the resultant force in z direction should be equal to zero. This condition can be achieved by superposing a uniform stress distribution $\sigma_z' = C_3$ so that the resultant force is zero.

For the solid cylinder, the resultant of σ_z from Eq. (30) is

$$\int_0^b 2\pi r\sigma_z\,dr = -\frac{2\pi\alpha E}{(1-v)}\int_0^b Tr\,dr + \frac{2VEC_1}{(1+v)(1-2v)}\pi b^2$$

The resultant of the superimposed uniform stress $\sigma_z' = C_3$ is $\pi b^2 C_3$. The value of C_3 to make the total force zero in z direction is, therefore, given by

$$C_3\pi b^2 = \frac{2\pi\alpha E}{(1-v)}\int_0^b Tr\,dr - \frac{2vEC_1}{(1+v)(1-2v)}\pi b^2$$

The resultant σ_z distribution is given by

$$\sigma_z = \frac{\alpha E}{(1-v)}\left(\frac{2}{b^2}\int_0^b Tr\,dr - T\right)$$

The u_z displacement is then given by Eq. (35) plus $(-vC_3r/E)$.

(ii) Hollow Cylinder with Inner Radius *a*: We shall write the solutions for this from the plane stress case results, Eqs (20) – (22), by putting E_1, V_1 σ_1 in place of E, V σ and then converting these according to Eq. (25). Accordingly, we get

$$\sigma_r = \frac{\alpha E}{(1-v)}\frac{1}{r^2}\left[\frac{r^2-a^2}{b^2-a^2}\int_a^b Tr\,dr - \int_a^b Tr\,dr\right] \tag{36}$$

$$\sigma_\theta = \frac{\alpha E}{(1-v)}\frac{1}{r^2}\left[\frac{r^2+a^2}{b^2-a^2}\int_a^b Tr\,dr + \int_a^b Tr\,dr - Tr^2\right] \tag{37}$$

and

$$u_r = \frac{(1+v)\alpha}{(1-v)r}\left[\int_a^r Tr\,dr + \frac{(1-2v)r^2+a^2}{(b^2+a^2)}\int_a^b Tr\,dr\right] \tag{38}$$

Also $$\sigma_z = \frac{\alpha E}{(1-v)}\left(\frac{2}{(b^2+a^2)}\int_a^b Tr\,dr - T\right) \tag{39}$$

THE PROBLEM OF A SPHERE

We shall now consider the problem of a sphere subjected to purely radial temperature variation, i.e. T is a function of r alone. Because of symmetry, the shear stresses are all zero and the normal stresses are such that $\sigma_\theta = \sigma_\phi$. The equation of equilibrium in the radial direction is, from Eq. (37),

$$\frac{1}{r^2}\frac{d}{dr}(r^2\sigma_r) - \frac{2}{r}\sigma_\phi = 0$$

or $$\frac{d}{dr}(r^2\sigma_r) - 2r\sigma_\phi = 0$$

The stress-strain relations are

$$\varepsilon_r \frac{1}{E}(\sigma_r - 2v\sigma_\phi) + \alpha T$$

$$\varepsilon_\phi = \varepsilon_\theta = \frac{1}{E}\left[\sigma_\phi - v(\sigma_r + \sigma_\phi)\right] + \alpha T$$

Solving the above equations for σ_θ and σ_ϕ

$$\sigma_r = \frac{E}{(1+v)(1-2v)}\left[(1-v)\varepsilon_r + 2v\varepsilon_\phi - (1+v)\alpha T\right] \tag{40}$$

$$\sigma_\phi = \frac{E}{(1+v)(1-2v)}\left[\varepsilon_\phi + v\varepsilon_r - (1+v)\alpha T\right] \tag{41}$$

From strain-displacement relations, we have

$$\varepsilon_r = \frac{du_r}{dr} \quad \text{and} \quad \varepsilon_\phi = \varepsilon_\theta = \frac{u_r}{r} \tag{42}$$

Substituting these in the expressions for σ_r and σ_θ and then substituting these in the equilibrium equation, we get

$$\frac{d^2u_r}{dr^2} + \frac{2}{r}\frac{du_r}{dr} - \frac{2u_r}{r^2} = \frac{1+v}{1-v}\alpha\frac{dT}{dr}$$

This can also be written as

$$\frac{d}{dr}\left[\frac{1}{r^2}\frac{d}{dr}(r^2u_r)\right] = \frac{1+v}{1-v}\alpha\frac{dT}{dr}$$

The solution is

$$u_r = \frac{1+v}{1-v}\alpha\frac{1}{r^2}\int_a^r Tr^2\,dr + C_1 r + \frac{C_2}{r^2} \tag{43}$$

where C_1 and C_2 are constants to be determined from boundary conditions. The lower limit of the integral in the above equation is zero if the sphere is solid or is equal to a, the inner radius, if the sphere is hollow. From the expression for u_r, the strain components ε_r and ε_ϕ can be determined from Eq. (42) and substituted in Eq. (40). The results are

$$\sigma_r = -\frac{2\alpha E}{1-\nu}\frac{1}{r^3}\int_a^r Tr^2\,dr + \frac{EC_1}{1-2\nu} - \frac{2EC_2}{1+\nu}\frac{1}{r^3} \tag{44}$$

$$\sigma_\theta = \sigma_\phi = \frac{\alpha E}{1-\nu}\frac{1}{r^3}\int_a^r Tr^2\,dr + \frac{EC_1}{1-2\nu} - \frac{EC_2}{1+\nu}\frac{1}{r^3} - \frac{ET}{1-\nu} \tag{45}$$

we shall consider two specific cases.

(i) Solid Sphere : In this case, the lower limit a in the integrals may be taken as zero. In Eq. (43), the limit

$$\lim_{r\to 0}\left(-\frac{1}{r^2}\int_0^r Tr^2\,dr\right) = 0$$

according to L' Hospital's rule. Consequently, the constant C_2 should be equal to zero, as otherwise, the displacement u_r would become infinite at $r = 0$. The remaining constant C_1 is determined from the condition that $\sigma_r = 0$ at $r = a$ and $r = b$. Hence, from Eq. (44),

$$-\frac{2\alpha E}{(1-\nu)}\frac{1}{b^3}\int_a^r Tr^2\,dr + \frac{EC_1}{1-2\nu} = 0$$

or $$C_1 = \frac{2(1-2\nu)}{(1-\nu)}\int_0^b Tr^2\,dr$$

Substituting this in Eqs (44) and (45)

$$\sigma_r = -\frac{2\alpha E}{1-\nu}\left(\frac{1}{b^3}\int_0^b Tr^2\,dr - \frac{1}{r^3}\int_0^r Tr^2\,dr\right) \tag{46}$$

$$\sigma_\theta = \sigma_\phi = \frac{\alpha E}{1-\nu}\left(\frac{2}{b^3}\int_0^b Tr^2\,dr + \frac{1}{r^3}\int_0^r Tr^2\,dr - T\right) \tag{47}$$

(ii) Hollow Sphere : Let a be the radius of the inner cavity and b the outer radius of the sphere. The boundary conditions are $\sigma_r = 0$ at $r = a$ and $r = b$. Hence, form Eq. (44).

$$\frac{EC_1}{1-2\nu} - \frac{2EC_2}{(1+\nu)}\cdot\frac{1}{a^3} = 0$$

$$-\frac{2\alpha E}{(1-\nu)}\frac{1}{b^3}\int_a^b Tr^2\,dr + \frac{EC_1}{1-2\nu} - \frac{2EC_2}{1+\nu}\cdot\frac{1}{b^3} = 0$$

The above equations can be solved for C_1 and C_2 and substituted in Eqs (44) and (45). The result is

$$\sigma_r = -\frac{2\alpha E}{1-\nu}\left[\frac{r^3 - a^3}{(b^3 - a^3)r^3}\int_a^b Tr^2\,dr - \frac{1}{r^3}\int_a^r Tr^2\,dr\right] \tag{48}$$

$$\sigma_\phi = \frac{2\alpha E}{1-\nu}\left[\frac{2r^3 + a^3}{2(b^3 - a^3)r^3}\int_a^b Tr^2\,dr + \frac{1}{2r^2}\int_a^r Tr^2\,dr - \frac{1}{2}T\right] \tag{49}$$

Therefore, the stress components can be calculated if the distribution of temperature is known.

NORMAL STRESSES IN STRAIGHT BEAMS DUE TO THERMAL LOADING

In this section, we shall develop an elementary formula for normal stresses in free beams subjected to thermal loadings. We shall make use of the Bernoulli-Euler assumption mentioned. According to this assumption, sections which are plane and perpendicular to the axis before loading remain so after loading and the effect of lateral contraction (due to Poisson effect) may be neglected. The beam is assumed to be statically determinate and free of external loads. The temperature variation is arbitrary and the cross-section of the beam is also arbitrary.

If the beam is prevented from bending and if warping is not allowed, then the displacement of any section in the axial direction due to temperature rise will be a function of the axial coordinate x. Let this be $f_0(x)$. If now, the beam is allowed to undergo bending with the plane section remaining plane, then the displacement in s direction of any point (y, z) in a plane will be a linear function of the coordinates y and z. This is equivalent to saying that the cross-section rotates about an axis. The section that was plane before bending will, therefore, remain plane after bending and axial displacement. Hence, the total axial displacement, according to the Euler-Bernoulli hypothesis, will be

$$u_x = f_0(x) + yf_1(x) + zf_2(x)$$

where f_1 and f_2 are functions of x alone. The axial strain ε_x is, therefore,

$$\varepsilon_x = \frac{\partial u_x}{\partial x} = f_0'(x) + yf_1'(x) + zf_2'(x) \tag{50}$$

The strain represented by the last two terms on the right-hand side is similar to the one expressed in Chapter 6. We can also assume that the section rotates about an axis, such as BB, and write the strain as $\varepsilon_x = f_0'(x) + ky'$, where y' is the perpendicular distance of a point from BB, which is inclined at β to the y-axis. This is what was done in Chapter 6. The unknowns k and

β are now replaced by $f_1'(x)$ and $f_2'(x)$. From Hooke's law, since ε_y and ε_z are assumed to be zero.

$$\sigma_x = E(\varepsilon_x - \alpha T)$$

Substituting for ε_x

$$\sigma_x = E[f_0'(x) + yf_1'(x) + zf_2'(x)\square\,\alpha T]$$

Since a free beam without external loading is considered, the conditions to be satisfied at any section are

$$\iint \sigma_x \, dA = 0; \quad \iint \sigma_x \, y \, dA = M_z = 0, \quad \iint \sigma_x \, z \, dA = M_y = 0$$

i.e. the resultant force over the section is zero and the moments about the y and z axes should individually vanish. Substituting the expression for $\sigma_{z'}$ the above conditions become

$$\begin{aligned} f_0' \iint dA + f_1' \iint y \, dA + f_2' \iint z \, dA &= \iint \alpha T \, dA \\ f_0' \iint y \, dA + f_1' \iint y^2 \, dA + f_2' \iint yz \, dA &= \iint \alpha Ty \, dA \\ f_0' \iint z \, dA + f_1' \iint yz \, dA + f_2' \iint z^2 \, dA &= \iint \alpha Tz \, dA \end{aligned} \tag{52}$$

The integrations extend over the entire cross-section. The expression

$$\iint y \, dA = \iint z \, dA = 0$$

because of the selection of the centroidal axes. Further,

$$\iint dA = A, \quad \iint y^2 dA = I_z, \quad \iint z^2 dA = I_y, \quad \iint yz \, dA = I_{yz}$$

Substituting these, Eq. (52) can be written as

$$\begin{aligned} Af_0'(x) &= \iint \alpha T \, dA \\ f_1' I_z + f_2' I_{yz} &= \iint \alpha Ty \, dA \\ f_1' I_{yz} + f_2' I_y &= \iint \alpha Tz \, dA \end{aligned} \tag{53}$$

Let $\quad E \iint \alpha T \, dA = p_t, \quad E \iint \alpha Ty \, dA = -M_{zt}, \quad E \iint \alpha Tz \, dA = M_{yt}$

A minus sign is used in the second expression in order to make the final result similar to the result of Chapter 6. The solutions for f_0', f_1' and f_2' are then given by

$$f_0' = \frac{p_t}{EA}, \quad f_1' = \frac{-I_y M_{zt} + I_{yz} M_{yt}}{E(I_y I_z - I^2_{yz})}, \quad f_2' = \frac{I_z M_{yt} + I_{yz} M_{zt}}{E(I_y I_z - I^2_{yz})} \tag{54}$$

Substituting these, the axial stress σ_x is, from Eq. (51),

$$\sigma_x = -\alpha ET + \frac{p_t}{A} - \frac{(I_y M_{zt} + I_{yz} M_{yt})}{(I_y I_z - I^2_{yz})} y + \frac{(I_z M_{yt} + I_{yz} M_{zt})}{(I_y I_z - I^2_{yz})}$$

or $$\sigma_x = -\alpha ET + \frac{p_t}{A} + \frac{M_{zt}(yI_y - zI_{yz}) + M_{yt}(yI_{yz} - zI_z)}{I^2_{yz} - I_y I_z} \tag{55}$$

Equation (55) bears a very close resemblance to Eq. (14) since the analyses in both cases have proceeded on similar lines.

If the axes chosen happen to be the principal axes of the section, then $I_{yz} = 0$ and Eq. (55) reduces to

$$\sigma_x = -\alpha ET + \frac{p_t}{A} - \frac{M_{zt}}{I_z} y + \frac{M_{yt}}{I_y} z \tag{56}$$

STRESSES IN CURVED BEAMS DUE TO THERMAL LOADING

An elementary analysis of the stresses developed in curved beams may be developed on the same basic assumptions as in the case of straight beams. Consider a free curved beam of arbitrary constant cross-section, the centre line of which is an arc of a circle (Fig. 2). It is assumed that this arc lies in one of the principal planes of the beam. Let the temperature vary as a function of r and θ, i.e. $T\ (r,\ \theta)$. In the isothermal case, the radius of curvature of the neutral surface is given by r_0 [Eq. (33)] such that

$$\iint \frac{y\, dA}{r_0 - y} = 0 \tag{57}$$

As in Sec, 7, the origin 0 lies on the neutral axis and y is measured towards the centre of curvature. A view of the deformed element is given in Fig. 3 Let the elementary length of an undeformed element enclose an angle $\Delta\theta$.

Because of thermal loading, the element deforms and it is assumed that sections which were plane before, remain plane after deformation. A fibre at a distance y from the chosen origin has a length $(r_0 - y)\,\Delta\theta$ before deformation. After deformation, the length of the same fibre becomes

$$\left[r_0{'} - y - \int_0^y \alpha T\, dy\right](\Delta\theta + \delta\Delta\theta) \tag{58}$$

The third term in the first bracket above represents the thermal expansion in y direction. The change in the length of the fibre is therefore

$$\left[r_0{'} - y - \int_0^y \alpha T\, dy\right](\Delta\theta + \delta\Delta\theta) - (r_0 - y)\,\Delta\theta$$

$$= \left[r_0{'} - r_0 - \int_0^y \alpha T\, dy\right]\Delta\theta + \left[r_0{'} - y - \int_0^y \alpha T\, dy\right](\delta\,\Delta\theta)$$

Hence, the strain is

$$\varepsilon_\theta = \frac{1}{(r_0 - y)}\left[\left(r_0' - r_0 - \int_0^y \alpha T\, dy\right) + \frac{\delta\Delta\theta}{\Delta\theta}\left(r_0' - y - \int_0^y \alpha T\, dy\right)\right]$$

We observe that

$$\int_0^y \alpha T\, dy << y$$

Hence,

$$\varepsilon_\theta = \frac{1}{(r_0 - y)}\left[r_0' - r_0 - \int_0^y \alpha T\, dy\right] + \frac{\delta\Delta\theta}{\Delta\theta}(r_0' - y) \tag{59}$$

From Hooke's law, taking only σ_θ into account

$$\sigma_\theta = E(\varepsilon_\theta - \alpha T)$$

Therefore,

$$\sigma_\theta = E\left\{\frac{1}{r_0 - y}\left[\left(r_0' - r_0 - \int_0^y \alpha T\, dy\right) + \frac{\delta\,\Delta\theta}{\Delta\theta}(r_0' - y)\right] - \alpha T\right\}$$

The two unknowns r_0' and $\frac{\delta\,\Delta\theta}{\Delta\theta}$ are determined from the boundary conditions of the beam. Since the beam is free of external loadings, we should have

$$\iint \sigma_\theta\, dA = 0; \qquad \iint \sigma_\theta\, y\, dA = 0$$

Beam with Rectangular Section

A somewhat more accurate result can be obtained for a curved beam with a rectangular cross-section and temperature independent of θ.

This is obtained by superposing the result for a thin circular disk subjected to radial thermal loading with the result for the bending of a curved beam subjected to pure bending moment.

$$\int_A \sigma_\theta\, dA = F_\theta = 0$$

$$\int_A \sigma_\theta\, y\, dA = F_m$$

where F_m is the moment about the median line.

If, on this curved beam, we apply an equal and opposite moment F_m, as shown in Fig. 4(c), then we get a free curved beam subjected to thermal loading only.

STATIONARY POTENTIAL ENERGY THEOREM

The energy method of analysing the problems of elastic stability is based on an extremum principle of mechanics. Consider an elastic body subjected to external surface and body forces. Let the body be in equilibrium. During the application of these forces, the body deforms and consequently, these forces do a certain amount of work W.

The internal forces which are set up inside the elastic body also do work during the deformation process and this is stored as elastic strain energy. When external forces are applied gradually and no dissipation of energy takes place due to friction etc. the work done by the external forces should be equal to the internal elastic energy U, i.e.

$$W = U \tag{1}$$

Let portions of the body be given small virtual displacements. These are small displacements that are consistent with the constraints imposed on the body. For example, if a point of the body is fixed, then the virtual displacement there is zero. If a point of the body is constrained to lie on the surface of another body, then the virtual displacement there should be tangential to the surface of the contacting body. These virtual displacements being very small, the changes necessary in the external forces to bring about these virtual displacements will also be very small and will vanish in the limit. The work done by external surface and body forces P_i during these virtual displacements is

$$\delta W = \Sigma P_i \delta\, \Delta_i + \text{higher order terms} \tag{2}$$

where $\delta\Delta_i$ are the work absorbing components of the virtual displacements. It is convenient to define a potential V of the external forces in such a manner that the work done during virtual displacements is equal to $-\delta V$, i.e. a decrease in potential energy in the form of an equation

$$-\delta V = \Sigma P_i \delta\Delta_i = \delta W \tag{3}$$

In the above equation, we have neglected the higher order terms of Eq. (2). If a part of the body is subjected to distributed external forces, then over that part, the summation must be replaced by a surface integral.

From Eq. (3)

$$-\delta V - \delta W = 0$$

Using Eq. (1), the above equation can be written as

$$\delta (U + V) = 0 \tag{4}$$

The expression $U + V$ is known as the total potential of the system. Consequently, Eq. (4) can be stated as follows:

The first–order change in the total potential energy must vanish for every first–order terms are considered. Terms of higher order, as in Eq. (2), are ignored. Equation (4) is also stated as that in which the quantity $U + V$ assumes a stationary value, i.e.

$$U + V = \text{stationary} \tag{5}$$

It is shown in books on elasticity that for stable equilibrium, any virtual displacement will cause a positive change in the total potential energy of the system, which means that for a system in stable equilibrium, the total potential energy is minimum.

ENERGY AND STABILITY CONSIDERATIONS

We have demonstrated the use of the theorem of stationary potential energy in solving a statically indeterminate problem. Now we shall show, with reference to a specific problem, how energy considerations can be used to analyse stability problems. Consider a vertical bar hinged at one end and supported at the other end by a spring. It is assumed that the bar is infinitely rigid. It carries a centrally applied load P.

Let the bar be displaced through a small angle α. Because of this displacement, the load P is lowered by the amount

$$L - L\cos\alpha = L(1-\cos\alpha) \approx L\frac{\alpha^2}{2}$$

The decrease in potential energy is equal to the work done by P, i.e. $\frac{1}{2}PL\alpha^2$. At the same tine, the spring elongates by an amount αL and the energy stored due to this is $\frac{1}{2}S(\alpha L)^2$ where S is the spring constant. If the decrease in the potential energy is greater than the energy stored in the spring, i.e. if

$$\frac{1}{2}PL\alpha^2 > \frac{1}{2}S\alpha^2L^2$$

then the system is unstable. On the other hand, if

$$\frac{1}{2}PL\alpha^2 < \frac{1}{2}S\alpha^2L^2$$

then the system is stable. If

$$\frac{1}{2}PL\alpha^2 = \frac{1}{2}S\alpha^2L^2, \quad \text{i.e.} \quad P_{cr} = SL$$

then we get the value of the critical load which keeps the column in equilibrium in a slightly displaced configuration.

The same conclusion can be obtained by applying the principles of statics. For equilibrium in the displaced position, the moment about A should be zero. The end B of the column is subjected to a vertical load P and a horizontal force $S\alpha L$. For moment equilibrium about A,

$$P\alpha L = S\alpha L^2 \quad \text{or} \quad P_{cr} = SL$$

An analysis of stability problems in column buckling, using the above concept, will be taken up again.

TIMOSHENKO'S CONCEPT OF SOLVING BUCKLING PROBLEMS

Consider a straight column subjected to an axial load P. If P is less than the critical load, then the column is in stable equilibrium, which means that if the column is slightly displaced from its straight equilibrium position by any transverse disturbing force, it will return to its vertical position as soon as the disturbing force is removed. In terms of energy, this means that when P is less than P_{cr}, in the slightly bent configuration, the elastic strain energy stored in the bent column is greater than the work done by the axial load in moving through a distance ΔL, i.e.

$$U - W > 0 \tag{1}$$

where U is the strain energy and $W = P\,\Delta L$. On the other hand, when P exceeds P_{cr}, if the column is slightly displaced, the work done by the external load P will exceed the strain energy in bending and the equilibrium becomes unstable. Consequently, the condition

$$U - W = 0 \tag{2}$$

characterises the state when the equilibrium configuration changes from stable to unstable.

Following the same procedure as in the Rayleigh–Ritz method, we can assume that the buckled column curve can be expressed as

$$y = a_1\phi_1 + a_2\phi_2 + \dots + a_n\phi_n$$

The ϕ terms are functions of x so that each term satisfies the boundary conditions of the column. The constants a_1, a_2,..., define the amplitudes of the terms. The strain energy is given by

$$U = \frac{1}{2}EI\int_0^L \left(\frac{d^2y}{dx^2}\right)^2 dx = F_1(a_1, a_2, \dots, a_n)$$

The work done by the external force during deformation is

$$W = \frac{1}{2}P\int_0^L \left(\frac{dy}{dx}\right)^2 dx = PF_2(a_1, a_2, \dots, a_n)$$

Using Eq. (2)

$$P = EI \frac{\int_0^L (d^2y/dx^2)^2 \, dx}{\int_0^L (dy/dx)^2 \, dx} = \frac{F_1(a_1, a_2, ..., a_n)}{F_2(a_1, a_2, ..., a_n)} \tag{3}$$

Observing that for pin–ended column or a column with one end free

$$M = -EI \frac{d^2y}{dx^2}$$

and that $M = Py$

and $$EI \int \left(\frac{d^2y}{dx^2} \right)^2 dx = \frac{P^2}{EI} \int y^2 \, dx$$

Eq. (3) can also be written as

$$P = \frac{EI \int_0^L (dy/dx)^2 \, dx}{\int_0^L y^2 \, dx} \tag{4}$$

Since we need the minimum value for the load P, the critical load is obtained when the expression in Eq. (3) or Eq. (4) is made a minimum. This requires that the derivatives of Eq. (3) or Eq. (4) with respect to each coefficient a_i must vanish. This yields

$$\frac{\partial F_1}{\partial a_i} - P \frac{\partial F_2}{\partial a_i} = 0, \quad (i = 1, 2, \ldots) \tag{5}$$

Since there are n homogeneous equations, a non–trivial solution exists when the determinant of the coefficients is equal to zero.

There is a fundamental difference between Eq. (3) and (4) though they appear to be equivalent. The elastic strain energy is obtained from the expression

$$U = \frac{1}{2EI} \int_0^L M^2 \, dx$$

If we take the deflection curve as $y = y(x)$, then M could be expressed in two ways

$$M = -EI \frac{d^2y}{dx^2} \tag{6}$$

or $M = Py$

where P is the axial force acting on a pin–ended column or a column with one end free end the other end fixed. If we use the first expression in Eq. (6), we get the strain energy for any assumed form of the column.

If we use the second expression, we take the external force also into account and consequently, the final result obtained for P_{cr} using Eq. (3) gives a slightly more accurate result. Equation (3) is generally referred to as the Rayleigh–Ritz formula and Eq. (4) as the Timoshenko formula.

COLUMNS WITH VARIABLE CROSS–SECTIONS

So for, in the examples considered, we have treated the moment of inertia I as independent of x. We shall consider a few problems where I varies with x. The energy method will be found to be very suitable to obtain fairly good solutions.

USE OF TRIGONOMETRIC SERIES

In many instances, it will be useful to represent the deflection curve in the form of a trigonometric series. The functions satisfying orthogonality conditions. The trigonometric series which we shall consider now is made up of such functions. Let the deflection curve be represented by

$$y = a_1 \sin\frac{\pi x}{L} + a_2 \sin\frac{2\pi x}{L} + \ldots + a_n \sin\frac{n\pi x}{L} + \ldots \tag{1}$$

By properly determining the coefficients a_1, a_2,..., the above series can be made to represent any deflection curve. These coefficients may be calculated by a consideration of the strain energy of the beam or the column. The strain energy is given by

$$U = \frac{1}{2} EI \int_0^L \left(\frac{d^2y}{dx^2}\right)^2 dx$$

Now, $$\frac{d^2y}{dx^2} = -a_1 \frac{\pi^2}{L^2}\sin\frac{\pi x}{L} - a_2 \frac{2^2\pi^2}{L^2}\sin\frac{2\pi x}{L} - a_3 \frac{3^2\pi^2}{L^2}\sin\frac{3\pi x}{L} - \ldots$$

Hence, the square of the above expression will involve terms of two kinds

$$a_n^2 \frac{n^4\pi^4}{L^4}\sin^2\frac{n\pi x}{L}$$

and $$2a_m a_n \frac{n^4 m^2 \pi^4}{L^4}\sin\frac{n\pi x}{L}\sin\frac{m\pi x}{L}$$

By direct integration it can be seen that

$$\int_0^L \sin^2\frac{n\pi x}{L}dx = \frac{L}{2}, \text{ and } \int_0^L \sin\frac{n\pi x}{L}\sin\frac{m\pi x}{L}dx = 0 \quad \text{for } n \neq m$$

In the expression for strain energy, terms containing products like $a_m\, a_n$ vanish and only terms like a_n^2 remain. Then

$$U = \frac{1}{2} EI\left[a_1^2 \frac{\pi^4}{L^4}\frac{L}{2} + a_2^2 \frac{2^4\pi^4}{L^4}\frac{L}{2} + a_3^2 \frac{3^4\pi^4}{L^4}\frac{L}{2} + \ldots \right]$$

$$= \frac{EI\pi^4}{4L^3}\left(1a_1^2 + 2^4 a_2^2 + 3^4 a_3^2 + \ldots\right)$$

$$= \frac{EI\pi^4}{4L^3}\sum_{n=1}^{\infty} n^4 a_n^2 \tag{2}$$

Similarly, if we consider the expression

$$\Delta L = \frac{1}{2}\int_0^L \left(\frac{dy}{dx}\right)^2 dx$$

we find the integrand to consist of two kinds of terms

$$a_n^2 \frac{n^2\pi^2}{L^2}\cos^2\frac{n\pi x}{L}$$

and $$2a_m a_n \frac{mn\pi^2}{L^2}\cos\frac{m\pi x}{L}\cos\frac{n\pi x}{L}$$

By direct integration it can be shown that

$$\int_0^L \cos^2\frac{n\pi x}{L} = \frac{L}{2}$$

and $$\int_0^L \cos^2\frac{n\pi x}{L}\cos\frac{m\pi x}{L}dx = 0 \quad \text{for } n \neq m$$

Consequently,

$$\Delta L = \frac{\pi^2}{4L}\sum_{n=1}^{\infty} n^2 a_n^2 \tag{3}$$

APPLICATION TO BUCKLING PROBLEMS

We shall now discuss the application of the minimum total energy principle to column buckling problems. Consider the column shown in Fig. 1 carrying an axial load P. Let the moment of inertia I_x be variable. In calculating the strain energy, we shall consider only the bending energy. From the straight equilibrium configuration, let the column be moved to a neighbouring bent configuration.

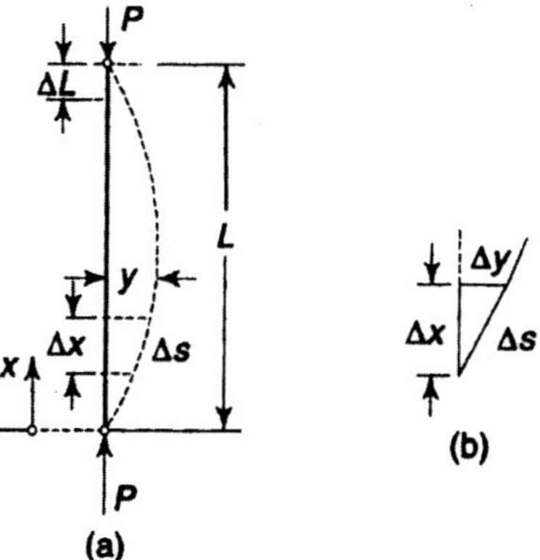

Fig. 1

Let the buckled form be expressed by $y = f(x)$. The elastic strain energy is

$$U = \frac{1}{2} E \int_0^L I_x \left(\frac{d^2 y}{dx^2} \right)^2 dx \tag{1}$$

Taking the undeflected position as datum, the potential energy in the buckled form is

$$V = -P\Delta L$$

To calculate ΔL, we observe,

$$\Delta L = \int_0^L (\Delta s - \Delta x)$$

But $\quad \Delta S = (\Delta x^2 + \Delta y^2)^{1/2} = \Delta x + \frac{1}{2}\left(\frac{\Delta y}{\Delta x}\right)^2 \Delta x$

Hence, $\quad P\Delta L = \frac{1}{2} P \int_0^L \left(\frac{dy}{dx}\right)^2 dx \qquad (2)$

The total potential energy is, therefore, given by

$$U + V = \frac{1}{2} E \int_0^L I_x \left(\frac{d^2 y}{dx^2} \right)^2 dx - \frac{1}{2} P \left(\frac{dy}{dx}\right)^2 dx \tag{3}$$

For equilibrium, the variation of the above quantity should vanish, i.e.

$$\delta(U + V) = \delta \left[\frac{1}{2} E \int_0^L I_x \left(\frac{d^2 y}{dx^2} \right)^2 dx - \frac{1}{2} P \left(\frac{dy}{dx}\right)^2 dx \right] = 0 \tag{4}$$

The above equation permits the determination of the function $y = f(x)$ by applying the technique of the calculus of variations. While the mathematical procedure will not be discussed here, it may be mentioned that the final result agrees. While this differential equation can be derived from static equilibrium considerations as was done in the derivation of the merit of the energy criterion for the solution of stability problems becomes evident in the Rayleigh–Ritz method discussed in the next section.

THE RAYLEIGH–RITZ METHOD

A direct solution to the extremum problem stated is obtained by the Rayleigh–Ritz method, revealing the importance of the energy criterion. We shall demonstrate the method with respect to the buckling problem discussed in the previous section. The deflection of the buckled column is expressed in the form of a finite series:

$$y = a_1\phi_1 + a_2\phi_2 + \dots + a_n\phi_n \tag{1}$$

The ϕ terms are a set of arbitrarily chosen functions of x, such that each term satisfies the prescribed boundary conditions of the column. These are called coordinate functions. The coefficients a correspond to a set of parameters, as yet undetermined. With the value of y as given by Eq. (1), the elastic strain energy and the potential energy. These lead to an expression involving the n parameters a, and having the form

$$U + V = F_1\,(a_1, a_2, \dots, a_n) - PF_2\,(a_1, a_2, \dots, a_n) \tag{2}$$

in which F_1 and F_2 are quadratic forms of the parameters a. If y, as given by Eq. (1), is to be a solution of the problem, then the parameters a must be chosen so as to make the total potential energy an extremum [Eq. (2)]. The problem has, therefore, been reduced to the familiar maximum–minimum problem involving the parameters $a_1, a_2, \dots, a_n$. Hence, the conditions become

$$\frac{\partial(U+V)}{\partial a_i} - 0, \quad (i = 1, 2, \dots, n) \tag{3}$$

i.e.

$$\frac{\partial F_1}{\partial a_1} - P\frac{\partial F_2}{\partial a_1} = 0$$

$$\frac{\partial F_1}{\partial a_2} - P\frac{\partial F_2}{\partial a_2} = 0 \tag{4}$$

$$\frac{\partial F_1}{\partial a_n} - P\frac{\partial F_2}{\partial a_n} = 0$$

The above set of equations involve only linear functions since these are derivatives of the quadratic expressions involved in $U + V$. Since Eq. (4) is a set of homogeneous equations, for the existence of a non–trivial solution, the determinant of the coefficients should be equal to zero, i.e,

$$\Delta = 0 \tag{5}$$

This is an equation of degree n in the unknown P and is the stability condition from which P can be determined. The smallest of the roots gives the critical load P_{cr}.

Introducing $P = P_{cr}$ in Eq. (4), a set of n linear homogeneous equations is obtained from which the ratios of the parameters a can be determined. Calling

$$\frac{a_2}{a_1} = \alpha_2, \quad \frac{a_3}{a_1} = \alpha_3, \quad \ldots, \quad \frac{a_n}{a_1} = \alpha_n$$

the buckling mode is obtained from Eq. (54) as

$$y = a_1(\phi_1\alpha_1 + \phi_2\alpha_2 + \ldots + \phi_n\alpha_n) \tag{6}$$

The importance of the Rayleigh–Ritz method lies in the fact that it offers a method of obtaining an approximate solution to the buckling problem. The method, in many cases, involves less labour than is involved in solving the differential equation and the associated eigenvalue. In the majority of cases, a few terms of the series in Eq. (4) give a sufficiently accurate result. Success or failure in applying the Rayleigh–Rize method to any problem depends largely on the proper choice of the coordinate functions. In the majority of cases, satisfactory results can be obtained only when the coordinate functions chosen form a system of orthogonal functions. This is the reason why Fourier Series play such an important role in the applications of the Rayleigh–Ritz method.

COMPARISON WITH THE PRINCIPLE OF CONSERVATION OF ENERGY

It is important to realise that the principle of the minimum total potential is different from the law of conservation of energy. The latter principle states that in an equilibrium condition, the work done by all external forces during the loading process is equal to the internal elastic strain energy stored, i.e.,

$$U - W = 0$$

If the loading is done gradually, the work done is equal to

$$W = \sum \frac{1}{2} P_i y_i$$

where y_i is the work absorbing component of the deflection at P_i. On the other hand, the virtual work done is

$$\delta W = \Sigma P_i \, \Delta y_i$$

There is no 1/2 factor in this case since the forces $p_i s$ are acting with full magnitude during the virtual displacements $\Delta y_i s$.

MISCELLANEOUS EXAMPLES

Example 1(a):

The inner surface of a hollow tube is at temperature T_i and the outer surface at zero temperature.

Assuming steady-state conditions, calculate the stresses. What are the values of σ_θ and σ_z near the inner and outer surfaces?

Solution:

Under steady heat flow conditions, the temperature at any distance r from the centre is given by the expression

$$T = \frac{T_i}{\log (b/a)} \log (b/r)$$

Substituting this in Eqs (36) - (38)

$$\sigma_r = \frac{\alpha E T_i}{2(1-\nu)\log (b/a)}\left[-\log\frac{b}{r} - \frac{a^2}{b^2-a^2}\left(1-\frac{b^2}{r^2}\right)\log\frac{b}{a}\right]$$

$$\sigma_\theta = \frac{\alpha E T_i}{2(1-\nu)\log (b/a)}\left[1-\log\frac{b}{r} - \frac{a^2}{b^2-a^2}\left(1+\frac{b^2}{r^2}\right)\log\frac{b}{a}\right]$$

$$\sigma_z = \frac{\alpha E T_i}{2(1-\nu)\log (b/a)}\left[1-2\log\frac{b}{r} - \frac{2a^2}{b^2-a^2}\log\frac{b}{a}\right]$$

$\sigma_r = 0$ at $r = 0$ and $r = b$. The stress components σ_θ and σ_z attain their maximum positive and negative values at $r = a$ and $r = b$. These values are

$$(\sigma_\theta)_{r=a} = (\sigma_z)_{r=a} = \frac{\alpha E T_i}{2(1-\nu)\log\frac{b}{a}}\left(1-\frac{2b^2}{b^2-a^2}\log\frac{b}{a}\right)$$

$$(\sigma_\theta)_{r=b} = (\sigma_z)_{r=b} = \frac{\alpha E T_i}{2(1-\nu)\log\frac{b}{a}}\left(1-\frac{2a^2}{b^2-a^2}\log\frac{b}{a}\right),$$

If T_i is positive, the radial stress is compressive at all points, whereas σ_θ and σ_z are compressive at the inner surface and tensile at the outer surface. These tensile stresses cause cracks in brittle materials such as stone, brick and concrete

Example 1(b):

Consider a column fixed at one end and free at the other end it is subjected to a compressive load P at the free end. Determine the approximate critical load assuming the deflection curve as

$$y = a_1\left(\frac{x}{L}\right)^2 + a_2\left(\frac{x}{L}\right)^3$$

The boundary conditions are

$$y = 0 \quad at\ x = 0, \qquad \frac{dy}{dx} = 0 \qquad at\ x = 0$$

and $$\frac{d^2y}{dx^2} = 0 \qquad at\ x = L$$

Solution:

(i) Let us ignore the last condition for the time being. The first two conditions are satisfied by the coordinate functions. The strain energy is equal to

$$U = \frac{1}{2}EI\int_0^L\left(\frac{d^2y}{dx^2}\right)^2 dx$$

$$= \frac{1}{2}EI\int_0^L\left(\frac{2a_1}{L^2} + \frac{6a_2x}{L^3}\right)^2 dx$$

$$= \frac{2EI}{L^3}\left(a_1^2 + 3a_1a_2 + 3a_2^2\right)$$

The potential energy is to equal to

$$V = -\frac{1}{2}P\int_0^L\left(\frac{dy}{dx}\right)^2 dx$$

$$= -\frac{1}{2}P\int_0^L\left(\frac{2a_1x}{L^2} + \frac{3a_2x^2}{L^3}\right)^2 dx$$

$$= \frac{P}{30L}\left(20a_1^2 + 45a_1a_2 + 27a_2^2\right)$$

Hence, the total potential energy is

$$U + V = a_1^2\left(\frac{2EI}{L^3} - \frac{2P}{3L}\right) + a_1a_2\left(\frac{6EI}{L^3} - \frac{3P}{2L}\right) + a_2^2\left(\frac{6EI}{L^3} - \frac{9P}{10L}\right)$$

For an extremum we should have

$$\frac{\partial(U+V)}{\partial a_1} = \left(\frac{4EI}{L^3} - \frac{4P}{3L}\right)a_1 + \left(\frac{6EI}{L^3} - \frac{3P}{2L}\right)a_2 = 0$$

For the existence of a non–trivial solution, the determinant D of the coefficients should be equal to zero. Hence,

$$\Delta = \left(\frac{4EI}{L^3} - \frac{4P}{3L}\right)\left(\frac{12EI}{L^3} - \frac{9P}{5L}\right) - \left(\frac{6EI}{L^3} - \frac{3P}{2L}\right) = 0$$

or $\quad 3P^2L^4 - 104PL^2EI + 240E^2I^2 = 0$

Therefore,

$$P = 2.49\frac{EI}{L^3} \quad \text{or} \quad 32.18\frac{EI}{L^2}$$

The smaller value is

$$P_{cr} = 2.49\frac{EI}{L^2}$$

Compared to the exact value $\dfrac{\pi^2 EI}{4L^2}$,

The error is only +0.92%.

(ii) In the above analysis, we have ignored the third boundary condition, i.e. at $x = L$, $\dfrac{d^2y}{dx^2} = 0$. If we use this condition

$$\frac{d^2y}{dx^2} \text{ at } x = L = \frac{2a_1}{L^2} + \frac{6a_2L}{L_3} = 0$$

or $\quad a_1 = -3a_2$

Using this

$$y = a_1\left(\frac{x}{L}\right)^2 - \frac{1}{3}a_1\left(\frac{x}{L}\right)^3$$

Substituting $a_1 = -3a_2$ in the expressions for U and V

$$U = \frac{2EI}{L^2}\left(a_1^2 - a_1^2 + \frac{a_1^2}{3}\right) = \frac{2}{3}\frac{EI}{L^3}a_1^2$$

and $\quad V = \dfrac{P}{L}\left(\dfrac{2}{3}a_1^2 - \dfrac{2}{3}a_1^2 + \dfrac{1}{10}a_1^2\right) = \dfrac{4}{15}\dfrac{P}{L}a_1^2$

Therefore,

$$U + V = \left(\frac{2}{3}\frac{EI}{L^3} - \frac{4}{15}\frac{P}{L}\right)a_1^2$$

For an extremum, the quantity inside the parentheses should be equal to zero, i.e.

$$\frac{4}{15}\frac{P}{L} = \frac{2}{3}\frac{EI}{L^3}$$

or $$P_{cr} = 2.5\frac{EI}{L^2}$$

which is almost identical with the previous solution but the solution has been obtained with comparative ease.

Example 2:

A beam column is subjected to an axial force P and a lateral force Q at x = c (Fig. 2). Determine the deflection curve using the energy method.

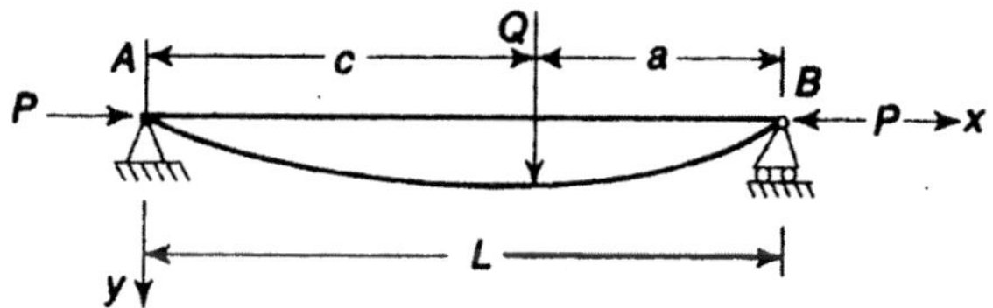

Fig. 2

Solution:

Let the deflection curve be

$$y = a_1 \sin\frac{\pi x}{L} + a_2 \sin\frac{2\pi x}{L} + \ldots$$

Let a virtual displacement δy_n be given. This virtual displacement is obtained by changing one of the terms a_n sin $(n\pi x/L)$ to $(a_n + \delta a_n)$ sin $(n\pi x/L)$.

In other words, the deflection curve δa_n sin $(n\pi x/L)$ is wuperimposed on the original deflection curve. The work done by the external forces Q and P is

$$\delta W = Q\,\delta a_n \sin\frac{n\pi c}{L} + P\delta\,(\Delta L)$$

$$\delta W = Q\,\delta a_n \sin\frac{n\pi c}{L} + P\frac{\pi^2}{4L}2n^2 a_n \delta_n$$

The increase in strain energy is

$$\delta U = \frac{\partial U}{\partial a_n}\delta a_n$$

From Eq. (67)

$$\delta U = \frac{EI\pi^4}{L^3}n^4\, a_n\, \delta a_n$$

Since the increase in strain energy should be equal to the work done, we have

$$Q\sin\frac{n\pi c}{L}\delta a_n + P\frac{\pi^2}{2L}n^2 a_n \delta a_n = \frac{EI\pi^4}{2L^3}n^4 a_n\ \delta a_n$$

from which,

$$a_n = \frac{2QL^3}{EI\pi^4}\frac{1}{\left(n^2 - \frac{PL^2}{EI\pi^2}\right)}\sin\frac{n\pi c}{L}$$

If we use the notation

$$\beta = \frac{PL^2}{EI\pi^2}$$

then, $$a_n = \frac{2QL^3}{EI\pi^4}\frac{1}{n^2(n^2-\beta)}\sin\frac{n\pi c}{L}$$

The deflection curve is, therefore, given by

$$y = \frac{2QL^3}{EI\pi^4}\sum_{n=1}^{\infty}\frac{1}{n^2(n^2-\beta)}\sin\frac{n\pi c}{L}\sin\frac{n\pi x}{L}$$

Example 3(a)

Let the inner surface of a hollow sphere be at temperature T_i and the outer surface at temperature zero. Let the system be in a steady heat flow condition. The temperature distribution is then given by

$$T = \frac{T_i a}{b-a}\left(\frac{b}{r} - 1\right)$$

Solution:

Determine the stress distribution.

Substituting the above expression for T in Eqs (48) and (49), we get

$$\sigma_r = -\frac{\alpha ET_i}{1-\nu}\frac{ab}{b^3-a^3}\left[a+b-\frac{1}{r}(b^2+ab+a^2)+\frac{a^2b^2}{r^3}\right]$$

$$\sigma_\theta = \sigma_\phi = \frac{\alpha ET_i}{1-\nu}\frac{ab}{b^3-a^3}\left[a+b-\frac{1}{2r}(b^2+ab+a^2)-\frac{a^2b^2}{2r^3}\right]$$

As can be seen $\sigma_r = 0$ at $r = a$ and $r = b$, according to the boundary conditions. Differentiating the expression for σ_r with respect to r and equating the resulting expression to zero, it is observed that σ_r is a maximum or a minimum when

$$r^2 = \frac{3a^2b^2}{b^2 + ab + a^2}$$

The expression for σ_ϕ shows that its value increases with r for T_i positive, and

$$(\sigma_r)_{r=a} = -\frac{\alpha ET_i}{2(1-v)} \frac{b(b-a)(a+2b)}{b^3 - a^3}$$

$$(\sigma_r)_{r=b} = -\frac{\alpha ET_i}{2(1-v)} \frac{a(b-a)(2a+b)}{b^3 - a^3}$$

Example 3(b):

Fig. 3 shows a three–bar truss, the point D of which is subjected to P units of force. Applying the principal of minimum potential energy, determine the vertical and horizontal displacements of D due to the load. The members have equal cross–sectional areas.

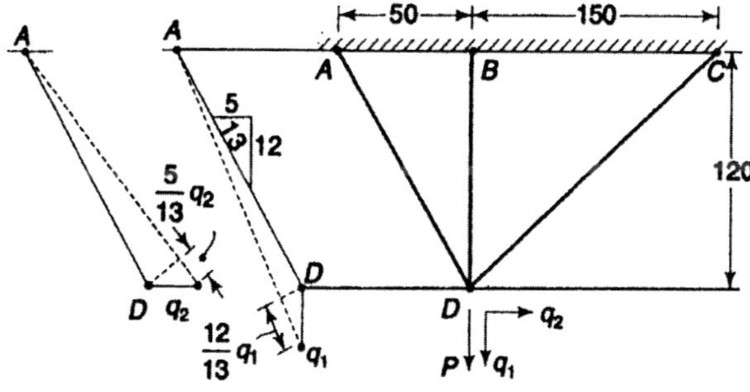

Fig. 3

Solution:

Let the point D have a vertical displacement q_1 and a horizontal displacement q_2. The elongation caused in each member due to these displacements can be calculated geometrically. This is shown in the figure for member AD. The total extension of each member is

member $AD \frac{12}{13} q1 + \frac{5}{13} q_2$

member $BD \quad q_1$

member $CD \quad \frac{4}{5} q_1 - \frac{3}{5} q_2$

If A is the cross–sectional area and E is the Young's modulus, then the elastic strain energy stored in a member of length L is

$$U = \frac{EA}{2L}\delta^2$$

where δ is the elongation. Hence, for the three members, the strain energies are

for AD $$U_1 = \frac{EA}{2(130)}\left(\frac{12}{13}q_1 + \frac{5}{13}q_2\right)^2$$

for BD $$U_2 = \frac{EA}{2(120)}q_1^2$$

for CD $$U_3 = \frac{EA}{2(150)}\left(\frac{4}{5}q_1 - \frac{3}{5}q_2\right)^2$$

The total elastic strain energy is the sum of the above three quantities.

Hence,

$$U = \left(958q_1^2 - 47q_1q_2 + 177q_2^2\right)\times 10^{-5}$$

Taking the undeformed position as the datum, the potential energy in the deformed configuration is

$$V = -Pq_1$$

Hence, the total potential energy is

$$U + V = EA\left(958q_1^2 - 47q_1q_2 + 177q_2^2\right)\times 10^{-5} - pq_1$$

For equilibrium position, the first–order variation of the above quantity should be equal to zero, i.e.

$$\delta(U + V) = EA\ [958(2q_1\ \delta q_1) - 47(q_1\delta q_2 + q_2\delta q_1) + 177(2q_2\ \delta q_2)\ ] \times 10^{-5} - P\delta q_1 = 0$$

or $$\left(1916q_1 - 47q_2 - \frac{P}{EA}\times 10^5\right)\delta q_1 + \left(-47q_1 + 354q_2\right)\delta q_2 = 0$$

Since δq_1 and δq_2 are arbitrary virtual displacements, the quantities inside the parentheses should vanish individually. Thus,

$$1916q_1 - 47q_2 - \frac{P}{EA}\times 10^5$$

$$-47q_1 + 354q_2 = 0$$

Solving these two equations, we obtain

$$q_1 = 0.0523 \quad \text{and} \quad q_2 = 0.00695$$

It should be observed that we have not made use of any equation of statics in solving the problem.

Example 4:

Using an infinite series, determine the deflection curve for the beam column shown in Fig. 4.

Solution:

In the solution of the previous example consider c to be very small. Then

$$\sin\frac{n\pi c}{L} \approx \frac{n\pi c}{L}$$

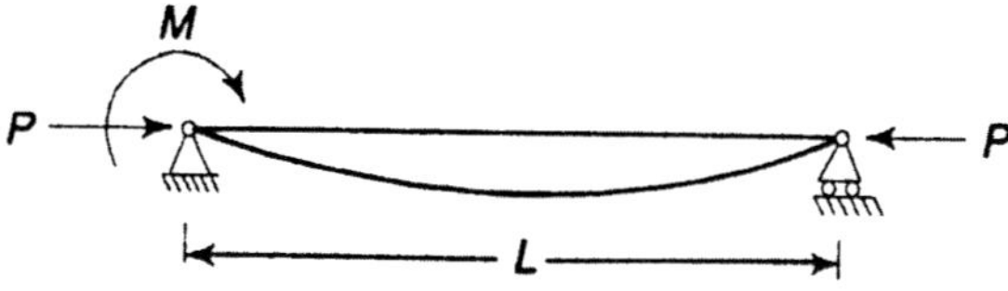

Fig. 4

and $$y = \frac{2L^3}{EI\pi^4}\frac{\pi}{L}Qc\sum_{n=1}^{\infty}\frac{n}{n^2(n^2-\beta)}\sin\frac{n\pi x}{L}$$

Let $c \to 0$ and $Q \to \infty$,

such that $Qc = M =$ constant.

Then, $$y = \frac{2ML^2}{EI\pi^3}\sum_{n=1}^{\infty}\frac{1}{n^2(n^2-\beta)}\sin\frac{n\pi x}{L}$$

Example 5:

Consider a column with the moment of inertia of the cross–sectional area varying according to the equation

$$I = I_0\left(1+\sin\frac{\pi x}{L}\right)$$

Solution:

The column is hinged at both ends. Assume that the deflection curve can be represented by the series

$$y = \sum a_n \sin\frac{n\pi x}{L}$$

Since the deflection curve must be symmetrical with respect to the middle point of the column (because the moment of inertia is symmetrical about the middle point), the even parameters in the above series vanish. The deflection equation then becomes

$$y = a_1 \sin\frac{\pi x}{L} + a_3 \sin\frac{3\pi x}{L} + a_5 \sin\frac{5\pi x}{L} + \ldots$$

We shall consider only two terms of the series. Thus

$$y = a_1 \sin\frac{\pi x}{L} + a_3 \sin\frac{3\pi x}{L}$$

$$U = \frac{1}{2}\int_0^L EI\left(\frac{d^2y}{dx^2}\right)^2 dx$$

$$= \frac{1}{2}\int_0^L EI_0\left(1+\sin\frac{\pi x}{L}\right)\left[-a_1\left(\frac{\pi}{L}\right)^2 \sin\frac{\pi x}{L} - a_3\left(\frac{3\pi}{L}\right)^2 \sin\frac{3\pi x}{L}\right]^2 dx$$

$$= \frac{1}{2}EI_0\left(\frac{\pi}{L}\right)^3\left[\left(\frac{4}{3}+\frac{\pi}{2}\right)a_1^2 - \frac{24}{5}a_1a_3 + \left(\frac{2916}{35}+\frac{81\pi}{2}\right)a_3^2\right]$$

and

$$\Delta L = EI\frac{1}{2}\int_0^L\left(\frac{dy}{dx}\right)^2 dx = \frac{1}{2}\left(\frac{L}{2}a_1^2 + \frac{L}{2}\times 9a_3^2\right)$$

Substituting

$$P = EI_0\frac{\pi}{L^2}\frac{(8/3+\pi)a_1^2 - (48/5)a_1a_3 + (5832/35) + 81\pi)a_3^2}{a_1^2 + 9a_3^2} = \frac{F_1}{F_2}$$

For minimum P, we should have

$$\frac{\partial}{\partial a_1}(F_1 - PF_2) = 0 \quad \text{and} \quad \frac{\partial}{\partial a_2}(F_1 - PF_2) = 0$$

Thus, $2\left(\frac{8}{3}+\pi-P^*\right)a_1 - \frac{48}{5}a_3 = 0$

and $-\frac{48}{5}a_1 + 2\left(\frac{5832}{35}+81\pi-9P^*\right)a_3 = 0$

where $P^* = \frac{PL^2}{\pi EI_0}$

For the existence of a non–trivial solution, the determinant of the coefficients should be equal to zero. This gives

$$\Delta = 9P^{*2} - \left(\frac{5832}{35}+81\pi+24+9\pi\right)P^*$$

$$+\left(\frac{8}{3}+\pi\right)\left(\frac{5832}{35}+81\pi\right) - \left(\frac{24}{5}\right)^2 = 0$$

Solving,

$$P^* = 5.746 \text{ or } P = 18.05\frac{EI_0}{L^2}$$

Example 6:

Consider a pin–ended column subjected to an axial compressive load P. Assume that the buckled shape is given by

$$y = a \sin \frac{\pi x}{L}$$

Solution:

Where a is an unknown parameter. The coordinate function chosen satisfies the boundary conditions which are

$$y = 0 \quad \text{at } x = 0 \quad \text{and} \quad \text{at } x = L$$

$$\frac{d^2 y}{dx^2} = 0 \quad \text{at } x = 0 \quad \text{and} \quad \text{at } x = L$$

The strain energy is obtained as

$$U = \frac{1}{2} EI \int_0^L \left(\frac{d^2 y}{dx^2} \right)^2 dx$$

$$= \frac{1}{2} EI \int_0^L a^2 \left(\frac{\pi}{L} \right)^4 \sin^2 \frac{\pi x}{L} dx$$

$$= \frac{1}{4} \pi^4 a^2 \left(\frac{EI}{L^3} \right)$$

The potential energy is obtained as

$$V = -\frac{1}{2} P \int_0^L \left(\frac{dy}{dx} \right)^2 dx$$

$$= -\frac{1}{2} P \int_0^L a^2 \left(\frac{\pi}{L} \right)^2 \cos^2 \frac{\pi x}{L} dx$$

$$= -\frac{1}{4} P \pi^2 \left(\frac{a^2}{L} \right)$$

Thus, the total potential energy is

$$U + V = \frac{1}{4} \pi^4 a^2 \frac{EI}{L^3} - \frac{1}{4} P \pi^2 \frac{a^2}{L}$$

For this to be an extremum, we must have

$$= \frac{1}{2} \pi^4 a \frac{EI}{L^3} - \frac{1}{2} P \pi^2 \frac{a}{L} = 0$$

or

$$\frac{1}{2} \pi^2 \frac{a}{L} \left(\pi^2 \frac{EI}{L^2} - P \right) = 0$$

The non–trivial solution is obtained when

$$P = P_{cr} = \frac{\pi^2 EI}{L^2}$$

We have been able to obtain an exact solution since the coordinate function we used happens to give the exact diflected shape for the column.

Example 7:

Show that the resultant circumferential force across any radial section of a hollow disk subjected to thermal loading is zero.

The value of the circumferential stress σ_θ is

$$\sigma_\theta = \frac{\alpha E}{r^2}\left[\frac{r^2+a^2}{b^2-a^2}\int_a^b Tr\,dr + \int_a^r Tr'\,dr' - Tr^2\right]$$

Solution:

Let the disk be of unit thickness perpendicular to the plane of the paper.

The resultant circumferential force across any section is

$$F_\theta = \int_a^b \sigma_\theta\,dr = \frac{\alpha E}{b^2-a^2}\left[\int_a^b dr \int_a^b Tr\,dr + \int_a^b \frac{a^2}{r^2}dr \int_a^r Tr\,dr\right]$$

$$+\alpha E\left[\frac{1}{r^2}dr \int_a^r Tr'\,dr' - \int_a^b T\,dr'\right]$$

Let $\int_a^b Tr\,dr = \beta$

Then, $F_\theta = \frac{\alpha E}{b^2-a^2}\left[\beta(b-a) - \beta a^2\left(\frac{1}{b}-\frac{1}{a}\right)\right]$

$$+\alpha E\left[\left(-\frac{1}{r}\int_a^r Tr'\,dr'\right)\Bigg|_a^b + \int_a^b T\,dr - \int_a^b T\,dr\right]$$

In the above expression, we have made use of the formula

$$\frac{d}{d\alpha}\int_{U(\alpha)}^{V(\alpha)} F(\alpha,x)dx = \int_{U(\alpha)}^{V(\alpha)} \frac{dF(\alpha,x)}{d\alpha}dx + F(V,\alpha)\frac{dV}{d\alpha} - F(U,\alpha)\frac{dU}{d\alpha}$$

Substituting the limits, it is observed that $F_\theta = 0$

There is no resultant circumferential force across any section.

Example 8:

Determine the bending moment due to the circumferential stress across a section of a thin hollow disk subjected to radial thermal variation.

Solution:

If ρ_0 is the radius of the median line and σ_θ the circumferential stress on a fibre at r from the centre of curvature, then the moment about the median line is

$$F_m = \int_a^b \sigma_\theta (r - \rho_0)\, dr$$

$$= \int_a^b \sigma_\theta\, r\, dr - \rho_0 \int_a^b \sigma_\theta\, dr$$

The second integral on the right-hand side is zero,

$$F_m = \frac{\alpha E}{b^2 - a^2}\left[\int_a^b r\, dr \int_a^b Tr\, dr + \int_a^b \frac{a^2}{r} dr \int_a^b Tr\, dr\right]$$

$$+\alpha E\left[\int_a^b \frac{1}{r} dr \int_a^r Tr'\, dr' - \int_a^b Tr\, dr\right]$$

Putting $\int_a^b Tr\, dr = \beta$, the above expression becomes

$$F_m = \frac{\alpha E}{b^2 - a^2}\left[\frac{\beta}{2}(b^2 - a^2) + a^2 \beta \log \frac{b}{a}\right]$$

$$+\alpha E\left[\log r \int_a^r Tr'\, dr'\right]\Bigg|_a^b - \alpha E \int_a^b (\log r)\, Tr\, dr - \alpha E \beta$$

$$= \frac{\alpha E \beta}{2(b^2 - a^2)}\left[b^2 + a^2\left(2 \log \frac{b}{a} - 1\right)\right] + \alpha E\left[\beta(\log b - 1) - \int_a^b (\log r) Tr\, dr\right]$$

EXERCISES

1. A thin hollow tube has its inner surface at temperature T_i and its outer surface at zero temperature. Assuming steady-state conditions, calculate the stresses. The inner radius is a and the thickness of the tube is t.

$$\left[\begin{aligned} Ans.\ (\sigma_\theta)_{r=a} = (\sigma_z)_{r=a} &= -\frac{\alpha E T_i}{1(1-\nu)}\left(1 + \frac{t}{3a}\right) \\ (\sigma_\theta)_{r=b} = (\sigma_z)_{r=b} &= -\frac{\alpha E T_i}{1(1-\nu)}\left(1 + \frac{t}{3a}\right) \end{aligned}\right]$$

2. A solid sphere of radius b is subjected to thermal loading $T = T(r)$. Show that the radial stress σ_r at any radius r is proportional to the difference between the mean temperature of the whole sphere and

the mean temperature of a sphere of radius r. Also, show that the circumferential stress at any point is equal to $\frac{2\alpha E}{3(1-v)}$ multiplied by the following expression:
[(mean temperature of the whole sphere)
+(1/2 the mean temperature within the sphere of radius r) – 3/2T]

3. For a column with one end built–in and the other end free and carrying an axial load P, it is assumed that the deflection curve has the form

$$y = \frac{\delta x^2}{L^2}$$

where L is the length of the column and x is measured form the fixed end. Using the energy method, determine the critical load.

4. The deflection curve for a pin–ended column is represented by a polynomial as

$$y = ax^4 + bx^3 + cx^2 + dx + e$$

Determine the critical load by the energy method.

5. A thin disk of inner radius a and outer radius b is subjected to a temperature variation which is symmetrical about the axis, i.e. $T = T(r)$. Consider a sectoral element, as shown in Fig. 4. Calculate the resultant moment due to σ_θ about the median line of the section across any radial section.

$$\left[\begin{aligned} \textit{Ans. } F_m = \frac{\alpha E \beta}{2(b^2 - a^2)}\left[b^2 + a^2\left(2\log\frac{b}{a} - 1\right)\right] + \alpha E\Big[\beta(\log b - 1) \\ - \int_a^b \log r\, Tr\, dr\Big] \\ \text{where } \beta = \int_a^b Tr\, dr \end{aligned} \right]$$

6. A thin, uniform disk of radius b is surrounded by a heavy ring of the same material. The assembly just fits when the disk and the ring are at a uniform temperature. The faces of the disk are kept at temperature T_i and the circumference is kept at temperature T_0. The temperature variation along r from the centre is given by

$$T = T_i - (T_i - T_0)\frac{r^2}{b^2}$$

The heavy ring is at temperature T_0 and its strain is assumed to be negligible. Show that the radial compressive stress in the disk at radius

$$\sigma_r = \frac{1}{4} E\alpha(T_i - T_0)\left(\frac{3-\nu}{1-\nu} - \frac{r^2}{b^2}\right)$$

7. The initial shape of a bar can be approximated by the series

$$y = \delta_1 \sin\frac{\pi x}{L} + \delta_2 \sin\frac{2\pi x}{L} + \ldots$$

If the bar is simply supported and subjected to axial force P only, show that the deflection curve due to P is given by

$$y_1 = \alpha\left(\frac{\delta_1}{1-\alpha}\sin\frac{\pi x}{L} + \frac{\delta_2}{2^2-\alpha}\sin\frac{2\pi x}{L} + \ldots\right), \text{ where } \alpha = \frac{k^2L^2}{\pi^2}$$

8. The temperature distribution in a long cylindrical conductor due to the passage of current is given by

$$T = \lambda(b^2 - r^2)$$

where λ is a constant. Determine the stresses due to thermal loading only.

$$\left[\begin{aligned} Ans.\ \sigma_r &= -\frac{E\alpha\lambda}{4(1-\nu)}(b^2 - r^2) \\ \sigma_\theta &= -\frac{E\alpha\lambda}{4(1-\nu)}(3r^2 - b^2) \\ \sigma_z &= \frac{E\alpha\lambda}{2(1-\nu)}(2r^2 - b^2) \end{aligned}\right]$$

4

Axisymmetric Problems

INTRODUCTION

Many problems of practical importance are concerned with solids of revolution which are deformed symmetrically with respect to the axis of revolution. Examples of such solids are circular cylinders subjected to uniform internal and external pressures, rotating circular disks, spherical shells subjected to uniform internal and external pressures, etc. In this chapter, a few of these problems will be investigated. Let the axis of revolution be the z-axis. The deformation being symmetrical with respect to the z-axis, it is convenient to use cylindrical coordinates. Since the deformation is symmetrical about the axis, the stress components do not depend on θ. Further, $\tau_{r\theta}$ and $\tau_{\theta z}$ do not exist. Consequently, the differential equations of equilibrium [Eqs (67) - (69)] can be reduced to our special case. However, it is instructive to derive the relevant equations applicable to axisymmetric body shown in Fig. 1. Let an elementary radial element be isolated. The stress vectors acting on its faces are as shown.

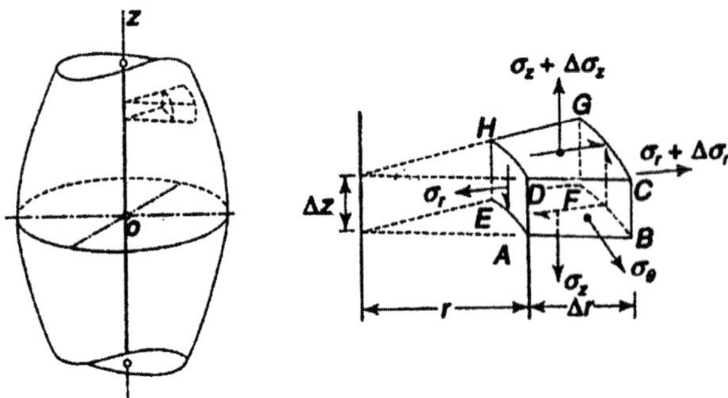

Fig. 1

On faces *ABCD* and *EFGH*, the normal stresses are σ_θ and there are no shear stresses. On face *ABFE*, the stresses are σ_z and τ_{zr}. On face *CDHG*, the normal and shear stresses are

$$\sigma_z + \Delta\sigma_z = \sigma_z + \frac{\partial \sigma_z}{\partial z}\Delta z$$

$$\tau_{rz} + \Delta\tau_{rz} = \tau_{rz} + \frac{\partial \tau_{rz}}{\partial z}\Delta z$$

On face *AEHD*, the normal and shear stresses are σ_r and τ_{rz}. On face *BCGF*, the stresses are $\sigma_r + \frac{\partial \sigma_r}{\partial r}\Delta r$ and $\tau_{rz} + \frac{\partial \tau_{rz}}{\partial r}\Delta r$

For equilibrium in z direction

$$\left(\sigma_z + \frac{\partial \sigma_z}{\partial z}\Delta z\right)\left(r + \frac{\Delta r}{2}\right)\Delta\theta\, \Delta r + \left(\tau_{rz} + \frac{\partial \tau_{rz}}{\partial r}\Delta r\right)(r + \Delta r)\Delta\theta\, \Delta z$$

$$-\tau_{rz} r\, \Delta\theta\, \Delta z - \sigma_z \left(r + \frac{\Delta r}{2}\right)\Delta\theta\, \Delta r + \gamma_z \left(r + \frac{\Delta r}{2}\right)\Delta\theta\, \Delta r\, \Delta z = 0$$

where γ_z is the body force per unit volume in z direction. Hence,

$$\frac{\partial \sigma_z}{\partial z}\left(r + \frac{\Delta r}{2}\right)\Delta r\, \Delta\theta\, \Delta z + \frac{\partial \tau_{rz}}{\partial r}(r + \Delta r)\Delta r\, \Delta\theta\, \Delta z$$

$$+\tau_{rz}\Delta r\, \Delta\theta\, \Delta z + \gamma_z \left(r + \frac{\Delta r}{2}\right)\Delta r\, \Delta\theta\, \Delta z = 0$$

Cancelling $\Delta r\, \Delta\theta\, \Delta z$ and going to limits

$$\frac{\partial \sigma_z}{\partial z} + \frac{\partial \tau_{rz}}{\partial r} + \frac{\tau_{rz}}{r} + \gamma_z = 0 \tag{1}$$

Similarly, for equilibrium in r direction we get

$$\frac{\partial \sigma_r}{\partial r} + \frac{\partial \tau_{rz}}{\partial z} + \frac{\sigma_r - \sigma_\theta}{r} + \gamma_r = 0 \tag{2}$$

where γ_r is the body force per unit volume in r direction. Since the stress components are independent of θ, the equilibrium equation for θ direction is identically satisfied.

For the problems that we are going to discuss in this chapter, we need expressions for the circumferential strain ε_θ and the radial strain ε_r.

Referring to Fig. 2(a), consider an arc *AE* at distance r, subtending an angle $\Delta\theta$ at the centre. The arc length is $r\Delta\theta$. The radial displacement is u_r. Consequently, the length of the arc becomes $(r + u)\, \Delta\theta$. Hence, the circumferential strain is

$$\varepsilon_\theta = \frac{(r+u_r)\Delta\theta - r\Delta\theta}{r\Delta\theta} = \frac{u_r}{r} \tag{3}$$

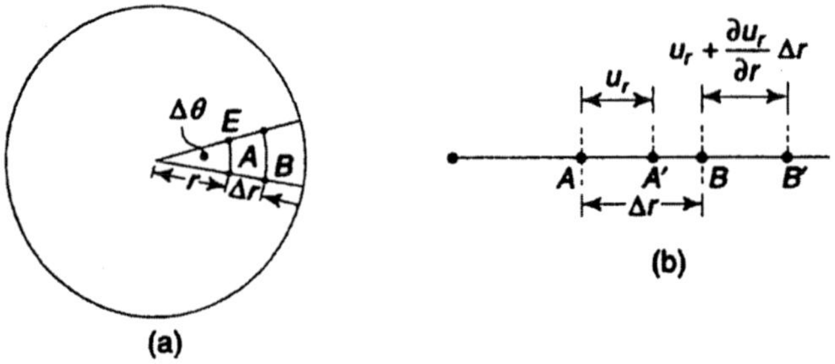

Fig. 2

The radial strain is, from Fig. 2(b),

$$\varepsilon_r = \frac{\partial u_r}{\partial r} \tag{4}$$

The axial strain is

$$\varepsilon_z = \frac{\partial u_z}{\partial z} \tag{5a}$$

where u_z is the axial displacement. In subsequent sections we shall consider the following problems:

Circular cylinder subjected to internal or external pressure

Sphere subjected to internal or external pressure

Sphere subjected to mutual gravitational attraction

Rotating disk of uniform thickness

Rotating disk of variable thickness

Rotating shaft and cylinder.

THICK-WALLED CYLINDER SUBJECTED TO INTERNAL AND EXTERNAL PRESSURE-LAME'S PROBLEM

Consider a cylinder of inner radius a and outer radius b (Fig. 3). Let the cylinder be subjected to an internal pressure P_a and an external pressure P_b. It is possible to treat this problem either as a plane stress case ($\sigma_z = 0$) or as a plane strain case ($\varepsilon_z = 0$). Appropriate solutions will be obtained for each case.

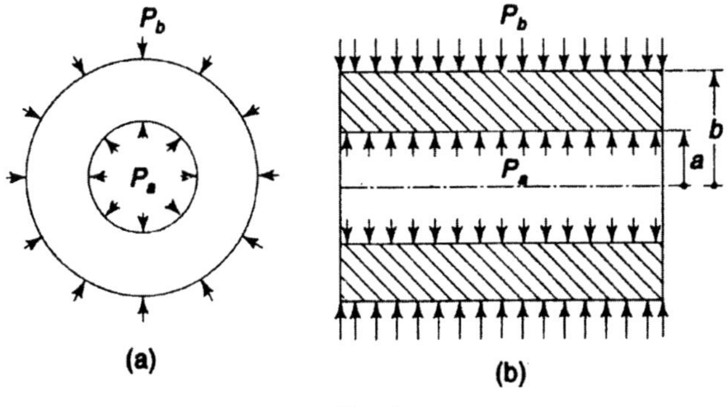

Fig. 3

Case (a) plane stress

Let the ends of the cylinder be free to expand. We shall assume that $\sigma_z = 0$ and our results will justify this assumption. Owing to uniform radial deformation, $\tau_{rz} = 0$. Neglecting body forces, Eq. (2) reduces to

$$\frac{\partial \sigma_r}{\partial r} + \frac{\sigma_r - \sigma_\theta}{r} = 0 \tag{5b}$$

Since r is the only independent variable, the above equation can be written as

$$\frac{d}{dr}(r\sigma_r) - \sigma_\theta = 0 \tag{5c}$$

Equation (1) is identically satisfied. From Hooke's law

$$\varepsilon_r = \frac{1}{E}(\sigma_r - V\sigma_\theta), \qquad \varepsilon_\theta = \frac{1}{E}(\sigma_\theta - V\sigma_r)$$

or the stresses in terms of strains are

$$\sigma_r = \frac{E}{1-V^2}(\varepsilon_r + V\varepsilon_\theta) \qquad \sigma_\theta = \frac{E}{1-V^2}(\varepsilon_\theta + V\varepsilon_r)$$

Substituting for ε_r and ε_θ from Eqs (3) and (4)

$$\sigma_r = \frac{E}{1-V^2}\left(\frac{du_r}{dr} + V\frac{u_r}{r}\right) \tag{6a}$$

$$\sigma_\theta = \frac{E}{1-V^2}\left(\frac{u_r}{r} + V\frac{du_r}{dr}\right) \tag{6b}$$

Substituting these in the equation of equilibrium given by Eq. (5a)

$$\frac{d}{dr}\left(r\frac{du_r}{dr}+Vu_r\right)-\left(\frac{u_r}{r}+V\frac{du_r}{dr}\right)=0$$

or $$\frac{du_r}{dr}+r\frac{d^2u_r}{dr^2}+V\frac{du_r}{dr}-\frac{u_r}{r}-V\frac{du_r}{dr}=0$$

i.e. $$\frac{d^2u_r}{dr^2}+\frac{1}{r}\frac{du_r}{dr}-\frac{u_r}{r^2}=0$$

This can be reduced to

$$\frac{d}{dr}\left(\frac{du_r}{dr}+\frac{u_r}{r}\right)=0$$

or $$\frac{d}{dr}\left[\frac{1}{r}\frac{d}{dr}(u_r r)\right]=0 \tag{7}$$

If the function u_r is found from this equation, the stresses are then determined from Eq. (6),

The solution to Eq. (7) is

$$u_r = C_1 r+\frac{C_2}{r} \tag{8}$$

where C_1 and C_2 are constants of integration. Substituting this function in Eq. (6)

$$\sigma_r = \frac{E}{1-V^2}\left[C_1(1+V)-C_2(1-V)\frac{1}{r^2}\right] \tag{9a}$$

$$\sigma_\theta = \frac{E}{1-V^2}\left[C_1(1+V)+C_2(1-V)\frac{1}{r^2}\right] \tag{9b}$$

The constants C_1 and C_2 are determined from the boundary conditions.

When $r = a,\quad \sigma_r = -P_a$

when $r = b,\quad \sigma_r = -P_b$

Hence,

$$\frac{E}{1-V^2}\left[C_1(1+V)-C_2(1-V)\frac{1}{a^2}\right]=-p_a$$

$$\frac{E}{1-V^2}\left[C_1(1+V)+C_2(1-V)\frac{1}{b^2}\right]=-p_b$$

whence, $C_1 = \frac{1-V}{E}\frac{p_a a^2 - p_b b^2}{b^2 - a^2}$

$$C_2 = \frac{1+V}{E}\frac{a^2 - b^2}{b^2 - a^2}(p_a - p_b)$$

Substituting these in Eqs (8) and (9) we get

$$u_r = \frac{1-V}{E}\frac{p_a a^2 - p_b b^2}{b^2 - a^2} r + \frac{1+V}{E}\frac{a^2 b^2}{r}\frac{p_a - p_b}{b^2 - a^2} \tag{10}$$

$$\sigma_r = \frac{p_a a^2 - p_b b^2}{b^2 - a^2} - \frac{a^2 b^2}{r^2}\frac{p_a - p_b}{b^2 - a^2} \tag{11}$$

$$\sigma_\theta = \frac{p_a a^2 - p_b b^2}{b^2 - a^2} + \frac{a^2 b^2}{r^2}\frac{p_a - p_b}{b^2 - a^2} \tag{12}$$

It is interesting to observe that the sum $\sigma_r + \sigma_\theta$ is constant through the thickness of the wall of the cylinder, i.e. independent or r. Hence, according to Hooke's Law, the stresses σ_r and σ_θ produce a uniform extension or contraction in z direction, and cross-sections perpendicular to the axis of the cylinder remain plane. If we consider two adjacent cross-sections, the deformation undergone by the element does not interfere with the deformation of the neighbouring element. Hence, the elements can be considered to be in a state of plane stress, i.e. $\sigma_z = 0$, as we assumed at the beginning of the discussion. It is important to note that in Eqs (10) - (12), P_c and P_b are the numerical values of the compressive pressures applied.

(i) **Cylinder Subjected to Internal Pressure** : In this case $P_b = 0$ and $P_a = P$. Then Eqs (11) and (12) become

$$\sigma_r = \frac{pa^2}{b^2 - a^2}\left(1 - \frac{b^2}{r^2}\right) \tag{13}$$

$$\sigma_\theta = \frac{pa^2}{b^2 - a^2}\left(1 + \frac{b^2}{r^2}\right) \tag{14}$$

These equations show that σ_r is always a compressive stress and σ_θ a tensile stress. Figure 4 shows the variation of radial and circumferential stresses across the thickness of the cylinder under internal pressure. The circumferential stress is greatest at the inner surface of the cylinder, where

$$(\sigma_\theta)_{max} = \frac{p(a^2 + b^2)}{b^2 - a^2} \tag{15}$$

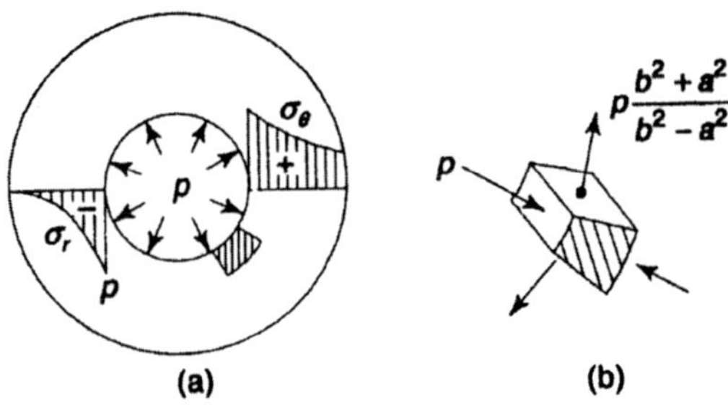

Fig. 4

Hence, $(\sigma_\theta)_{max}$ is always greater than the internal pressure and approaches this value as b increases so that it can never be reduced below P_a irrespective of the amount of material added on the outside.

(ii) Cylinder Subjected to External Pressure : In this case $P_a = 0$ and $P_b = p$. Equations (11) and (12) reduce to

$$\sigma_r = -\frac{pb^2}{b^2 - a^2}\left(1 - \frac{a^2}{r^2}\right) \tag{16}$$

$$\sigma_\theta = -\frac{pb^2}{b^2 - a^2}\left(1 + \frac{a^2}{r^2}\right) \tag{17}$$

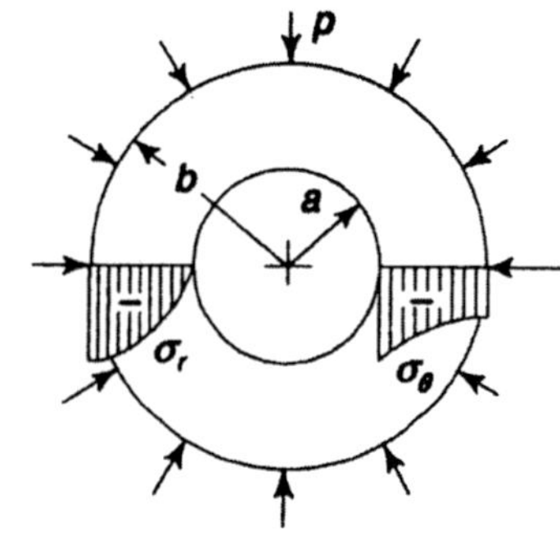

Fig. 5.

The variations of these stresses across the thickness are shown in Fig. 5. If there is no inner hole, i.e. if $a = 0$, the stresses are uniformly distributed in the cylinder with $\sigma_r = \sigma_\theta = -p$.

Example 1:

Select the outer radius b for a cylinder subjected to an internal pressure p = 500 atm with a factor of safety 2. The yield point for the material (in tension as well as in compression) is σ_{yp} = 5000 kgf/cm² (490000 kPa). The *inner radius is 5 cm. Assume that the ends of the cylinder are closed. The critical point lies on the inner surface of the cylinder, where*

$$\sigma_r = -p, \quad \sigma_\theta = p\frac{b^2+a^2}{b^2-a^2}, \quad \sigma_z = p\frac{a^2}{b^2-a^2}$$

Solution:

In the above expressions, it is assumed that away from the ends, σ_z caused by p is uniformly distributed across the thickness. The maximum and minimum principal stresses are $\sigma_1 = \sigma_\theta$ and $\sigma_3 = \sigma_r$. Hence,

$$\tau_{max} = \frac{1}{2}(\sigma_1 - \sigma_3) = p\frac{b^2}{b^2-a^2}$$

Substituting the numerical values (1 atm = 98.07 kPa),

$$b = \sqrt{\frac{5}{3}}a = 6.45 \text{ cm}$$

Example 2:

A thick-walled steel cylinder with radii a = 5 cm and b = 10 cm is subjected to an internal pressure p. The yield stress in tension for the material is 350 MPa. Using a factor of safety of 1.5, determine the maximum working pressure p according to the major theories of failure. $E = 207 \times 10^6$ kPa, $\nu = 0.25$.

Solution:

(i) *Maximum normal stress theory:*

Maximum normal stress $= \sigma_\theta$ at $r = a$

$$= p\frac{(b^2+a^2)}{(b^2-a^2)}$$

$$\therefore \quad p\frac{(b^2+a^2)}{(b^2-a^2)} = \frac{\sigma_y}{N}$$

or

$$p = \frac{350\times10^6}{1.5}\times\frac{100-25}{100+25} = 140\times10^3 \text{ kPa}$$

(ii) *Maximum shear stress theory:*

Maximum shear stress $= \frac{1}{2}(\sigma_\theta - \sigma_r)$ at $r = a$

$$= \frac{1}{2}p\left(\frac{2b^2}{b^2-a^2}\right)$$

$$\therefore \qquad p\frac{b^2}{b^2-a^2}=\frac{1}{2}\frac{\sigma_y}{N}$$

$$\text{or} \qquad p=\frac{350\times10^6}{3}\times\frac{100-25}{100}=87.5\times10^3 \text{ kPa}$$

(iii) *Maximum strain theory:*

$$\text{Maximum strain } \varepsilon_\theta=\frac{1}{E}(\sigma_\theta-v\,\sigma_r)\,r=a$$

$$=\frac{P}{E}\frac{a^2}{(b^2-a^2)}\left[\left(1+\frac{b^2}{a^2}\right)-v\left(1-\frac{b^2}{a^2}\right)\right]$$

$$\therefore \qquad \frac{P}{E(b^2-a^2)}[(a^2+b^2)-v(a^2-b^2)]=\frac{\sigma_y}{N}$$

$$\text{or} \qquad \frac{P}{(100-25)}[125+(0.25\times75)]=\frac{350\times10^6}{1.5}$$

$$\therefore \qquad p=\frac{350\times10^6}{1.5\times143.75}=121.7\times10^3 \text{ kPa}$$

(iv) *Octahedral shear stress theory:*

$$\tau_{oct}=\frac{1}{3}[(\sigma_\theta^2+\sigma_r^2+(\sigma_r-\sigma_\theta)^2]^{1/2} \quad \text{at } r=a$$

$$=\frac{1}{3}[2((\sigma_r-\sigma_\theta)^2+2\sigma_r\sigma_\theta]^{1/2}$$

$$=\frac{\sqrt{2}}{3}\left\{\left[-p-\frac{p(b^2+a^2)}{(b^2-a^2)}\right]^2-p^2\frac{(b^2+a^2)}{(b^2-a^2)}\right\}^{1/2}$$

$$=\frac{\sqrt{2}}{3}\frac{\sigma_y}{N}$$

$$\therefore \qquad \frac{\sqrt{2}}{3}p\left[\frac{4b^4}{(b^2-a^2)^2}-\frac{(b^2+a^2)}{(b^2-a^2)}\right]^{1/2}=\frac{\sqrt{2}}{3}\frac{\sigma_y}{N}$$

$$\text{or} \qquad p\left(\frac{40000}{5625}-\frac{125}{75}\right)^{1/2}=\frac{350}{1.5}$$

$$\therefore \qquad p=100\times10^3 \text{ kPa.}$$

(v) *Energy of distortion theory:*

This will give a value identical to that obtained based on octahedral shear stress theory, i.e. $p = 100 \times 10^3$ kPa.

Example 3:

A pipe made of steel has a tensile elastic limit σ_y = 275 MPa and E = 207 × 10^6 kPa. If the pipe has an internal radius a = 5 cm and is subjected to an internal pressure p = 70 × 10^3 kPa, determine the proper thickness for the pipe wall according to the major theories of failure. Use a factor of safety $N = \dfrac{4}{3}$.

Solution:

(i) *Maximum principal stress theory:*

Maximum principal stress = σ_θ at $r = a$

$$= p\frac{(b^2+a^2)}{(b^2-a^2)} = \frac{\sigma_y}{N}$$

$$\therefore \qquad \frac{70\times10^6[(25\times10^{-4})+b^2]}{[b^2-(25\times10^{-4})]} = \frac{275\times10^6\times3}{4}$$

or $\quad 1750 \times 10^{-4} + 70b^2 = 825b^2 - \dfrac{20625}{4}\times10^{-4}$

or $\quad 136.25b^2 = 6906.25 \times 10^{-4}$

$\therefore \quad b = 7.12\times 10^{-2}$ m = 7.12 cm

$\therefore \quad$ Wall thickness t = 2.12 cm

(ii) *Maximum shear stress theory:*

$$\tau_{max} = \frac{1}{2}(\sigma_\theta - \sigma_r) \text{ at } r = a$$

$$= \frac{pb^2}{(b^2-a^2)}\frac{\sigma_y}{N}$$

$$\therefore \qquad \frac{70\times10^6 b^2}{[b^2-(25\times10^{-4})]} = \frac{3}{8}\times275\times10^6$$

or $\quad 70b^2 = 103.13 \quad - 2578.1 \times 10^{-4}$

$\therefore \quad b = 8.82 \times 10^{-2}$ m = 8.82 cm

$\therefore \quad$ Wall thickness t = 3.82 cm

(iii) *Maximum strain theory (with* ν = 0.25)*:*

$$\varepsilon_{max} = \frac{1}{E}(\sigma_\theta - \nu\sigma_r \quad \text{at } r = a$$

$$= \frac{p}{E(b^2-a^2)}[(a^2+b^2)-\nu(a^2-b^2)] = \frac{\sigma_y}{NE}$$

$$\therefore \quad \frac{70\times10^6}{[b^2-(25\times10^{-4})]}[(0.75\times25\times10^{-4})+$$

$$(1.25\times b^2)] = \frac{3}{4}\times275\times10^6$$

or $\quad 1312.5 \times 10^{-4} + 87.5b^2 = 206.25b^2 - 5156.25 \times 10^{-4}$

$\therefore \quad b = 7.38 \times 10^{-2}\text{m} = 7.38$ cm

$\therefore$ Wall thickness $t = 2.38$ cm

(iv) *Maximum distortion energy theory:*

From Eq. (12) with

$$\sigma_1 = \sigma_\theta, \quad \sigma_2 = 0, \quad \sigma_3 = \sigma_r = -p$$

$$U^* = \frac{1}{12G}[\sigma_\theta^2+\sigma_r^2+(\sigma_r-\sigma_\theta)^2]$$

$$= \frac{(1+V)}{3E}(\sigma_\theta^2+\sigma_r^2-\sigma_r\sigma_\theta) = \frac{1+V}{E}\frac{\sigma_y^2}{N^2}$$

$$\therefore \quad \sigma_\theta^2+\sigma_r^2-\sigma_r\sigma_\theta)\frac{\sigma_y^2}{N^2}$$

i.e. $$p^2\left[\frac{(b^2+a^2)^2}{(b^2-a^2)^2}+1+\frac{(b^2-a^2)}{(b^2+a^2)}\right] = \frac{\sigma_y^2}{N^2}$$

putting $\left(\frac{\sigma_y}{pN}\right) = f_y$ and simplifying one gets

$$(3-f_y^2)b^4+2a^2f_y^2b^2+(1-f_y^2)a^4 = 0$$

$$\therefore \quad b^2 = \frac{-2a^2f_y^2 \pm \sqrt{[4a^4f_y^4-4a^4(1-f_y^2)(3-f_y^2)]}}{2(3-f_y^2)}$$

$$= \frac{a^2\left[-f_y^2 \pm \sqrt{(4f_y^4-3)}\right]}{2(3-f_y^2)}$$

with $a = 5 \times 10^{-2}$

$$fy = \frac{275\times10^6\times3}{70\times10^6\times4} = 2.946$$

$\therefore \quad b^2 = 63$ or $13.4)\ 10^{-4}$ or $b = 7.9 \times 10^{-2}$ m $= 7.9$ cm

Wall thickness $t = 2.9$ cm

Case (b) Plane Strain

When the cylinder is fairly long, sections that are far from the ends can be considered to be in a state of plane strain and we can assume that σ_z does not vary along the z-axis. As in the case of plane stress, the equation is

$$\frac{d}{dr}(r\sigma_r) - \sigma_\theta = 0$$

From Hooke's law

$$\varepsilon_r = \frac{1}{E}[\sigma_r - V(\sigma_\theta + \sigma_z)]$$

$$\varepsilon_\theta = \frac{1}{E}[\sigma_\theta - V(\sigma_r + \sigma_z)]$$

$$\varepsilon_z = \frac{1}{E}[\sigma_z - V(\sigma_r + \sigma_\theta)]$$

Since $\varepsilon_z = 0$ in this case, one has from the last equation

$$\sigma_z = \nu(\sigma_r - \sigma_\theta)$$

$$\varepsilon_r = \frac{1+V}{E}[(1-V)\sigma_r - V\sigma_\theta] \tag{18}$$

$$\varepsilon_\theta = \frac{1+V}{E}[(1-V)\sigma_\theta - V\sigma_r]$$

Solving for σ_θ and σ_r

$$\sigma_\theta = \frac{E}{(1-2V)(1+V)}[V\varepsilon_r + (1-V)\varepsilon_\theta] \tag{19a}$$

$$\sigma_r = \frac{E}{(1-2V)(1+V)}[(1-V)\varepsilon_r + V\varepsilon_\theta] \tag{19b}$$

On substituting for ε_r and ε_θ from Eqs (3) and (4), the above equations become

$$\sigma_\theta = \frac{E}{(1-2V)(1+V)}\left[V\frac{du_r}{dr} + (1-V)\frac{u_r}{r}\right] \tag{20}$$

$$\sigma_r = \frac{E}{(1-2V)(1+V)}\left[(1-V)\frac{du_r}{dr} + V\frac{u_r}{r}\right] \tag{21}$$

Substituting these in the equation of equilibrium

$$\frac{d}{dr}\left[(1-V)r\frac{du_r}{dr} + Vu_r\right] - V\frac{du_r}{dr} - (1-V)\frac{u_r}{r} = 0$$

or $$\frac{du_r}{dr} + r\frac{d^2u_r}{dr^2} - \frac{u_r}{r} = 0$$

i.e. $$\frac{d}{dr}\left(\frac{du}{dr} + \frac{u_r}{r}\right) = 0$$

This is the same as Eq. (7) for the plane stress case. The solution is the same as in Eq. (8).

$$u_r = C_1 r + \frac{C_2}{r}$$

where C_1 and C_2 are constants of integration. From Eqs (20) and (21)

$$\sigma_\theta = \frac{E}{(1-2V)(1+V)}\left[C_1 + (1-2V)\frac{C_2}{r^2}\right] \tag{22a}$$

$$\sigma_r = \frac{E}{(1-2V)(1+V)}\left[C_1 - (1-2V)\frac{C_2}{r^2}\right] \tag{22b}$$

Once again, we observe that $\sigma_r + \sigma_\theta$ is a constant independent of r. Further, the axial stress from Eq. (18) is

$$\sigma_z = -\frac{2VE}{(1-2V)(1+V)}C_1 \tag{22c}$$

Applying the boundar conditions

$$\sigma_r = -p_a \quad \text{when } r = a, \qquad \sigma_r = -p_b \quad \text{when } r = b$$

$$\frac{E}{(1-2V)(1+V)}\left[C_1 - (1-2V)\frac{C_2}{a^2}\right] = -p_a$$

$$\frac{E}{(1-2V)(1+V)}\left[C_1 - (1-2V)\frac{C_2}{b^2}\right] = -p_b$$

Solving, $C_1 = \dfrac{(1-2V)(1+V)}{E}\dfrac{p_b b^2 - p_a a^2}{a^2 - b^2}$

and $C_2 = \dfrac{1+V}{E}\dfrac{(p_b - p_a)a^2b^2}{a^2 - b^2}$

Substituting these, the stress components become

$$\sigma_r = \frac{p_a a^2 - p_b b^2}{b^2 - a^2} - \frac{p_a - p_b}{b^2 - a^2}\frac{a^2b^2}{r^2} \tag{23}$$

$$\sigma_\theta = \frac{p_a a^2 - p_b b^2}{b^2 - a^2} + \frac{p_a - p_b}{b^2 - a^2}\frac{a^2b^2}{r^2} \tag{24}$$

$$\sigma_z = 2V\frac{p_b a^2 - p_a b^2}{b^2 - a^2} \tag{25}$$

It is observed that the values of σ_r and σ_θ are identical to those of the plane stress case. But in the plane stress case, $\sigma_z = 0$, whereas in the plane strain case, σ_z has a constant value given by Eq. (25).

STRESSES IN COMPOSITE TUBES-SHRINK FITS

The problem which will be considered now, involves two cylinders made of two different materials and fitted one inside the other. Before assembling, the inner cylinder has an internal radius a and an external radius c. The internal radius of the outer cylinder is less than c by Δ, i.e. its internal radius is $c - \Delta$. Its external radius is b. If the inner cylinder is cooled and the outer cylinder is heated, then the two cylinders can be assembled, one fitting inside the other. When the cylinders come to room temperature, a shrink fit is obtained. The problem lies in determining the contact pressure p_c between the two cylinders.

The above construction is often used to obtain thick-walled vessels to withstand high pressures. For example, if we need a vessel to withstand a pressure of say 15000 atm, the yield point of the material must be at least 30000 kgf/cm^2 (2940000 kPa). Since no such high-strength material exists, shrink-fitted composite tubes are designed.

The contact pressure p_c acting on the outer surface of the inner cylinder reduces its outer radius by u_1. On the other hand, the same contact pressure increases the inner radius of the outer cylinder by u_2. The sum of these two quantities, i.e. $(-u_1 + u_2)$ must be equal to Δ, the difference in the radii of the cylinders. To determine u_1 and u_2, we make use of Eq. (10), assuming a plane stress case.

For the inner tube

$$u_1 = \frac{1-V_1}{E_1}\left(-p_c \frac{c^2}{c^2 - a^2}\right)c + \frac{1-V_1}{E_1}\frac{a^2c^2}{c}\left(-\frac{p_c}{c^2 - a^2}\right)$$

or $$u_1 = -\frac{cp_c}{E_1(c^2 - a^2)}[(1-V_1)c^2 + (1-V_1)a^2]$$

For the outer tube

$$u_2 = \frac{1-V_2}{E_2}\left(p_c \frac{c^2}{b^2 - c^2}\right)c + \frac{1+V_2}{E_2}\frac{c^2b^2}{c}\left(\frac{p_c}{b^2 - c^2}\right)$$

or $$u_2 = -\frac{cp_c}{E_2(b^2 - c^2)}[(1-V_2)c^2 + (1+V_2)b^2]$$

In calculating u_2, we have neglected Δ since it is very small as compared to c. Noting that u_1 is negative and u_2 is positive, we should have

$$-u_1 + u_2 = \Delta$$

i.e.
$$\frac{cp_c}{E_1(c^2-a^2)}[(1-V_1)c^2+(1+V_1)a^2]+$$

$$\frac{cp_c}{E_2(b^2-c^2)}[(1-V_2)c^2+(1+V_2)b^2]=\Delta \qquad (26a)$$

Regrouping, the contact pressure p_c is given by

$$p_c = \frac{E\Delta}{2c^2}\frac{(c^2-a^2)(b^2-c^2)}{(b^2-a^2)} \qquad (27)$$

It is important to note that in Eqs (26) and (27), Δ is the difference in radii between the inner cylinder and the outer jacket. Because of shrink fitting, therefore, the inner cylinder is under external pressure p_c. The stress distribution in the assembled cylinders is shown in Fig. 6.

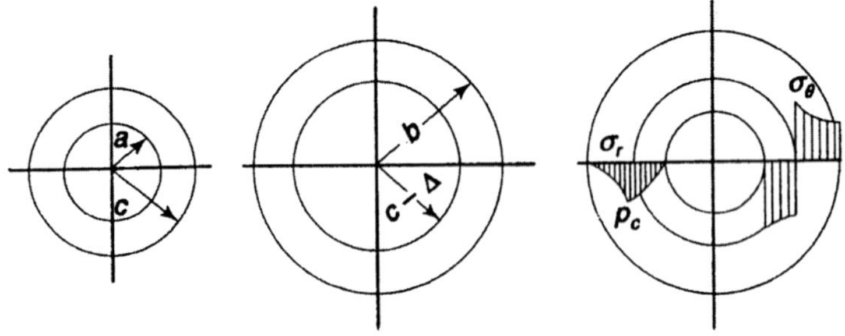

Fig. 6

If the composite cylinder made up of the same material is now subjected to an internal pressure p, then the two parts will act as a single unit and the additional stresses induced in the composite can be determined from Eqs (13) and (14). At the inner surface of the inner cylinder, the internal pressure p causes a tensile tangential stress σ_θ, Eq. (14), but, the contact pressure p_c cause at the same points a compressive tengential stress, Eq (17). Hence, a composite cylinder can support greater internal pressure than an ordinary one. However, at the inner points of the jacket or the outer cylinder, the internal pressure p and the contact pressure p_c both will induce tensile tangential (i.e. circumferential) stresses σ_θ. For design purposes, one can choose the shrind-fit allowance Δ such that the strengths of the two cylinders are equal. To determine this value of Δ, one can proceed as follows.

Let a and c be the radii of the inner cylinder, and c and b the radii of the jacket (see Fig. 7) c is the common radius of the two cylinders at the contact surface when the composite cylinder is experiencing an internal pressure p and the shrink-fit pressure p_c.

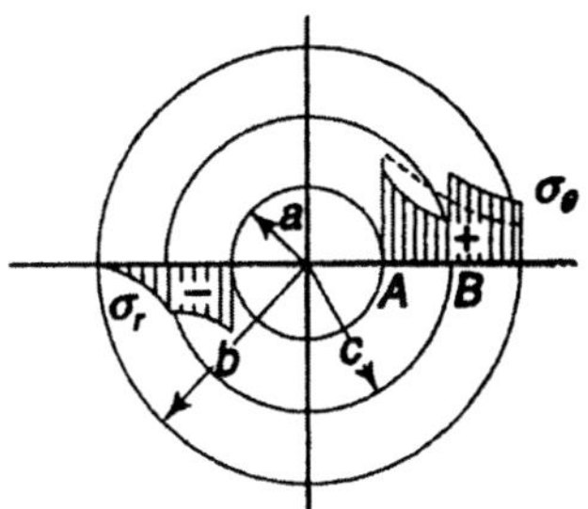

Fig. 7

If the strengths of the two cylinders are the same, then according to the maximum shear stress theory, $(\sigma_1 - \sigma_3)$ at point A of the inner cylinder should be equal to $(\sigma_1 - \sigma_3)$ at point B of the outer cylinder. σ_1 and σ_3 are the maximum and minimum normal stresses, which are respectively equal to σ_θ and σ_r.

At point A, due to internal pressure p, from Eqs (13) and (14),

$$(\sigma_\theta - \sigma_r)_A = p\frac{b^2 + a^2}{b^2 - a^2} - (-p)$$

$$= 2p\frac{b^2}{b^2 - a^2}$$

Because of shrink-fitting pressure p_c at the same point, from Eqs (8.16) and (17),

$$(\sigma_\theta - \sigma_r)_A = -2p_c\frac{c^2}{c^2 - a^2}$$

Hence, the resultant value of $(\sigma_\theta - \sigma_r)$ at A is

$$(\sigma_\theta - \sigma_r)_A = -2p\frac{b^2}{b^2 - a^2} - 2p_c\frac{c^2}{c^2 - a^2} \tag{28}$$

At point B of the outer cylinder, since the composite involves the same material, due to the pressure p, form Eqs (30 and (14), and observing that $r = c$ in these equations,

$$(\sigma_\theta - \sigma_r)_B = p\left[\frac{a^2(c^2 + b^2)}{c^2(b^2 - a^2)} - \frac{a^2(c^2 - b^2)}{c^2(b^2 - a^2)}\right]$$

$$= 2p\frac{a^2b^2}{c^2(b^2-a^2)}$$

At the same point B, due to the contact pressure p_c, form Eqs (13) and (14), with internal radius equal to c and external radius b,

$$(\sigma_\theta - \sigma_r)_B = p_c\left[\frac{c^2+b^2}{b^2-c^2} - \frac{c^2-b^2}{b^2-c^2}\right]$$

$$= 2p_c\frac{b^2}{b^2-c^2}$$

The resultant value of $(\sigma_\theta - \sigma_r)$ at B is therefore

$$(\sigma_\theta - \sigma_r)_B = 2p\frac{a^2b^2}{c^2(b^2-a^2)} + 2p_c\frac{b^2}{(b^2-c^2)} \tag{29}$$

For equal strength, equating Eqs (28) and (29)

$$2p\frac{b^2}{(b^2-a^2)} - 2p_c\frac{c^2}{(c^2-a^2)} = 2p\frac{a^2b^2}{c^2(b^2-a^2)} + 2p_c\frac{b^2}{(b^2-c^2)}$$

or $$p_c\left[\frac{b^2}{b^2-c^2} + \frac{c^2}{(c^2-a^2)}\right] = p\left[\frac{b^2}{(b^2-a^2)} - \frac{a^2b^2}{c^2(b^2-a^2)}\right] \tag{30}$$

The shrink-fitting pressure p_c is related to the negative allowance Δ through Eq. (27) and it is this value of Δ that is required now for equal strength. Hence, substituting for p_c from Eq. (27), Eq. (30) becomes

$$\frac{\Delta E}{2c^2}\frac{(c^2-a^2)(b^2-c^2)}{(b^2-a^2)}\left[\frac{b^2}{(b^2-c^2)} + \frac{c^2}{(c^2-a^2)}\right] = p\left[\frac{b^2}{(b^2-a^2)} - \frac{a^2b^2}{c^2(b^2-a^2)}\right]$$

or $$\frac{\Delta E}{2c^3}\frac{(2b^2c^2-b^2a^2-c^4)}{(b^2-a^2)} = \frac{pb^2(c^2-a^2)}{c^2(b^2-a^2)}$$

or $$\Delta = \frac{2p}{E}\frac{b^2c(c^2-a^2)}{b^2(c^2-a^2)-c^2(b^2-c^2)} \tag{31a}$$

Also, from Eq. (30),

$$p_c = p\frac{b^2(c^2-a^2)^2(b^2-c^2)}{c^2(b^2-a^2)[b^2(c^2-a^2)+c^2(b^2-c^2)]} \tag{31b}$$

The value of $(\sigma_\theta - \sigma_r)$ either at A or at B, from Eqs (28) and (31), is

$$\sigma_\theta - \sigma_r = p\frac{2b^2}{(b^2-a^2)}\left[1 - \frac{(c^2-a^2)(b^2-c^2)}{b^2(c^2-a^2)+c^2(b^2-c^2)}\right]$$

or $$\sigma_\theta - \sigma_r = p\frac{2b^2}{(b^2-a^2)}\left[1-\frac{1}{\dfrac{b^2}{b^2-c^2}+\dfrac{c^2}{c^2-a^2}}\right] \qquad (32)$$

Therefore, for composites made of the same material, in order to have equal strength according to the shear stress theory, the shrink-fit allowance Δ that is necessary is given by Eq. (31a), and this depends on the internal pressure p. Further, Δ depends upon the difference between the external radius of the inner cylinder and the internal radius of the jacked. In other words, this depends on $c(+)$ and $c(-)$. With a, b and p fixed, one can determine the minimum value of $(\sigma_\theta - \sigma_r)$ at A and B. From Eq. (32), the minimum value of $(\sigma_\theta - \sigma_r)$ is obtained when the denominator of the second expression within the square brackets is a maximum, i.e. when D is a maximum, where

$$D = \frac{b^2}{b^2-c^2}+\frac{c^2}{c^2-a^2}$$

Differentiating with respect to c and equating the differential to zero,

$$\frac{dD}{dc} = \frac{2cb^2}{(b^2-c^2)}+\frac{2c(c^2-a^2)-2c^3}{(c^2-a^2)^2} = 0$$

Simplifying, one gets

$$c = \sqrt{ab}$$

The corresponding value of $(\sigma_\theta - \sigma_r)$, form Eq. (32), is

$$(\sigma_\theta - \sigma_r)_{\min} = p\frac{2b^2}{(b^2-a^2)}\left[1-\frac{1}{\dfrac{b^2}{b(b-a)}+\dfrac{ab}{a(b-a)}}\right]$$

$$= p\frac{2b^2}{(b^2-a^2)}\left[1-\frac{(b-a)}{2b}\right]$$

or $$(\sigma_\theta - \sigma_r)_{\min} = p\frac{b}{(b-a)} \qquad (33)$$

Also, the optimum value of Δ is from Eq. (31a),

$$\Delta_{\text{opt}} = \frac{1}{E}pc = \frac{p}{E}\sqrt{ab} \qquad (34)$$

Example 4:

Determine the diameters 2c and 2b and negative allowance Δ for a two-layer barrel of inner diameter 2a = 100 mm. The maximum pressure the *barrel is to withstand is p_{max} = 2000 kgf/cm² (196000 kPa). The material is steel with E = $2(10)^6$ kgf/cm² (196 × 10^5 kPa); σ_{yp} in tension or compression is 6000 kgf/cm² (588 × 10^3 kPa). The factor of safety is 2.*

Solution:

From Eq. (33),

$$\frac{6000}{2} = 2000\frac{b}{b-a}$$

Therefore, $b = 3a$

Since $c = \sqrt{ab}.c = \sqrt{3}a.$ The numerical values are therefore, $2a$ = 100 mm, $2b$ = 300 mm, $2c$ = 173 mm. With $c = ab$, the value of Δ is, from Eq. (8.34),

$$\Delta = \frac{P}{E}\sqrt{ab} = \frac{2000}{2\times10^6}\sqrt{(50\times150)} = 0.866 \text{ mm}$$

Example 5:

A steel shaft of 10 cm diameter is shrunk inside a bronze cylinder of 25 cm outer diameter. The shrink allowance is 1 part per 1000 (i.e. 0.005 cm difference between the radii). Find the tangential stress in the bronze cylinder at the inside and outer radii and the stress in the shaft.

E_{steel} = *2.18 × 10^6 kgf/cm² (214 × 10^6 kPa)*

E_{bronze} = *1.09 × 10^6 kgf/cm² (107 × 10^6 kPa)*

and V = *0.3 for both metals.*

Solution:

In Eq. (26),

$$a = 0, \quad c = 5, \quad b = 12.5, \ \Delta = 0.005, \ V_1 = V_2 = 0.3$$

Substituting in Eq. (26a),

$$\frac{5P_c}{2.18\times10^6\times25}(0.7\times25)+\frac{5P_c}{1.09\times10^6\times(156.25-25)}$$

$$\times (0.7 \times 25 + 1.3 \times 156.25) = 0.005$$

or, P_c = 610 kgf/cm² (59780 kPa)

For the bronze tube, the circumferential stress is, from Eq. (14),

$$\sigma_\theta = \frac{610\times25}{(156.25-25)}\left(1+\frac{156.25}{r^2}\right)$$

when r = 5 cm and r = 12.5 cm

σ_θ = 842.4 kgf/cm² (82555 kPa)

σ_θ = 232.4 kgf/cm² (22775 kPa)

The shaft experiences equal σ_r and σ_θ at every point, from Eqs (16) and (17). Hence,

$$\sigma_r = \sigma_\theta = -610 \text{ kgf/cm}^2 \text{ (59780 kPa)}$$

Example 6:

A compound cylinder made of copper inner tube of radii a = 10 cm and c = 20 cm is snug fitted (Δ = 0) inside a steel jacket of external radius b = 40 cm. If the compound cylinder is subjected to an internal pressure p = 1500 kgf/cm² (147009 kPa), determine the contact pressure p_c and the values of σ_r and σ_θ at the inner and external points of the inner cylinder and of the jacket. Use the following data:

$$E_{st} = 2 \times 10^6 \text{ kgf/cm}^2 \text{ } (196 \times 10^6 \text{ kPa}),$$

$$E_{cu} = 1\times 10^6 \text{ kgf/cm}^2 \text{ } (98 \times 10^6 \text{ kPa}), \quad V_{st} = 0.3, \quad V_{cu} = 0.34$$

Solution:

Since the initial shrink-fit allowance Δ is zero, the initial contact pressure is zero. When the compound cylinder is subjected to an internal pressure p, the increase in the external radius of the upper cylinder under p and contact pressure p_c should be equal to the increase in the internal radius of the jacker under the contact pressure p_c,

i.e. $[(u_r)_p + (u_r)_{pc} \text{ at } r = c]_{cu} = [(u_r)_{pc} \text{ at } r = c]_{st}$

For copper cylinder, from Eq. (10),

$$(u_r)_p = \frac{1-V_{cu}}{E_{cu}}\frac{pa^2c}{(c^2-a^2)} + \frac{1+V_{cu}}{E_{cu}}\frac{a^2c^2}{c}\frac{p}{(c^2-a^2)}$$

$$(u)_{pc} = -\frac{1-V_{cu}}{E_{cu}}\frac{p_c c^3}{(c^2-a^2)} - \frac{1+V_{cu}}{E_{cu}}\frac{a^2c^2}{c}\frac{p_c}{(c^2-a^2)}$$

$$(u_r)_{\text{total}} = \frac{2pa^2c}{E_{cu}(c^2-a^2)} - \frac{p_c c}{E_{cu}(c^2-a^2)}\left[c^2(1-V_{cu}) + a^2(1+V_{cu})\right]$$

For steel jacket, from Eq. (10),

$$(u)_{p_c} = \frac{1-V_{st}}{E_{st}}\frac{p_c c^3}{(b^2-c^2)} + \frac{1+V_{st}}{E_{st}}\frac{c^2b^2}{c}\frac{p_c}{(b^2-c^2)}$$

$$= \frac{p_c c}{E_{st}(b^2-c^2)}\left[c^2(1-V_{st}) + b^2(1+V_{st})\right]$$

Equating the (u_r)s

$$\frac{2pa^2c}{E_{cu}(c^2-a^2)}-\frac{p_c c}{E_{cu}(c^2-a^2)}\left[c^2(1-V_{cu})+a^2(1+V_{cu})\right]$$

$$=\frac{p_c c}{E_{st}(b^2-c^2)}\left[c^2(1-V_{st})+b^2(1+V_{st})\right]$$

or $$p_c=\frac{(c^2+a^2)-V_{cu}(c^2-a^2)}{E_{cu}\,(c^2-a^2)}+\frac{(b^2+c^2)+V_{st}(b^2-c^2)}{E_{st}(b^2-c^2)}$$

$$=p\frac{2a^2}{E_{cu}(c^2-a^2)}$$

With p = 1500 kgf/cm^2,

a = 10, c = 20,

b = 40, V_{st} = 0.3 V_{cu} = 0.34,

$$p_c=\left[\frac{500-300\times 0.34}{300\times 10^6}+\frac{2000+1200\times 0.3}{2\times 1200\times 10^6}\right]=\frac{3000\times 100}{300\times 10^6}$$

$\therefore$ p_c = 433 kgf/cm^2 (42453 kPa)

Now, p_c will act as an external pressure on the copper tube and as an internal pressure on the steel jacket. For copper tube, from Eqs (11) and (12),

(i) Inner surface:,

σ_r at $r = a$ is – 1500 kgf/cm^2 and,

σ_θ at $r = a$ is

$$=\frac{(1500\times 100)-(433\times 400)}{(400-100)}+\frac{100\times 400}{100}\times\frac{(1500-433)}{(400-100)}$$

= 1357 kgf/cm^2 (compressive)

(ii) Outer surface:

σ_r at $r = c$ is – 433 kgf/cm^2 and,

σ_θ at $r = c$ is

$$=\frac{(1500\times 100)-(433\times 400)}{(400-100)}+\frac{100\times 400}{400}\times\frac{(1500-433)}{(400-100)}$$

= 279 kgf/cm^2

For steel jacket, from Eqs (11) and (12),

(i) Inner surface:

σ_r at $r = c$ is – 433 kgf/cm^2 and,

σ_θ at $r = c$ is

$$= \frac{(433 \times 400)}{(1600 - 400)} = \frac{400 \times 1600}{400} \times \frac{433}{(1600 - 400)}$$

$$= 722 \text{ kgf/cm}^2$$

(ii) Outer surface:

σ_r at $r = b$ is zero and,

σ_θ at $r = b$ is

$$= \frac{(433 \times 400)}{(1600 - 400)} + \frac{400 \times 1600}{1600} \times \frac{433}{(1600 - 400)}$$

$$= 289 \text{ kgf/cm}^2$$

SPHERE WITH PURELY RADIAL DISPLACEMENTS

Consider a uniform sphere or spherical shell subjected to radial forces only, such as internal or external pressures. The sphere or the spherical shell will then undergo radial displacements only. Consider a particle situated at radius r before deformation. After deformation, the spherical surface of radius r becomes a surface of radius $(r + u_r)$ and the particle undergoes a displacement u_r. Similarly, another particle at distance $(r + \Delta r)$ along the same radial line will undergo a displacement

$$\left(u_r + \frac{\partial u_r}{\partial r} \Delta r \right)$$

Hence, the radial strain is

$$e_r = \frac{\partial u_r}{\partial r}$$

Before deformation, the circumference of any great circle on the surface of radius r is $2\pi r$. After deformation, the radius becomes $(r + u_r)$ and the circumference of the great circle is $2\pi\,(r + u_r)$ Hence, the circumferential strain is

$$\varepsilon_\phi = \frac{2\pi(r + u_r) - 2\pi r}{2\pi r} = \frac{u_r}{r}$$

This is the strain in every direction perpendicular to the radius r. Because of complete symmetry, we can choose a frame of reference, as shown in Fig. 8.

This the three extensional strains along the three axes are

$$\varepsilon_r = \frac{\partial u_r}{\partial r}, \quad \varepsilon_\theta = \frac{u_r}{r}, \quad \varepsilon_\phi = \frac{u_r}{r} \tag{35}$$

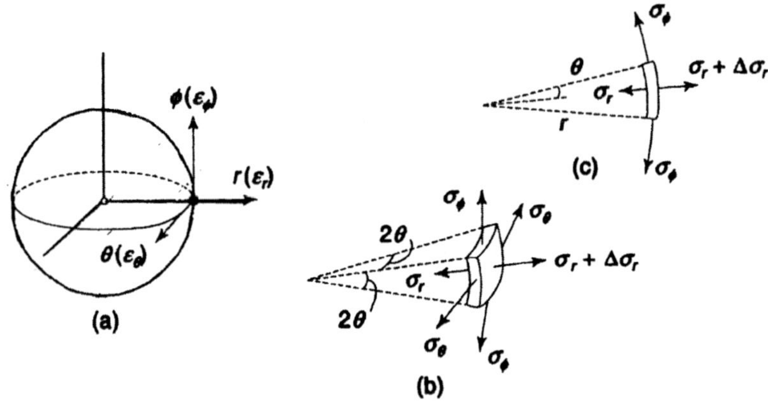

Fig. 8

Because of symmetry, there are no shear stresses and shear strains. Let γ_r be the body force per unit volume in the radial direction.

The stress equations of equilibrium can also be derived easily. Consider a spherical element of thickness Δr at distance r, subtending a small angle 2θ at the centre, Because of spherical symmetry, $\sigma_\theta = \sigma_\phi$. For equilibrium in the radial direction,

$$-\sigma_r(2\theta r)(2\theta r) + (\sigma_r + \Delta\sigma_r)(r + \Delta r)2\theta\,(r + \Delta r)2\theta -$$

$$2 = \left(r + \frac{\Delta r}{2}\right)2\theta\Delta r\sigma_\phi \sin\theta = -2\left(r + \frac{\Delta r}{2}\right)2\theta\,\Delta r\sigma_\theta \sin\theta + \gamma_r 4\theta^2 r^2 \Delta r = 0$$

putting $\sigma_\theta = \sigma_\phi$ and $\Delta\sigma_r = \frac{\partial\sigma_r}{\partial r}\Delta r$, the sbove equation reduces in the limit to $r^2\frac{\partial\sigma_r}{\partial r} + 2r\sigma_r - 2r\sigma_\phi + r^2\gamma_r = 0$

Since r is the only independent variable, the above equation can be rewritten as

$$\frac{d}{dr}(r^2\sigma_r) - 2r\sigma_\phi + r^2\gamma_r = 0 \tag{36}$$

If body force is ignored,

$$\frac{1}{r^2}\frac{d}{dr}(r^2\sigma_r) = \frac{2}{r}\sigma_\phi \tag{37}$$

From Hooke's law

$$\varepsilon_r = \frac{1}{E}\left[\sigma_r - V(\sigma_\theta + \sigma_\phi)\right]$$

or $$\frac{du_r}{dr} = \frac{1}{E}\left[(\sigma_r - 2V\sigma_\phi)\right] \tag{38}$$

and

$$\varepsilon_\phi = \frac{1}{E}\left[\sigma_\phi - V(\sigma_\phi + \sigma_r)\right]$$

or

$$\frac{u_r}{r} = \frac{1}{E}\left[(1-V)\sigma_\phi + V\sigma_r\right] \tag{39}$$

Equations (37) – (39) can be solved. From Eq. (39)

$$u_r = \frac{1}{E}\left[(1-V)r\sigma_\phi + Vr\sigma_r\right]$$

Differentiating with respect to r

$$\frac{du_r}{dr} = \frac{1}{E}\left[(1-V)\frac{\partial(r\sigma_\phi)}{\partial r} - V\frac{\partial(r\sigma_r)}{\partial r}\right]$$

Subtracting the above equation from Eq. (38)

$$0 = -(1-V)\frac{d(r\sigma_\phi)}{dr} + V\frac{d(r\sigma_r)}{dr} + \sigma_r - 2v\sigma_\phi$$

Substituting for σ_ϕ from Eq. (36)

$$\frac{1}{2}(1-V)\frac{d^2(r^2\sigma_r)}{dr^2} - V\frac{d(r\sigma_r)}{dr} - \sigma_r + \frac{V}{r}\frac{d(r^2\sigma_r)}{dr} = 0 \tag{40}$$

If $r^2\ \sigma_r = y$,

$$\frac{d}{dr}(r\sigma_r) = \frac{d}{dr}\left(\frac{y}{r}\right) = \frac{1}{r}\frac{dy}{dr} - \frac{1}{r^2}y$$

Therefore, Eq. (40) becomes

$$\frac{1}{2}(1-V)\frac{d^2y}{dr^2} - \frac{V}{r}\frac{dy}{dr} + \frac{V}{r^2}\frac{y}{} - \frac{y}{r^2} + \frac{V}{r}\frac{dy}{dr} = 0$$

or $$\frac{d^2y}{dr^2} - 2\frac{y}{r^2} = 0 \tag{41}$$

This is a homogeneous linear equation with the solution

$$y = Ar^2 + \frac{B}{r}$$

where A and B are constants. Hence,

$$\sigma_r = A + \frac{B}{r^3} \tag{8.42}$$

And from Eq. (37)

$$\sigma_\phi = \frac{1}{2r}\frac{d}{dr}\left(Ar^2 + \frac{B}{r}\right) = A - \frac{B}{2r^3} \tag{43}$$

The constants A and B are determined from the boundary conditions.

Problem of Thick Hollwo Sphere

Consider a spherical body formed by the boundaries of two spherical surfaces of radii a and b respectively. Let the hollow sphere be subjected to an internal pressure p_a and an external pressure p_b. The boundary conditions are therefore

$\sigma_r = -p_a$ when $r = a$, and $\sigma_r = -p_b$ when $r = b$

$$-p_a = A + \frac{B}{a^3} \text{ and } -p_b = A + \frac{B}{b^3}$$

Solving,

$$A = -\frac{b^3 p_b - a^3 p_a}{b^3 - a^3}, \quad B = \frac{a^3 b^3}{b^3 - a^3}(p_b - p_a)$$

Thus, the general expressions for σ_r and σ_ϕ are

$$\sigma_r = \frac{1}{b^3 - a^3}\left[-b^3 p_b + a^3 p_a + \frac{a^3 b^3}{r^3}(p_b - p_a)\right] \tag{44}$$

$$\sigma_\phi = \sigma_\theta = \frac{1}{b^3 - a^3}\left[-b^3 p_b + a^3 p_a - \frac{a^3 b^3}{2r^3}(p_b - p_a)\right] \tag{45}$$

If the sphere is subjected to internal pressure only, $p_b = 0$, and

$$\sigma_r = p_a \frac{a^3}{b^3 - a^3}\left(1 - \frac{b^3}{r^3}\right) \tag{46}$$

$$\sigma_\phi = \sigma_\theta = p_a \frac{a^3}{b^3 - a^3}\left(1 + \frac{b^3}{2r^3}\right) \tag{8.47}$$

The above two equations can also be written as

$$\sigma_r = p_a \frac{a^3}{1 - (a^3/b^3)}\left(\frac{1}{b^3} - \frac{1}{r^3}\right)$$

$$\sigma_\phi = \sigma_\theta = p_a \frac{a^3}{1 - (a^3/b^3)}\left(\frac{1}{b^3} + \frac{1}{2r^3}\right)$$

In the case of a cavity inside an infinite or a large medium, $b \to \infty$ and the above equations reduce to

$$\sigma_r = -p_a \frac{a^3}{r^3} \tag{48}$$

$$\sigma_\phi = \sigma_\theta = +p_a \frac{a^3}{2r^3} \tag{49}$$

The above equations can also be used to calculate stresses in a body of any shape with a spherical hole under an internal pressure p_a, provided the outer surface of the body is free from pressure and provided that every point of this outer surface is at a distance greater than four or five times the diameter of the hole from its centre.

Example 7:

Calculate the thickness of the shell of a bomb calorimeter of spherical form of 10 cm inside diameter if the working stress is σ kgf/cm² (98 σ kPa) and the internal pressure is $\sigma/2$ kgf/cm² (49 σ kPa).

Solution:

From Eqs (46) and (47) , the maximum tensile stress is due σ_ϕ, which occurs at $r = a$. Hence,

$$\sigma_\phi = \frac{\sigma}{2}\frac{5^3}{b^3 - 5^3}\left(1 + \frac{b^3}{2\times 5^3}\right)$$

Equating this to the working stress σ

$$\frac{5^3}{b^3 - 5^3}\left(1 + \frac{b^3}{2\times 5^3}\right) = 2$$

$\therefore \qquad b \approx 6.3$ cm

Hence, the thickness of the shell is 1.3 cm.

Example 8:

Express the stress equation of equilibrium, i.e.

$$\frac{1}{r^2}\frac{d}{dr}(r^2\sigma_r) - \frac{2}{r}\sigma_\phi = 0$$

given by Eq. (37), in terms of the displacement component u_r using Hooke's law and strain-displacement relations.

Solution:

We have $\quad \varepsilon_r = \dfrac{1}{E}\left[\sigma_r - 2V\sigma_\phi)\right]$

$$\varepsilon_\phi = \frac{1}{E}\left[(1 - V)\sigma_\phi - V\sigma_r)\right]$$

Solving for σ_r and σ_ϕ,

$$\sigma_r = \frac{E}{(1+V)(1-2V)}\left[\varepsilon_r(1+V)+2V\varepsilon_\phi\right] \tag{50}$$

$$\sigma_\phi = \frac{E}{(1+V)(1-2V)}(V\,\varepsilon_r+\varepsilon_\phi) \tag{51}$$

Using the strain-displacement relations

$$\varepsilon_r = \frac{du_r}{dr} = \quad \text{and} \quad \varepsilon_\phi = \frac{u_r}{r}$$

and substituting in the equilibrium equation, we get

$$\frac{E(1-V)}{(1+V)(1-2V)}\left[\frac{d^2u_r}{dr^2}+\frac{2}{r}\frac{du_r}{dr}-\frac{2}{r^2}u_r\right]=0$$

or

$$\frac{E(1-V)}{(1+V)(1-2V)}\frac{d}{dr}\left(\frac{du_r}{dr}+2\frac{u_r}{r}\right)=0 \tag{52}$$

STRESSES DUE TO GRAVITATION

When body forces are operat:ve, the stress equation of equilibrium is, from Eq. (36),

$$\frac{1}{r^2}\frac{d}{dr}\left(r^2\sigma_r\right)-\frac{2}{r}\sigma_\phi+\gamma_r=0 \tag{53}$$

where γ_r is the body force per unit volume. The problem of a sphere strained by the mutual gravitation of its parts will now be considered. It is known from the theory of attractions

$$\gamma_r = -\rho g\frac{r}{a}$$

where a is the radius of the sphere, ρ is the mass density, r is the radius of any point from the centre and g is the acceleration due to gravity. Expressing the equations of equilibrium in terms of stresses [Eq. (52)], we have

$$\frac{E(1-V)}{(1+V)(1-2V)}\frac{d}{dr}\left(\frac{du_r}{dr}+2\frac{u_r}{r}\right)-\rho g\frac{r}{a}=0 \tag{54}$$

The complementary solution is

$$u_r = C_r + \frac{C_1}{r}$$

and the particular solution is

$$u_r = \frac{1}{10}\frac{(1+V)(1-2V)}{E(1-V)a}\rho g r^3$$

Hence, the complete solution is

$$u_r = C_r + \frac{C_1}{r} + \frac{1}{10}\frac{(1+V)(1-2V)}{E(1-V)a}\rho g r^3$$

For a solid sphere, C_1 should be equal to zero as otherwise the displacement will become infinite at $r = 0$. The remaining constant is determined from the boundary condition $\sigma_r = 0$ at $r = a$. From the general solution

$$\frac{du_r}{dr} = C + \frac{3(1+V)(1-2V)}{10E(1-V)a}\rho g r^3, \quad \frac{u_r}{r} = C + \frac{(1+V)(1-2V)}{10E(1-V)a}\rho g r^2$$

and from Eq. (50)

$$\sigma_r = \frac{E}{(1+V)(1-2V)}\left[\varepsilon_r(1-V) + 2V\varepsilon_\phi\right]$$

$$= \frac{E}{(1+V)(1-2V)}\left[\frac{du_r}{dr}(1-V) + 2V\frac{u_r}{r}\right]$$

$$= \frac{E}{(1+V)(1-2V)}\left[C(1-V) + \frac{3(1+V)(1-2V)}{10Ea}\rho g r^2 + 2VC + \frac{V(1+V)(1-2V)}{5E(1-V)a}\rho g r^2\right]$$

$$= \frac{E}{(1+V)(1-2V)}\left[C(1+V) + \frac{(3-V)(1+V)(1-2V)}{10Ea(1-V)}\rho g r^2\right]$$

Form the boundary condition $\sigma_r = 0$ at $r = a$,

$$C = \frac{(3-V)(1-2V)}{10E(1-V)}\rho g a$$

Hence, $$\sigma_r = -\frac{1}{10}\frac{(3-V)}{(1-V)}(a^2 - r^2)\frac{\rho g}{a} \tag{8.55}$$

and from Eq. (53)

$$\sigma_\phi = \sigma_\theta = -\frac{1}{10}\frac{(3-V)a^2 - (1+2V)r^2}{(1-V)}\frac{\rho g}{a} \tag{56}$$

It will be observed that both stress components σ_r and σ_θ are compressive at every point. At the centre ($r = 0$), they are equal and have a magnitude

$$\sigma_r = \sigma_\phi = \sigma_\theta = \frac{1}{10}\frac{3-V}{1-V}\rho g a \text{ (compressive)}$$

Further,

$$\frac{du_r}{dr} = C + \frac{1}{10}\frac{3(1+V)(1-2V)}{E(1-V)a}\rho g r^2$$

$$= -\frac{1}{10}\frac{(3-V)(1-2V)}{E(1-V)}\rho ga + \frac{1}{10}\frac{3(1+V)(1-2V)}{E(1-V)a}\rho gr^2$$

$$= \frac{1}{10}\frac{(1+V)(1-2V)}{E(1-V)}\frac{\rho g}{a}\left[3r^2 - \frac{(3-V)}{(1+V)}a^2\right]$$

The above value is zero when

$$r^2 = \frac{(3-V)}{3(1+V)}a^2$$

Hence, if V is positive (which is true for all known materials), there is a definite surface outside which the radial strain is an extension. In other words, for

$$r > a\left[\frac{(1+V)}{3(1+V)}\right]^{1/2}$$

the radial strain ε_r is positive though the radial stress σ_r is compressive everywhere. This result is due, of course, to the 'Poisson effect' of the large circumferential stress, i.e. hoop stress, which is compressive.

ROTATING DISKS OF UNIFORM THICKNESS

We shall now consider the stress distribution in rotating circular disks which are thin. We assume that over the thickness, the radial and circumferential stresses do not vary and that the stress σ_z in the axial direction is zero. The equation of equilibrium given by Eq. (5b) can be used, provided we add the inertia force term $\rho\omega^2 r$, i.e. in the general equation of equilibrium [Eq. (2)] we put the body force term equal to the inertia term $\rho\omega^2 r$, where ω is the angular velocity of the rotating disk and ρ is the density of the disk material. The z-axis is the axis of rotation. Then;

$$\frac{\partial\sigma_r}{\partial r} + \frac{\sigma_r - \sigma_\theta}{r}\rho\omega^2 r = 0 \tag{57a}$$

or

$$\frac{d}{dr}(r\sigma_r) - \sigma_\theta + \rho\omega^2 r^2 = 0 \tag{57b}$$

The strain components are, as before,

$$\varepsilon_r = \frac{du_r}{dr} \quad \text{and} \quad \varepsilon_\theta = \frac{u_r}{r} \tag{58}$$

From Hooke's law, with $\sigma_z = 0$,

$$\varepsilon_r = \frac{1}{E}(\sigma_r - V\sigma_\theta)$$

$$\varepsilon_\theta = \frac{1}{E}(\sigma_\theta - V\sigma_r)$$

From Eq. (58)

$$\varepsilon_r = \frac{d}{dr}(r\varepsilon_\theta)$$

Using Hooke's law

$$\frac{1}{E}(\sigma_r - V\sigma_\theta) = \frac{1}{E}\frac{d}{dr}(r\sigma_\theta - Vr\sigma_r) \tag{59}$$

Let $\quad r\sigma_r = y \qquad$ (60a)

Then, from Eq. (57b)

$$\sigma_\theta = \frac{dy}{dr} + \rho\omega^2 r^2 \tag{60b}$$

Substituting these in Eq. (59)

$$r^2\frac{d^2y}{dr^2} + r\frac{dy}{dr} - y + (3+V)\rho\omega^2 r^3 = 0 \tag{61}$$

The solution of the above differential equation is

$$y = Cr + C_1\frac{1}{r} - \frac{(3+V)}{8}\rho\omega^2 r^3 \tag{62}$$

From Eq. (60)

$$\sigma_r = C + C_1\frac{1}{r^2} - \frac{(3+V)}{8}\rho\omega^2 r^3 \tag{63}$$

$$\sigma_\theta = C - C_1\frac{1}{r^2} - \frac{(1+3V)}{8}\rho\omega^2 r^3 \tag{64}$$

The integration constants are determined from boundary conditions.

Solid Disk

For a solid disk, we must take $C_1 = 0$, since otherwise the stresses σ_r and σ_θ become infinite at the centre. The constant C is determined from the condition at the periphery ($r = b$) of the disk. If there are no forces applied there, then,

$$(\sigma_r)_{r=b} = C - \frac{3+V}{8}\rho\omega^2 b^2 = 0$$

Hence,

$$C = \frac{3+V}{8}\rho\omega^2 b^2$$

and the stress components become

$$\sigma_r = \frac{3+V}{8}\rho\omega^2(b^2 - r^2) \tag{65a}$$

$$\sigma_\theta = \frac{3+V}{8}\rho\omega^2 b^2 - \frac{1+3V}{8}\rho\omega^2 r^2 \tag{65b}$$

These stresses attain their maximum values at the centre of the disk, where

$$\sigma_r = \sigma_\theta = \frac{3+V}{8}\rho\omega^2 b^2 \tag{66}$$

Circular Disk with a Hole of Radius *a*

If there are no forces applied at the boundaries a and b, then

$$(\sigma_r)_{r=a} = 0, \quad (\sigma_r)_{r=b} = 0$$

from which we find that

$$C = \frac{3+V}{8}\rho\omega^2\left(b^2 + a^2\right), \quad C_1 = -\frac{3+V}{8}\rho\omega^2 a^2 b^2$$

Substituting these in Eqs (63) and (64)

$$\sigma_r = \frac{3+V}{8}\rho\omega^2\left(b^2 + a^2 - \frac{a^2b^2}{r^2} - r^2\right) \tag{67}$$

$$\sigma_\theta = \frac{3+V}{8}\rho\omega^2\left(b^2 + a^2 + \frac{a^2b^2}{r^2} - \frac{1+3V}{3+V}r^2\right) \tag{68}$$

The radial stress σ_r reaches its maximum at $r = \sqrt{ab}$ where

$$(\sigma_r)_{max} = \frac{3+V}{8}\rho\omega^2(b-a)^2 \tag{69}$$

The maximum circumferential stress is at the inner boundary, where

$$(\sigma_r)_{max} = \frac{3+V}{8}\rho\omega^2\left(b^2 + \frac{1-V}{3+V}a^2\right) \tag{70}$$

It can be seen that $(\sigma_\theta)_{max}$ is greater than $(\sigma_r)_{max}$.

When the radius a of the hole approaches zero, the maximum circumferential stress approaches a value twice as great as that for a solid disk [Eq. (66)]. In other words, by making a small circular hole at the centre of a solid rotating disk, we double the maximum stress.

The displacement u_r for all the cases considered above can be calculated from Eq. (58), i.e.

$$u_r = r\varepsilon_\theta = \frac{r}{E}(\sigma_\theta - V\sigma_r) \tag{71}$$

Example 9:

A flat steel disk of 75 cm outside diameter with a 15 cm diameter hole is shrunk around a solid steel shaft. The shrink-fit allowance is 1 part in 1000 (i.e. an allowance of 0.0075 cm in radius). E = 2.18 × 10⁶ kgf/cm² (214 × 10⁶ kPa).

(i) What are the stresses due to shrink-fit?

(ii) At what rpm will the shrink-fit loosen up as a result of rotation?

(iii) What is the circumferential stress in the disk when spinning at the above speed?

Solution:

Assume that the same equations as for the disk are applicable to the solid rotating shaft also.

(i) To calculate the shrink-fit pressure, we have from Eq. (27)

$$P_c = \frac{2.18\times10^6\times0.0075}{2\times7.5^3}\times\frac{(7.5^2-0)(37.5^2-7.5^2)}{(37.5^3-0)}$$

or $\qquad P_c$ = 1044 kgf/cm² (102312 kPa)

The tangential stress at the hole will be the largest stress in the system and from Eq. (24)

$$\sigma_\theta = \frac{1044\times7.5^2}{(37.5^2-7.5^2)}\left(1+\frac{37.5^2}{7.5^2}\right)$$

(ii) When the shrink-fit loosens up as a result of rotation, there will no radial pressure on any boundary. When the shaft and the disk are rotating, the radial displacement of the disk at the hole will be greater than the radial displacement of the shaft at its boundary. The difference between these two radial displacements should equal $\Delta = 0.0075$ cm at 7.5 cm radius. From Eqs (71), (67) and (68)

$$u_{\text{disk}} = \frac{r}{E}(\sigma_\theta - V\sigma_r)$$

$$= \frac{r}{E}\frac{3+V}{8}\rho\omega^2\left[b^2+a^2+\frac{a^2b^2}{r^2}-\frac{1+3V}{3+V}r^2\right.$$

$$\left.-V\left(b^2+a^2-\frac{a^2b^2}{r^2}-r^2\right)\right]$$

$$= \frac{r}{E}\frac{(3+V)(1-V)}{8}\rho\omega^2\left(b^2+a^2+\frac{1+V}{1-V}\frac{a^2b^2}{r^2}-\frac{1+V}{3+V}r^2\right)$$

$$= \frac{7.5^2}{2.18\times10^6}\times\frac{3.3\times0.7}{8}\rho\omega^2$$

$$\times\left(37.5^2+7.5^2+\frac{1.3}{0.7}\times\frac{37.5^2\times7.5^2}{7.5^2}-\frac{1.3}{3.3}\times7.5^2\right)$$

$= 4052 \times 10^{-6}\ rw^2$

From Eqs (71), (65a) and (65b)

$$u_{\text{shaft}} = \frac{r}{E}(\sigma_\theta - V\sigma_r)$$

$$= \frac{1-V}{8}\rho\omega^2\ r\left[(3+V)b^2-(1+V)r^2\right]$$

$$= \frac{0.7}{8\times2.18\times10^6}\rho\omega^2\times7.5(3.3\times7.5^2-1.3\times7.5^2)$$

$= 434 \times 10^6\ rw^2$

Therefore,

$= (4052–34) \times 10^{-6}\ rw^2 = 0.0075$

or

$$\omega^2 = 0.0075\times10^6\times\frac{1}{4018}\times\frac{981}{0.0081}$$

$= 226066\ (\text{rad/s})^2$

Therefore,

$\omega = 475$ rad/s or 7536 rpm

(iii) The stresses in the disk can be calculated from Eq. (68)

$$\sigma_r = \frac{3.3}{8}\rho\omega^2\left(37.5^2+7.5^2+37.5^2-\frac{1.9}{3.3}\times7.5^2\right)$$

$= 1170\ rw^2$

$$= 1170\times\frac{0.0081}{981}\times226066$$

$= 2184$ kgf/cm^2 (214024 kPa)

Example 10:

A flat steel turbine disk of 75 cm outside diameter and 15 cm inside diameter rotates at 3000 rpm, at which speed the blades and shrouding cause a tensile rim loading of 44kgf/cm² (4312 kPa). The maximum stress at this speed is to be 1164 kgf/cm² (114072 kPa). Find the maximum shrinkage allowance on the diameter when the disk and the shift are rotating.

Solution:

Let c be the radius of the shaft and b that of the disk. From Eq. (70), the maximum circumferential stress due to rotation alone is

$$(\sigma_\theta)_1 = \frac{3+V}{4}\rho\omega^2\left(b^2 + \frac{1-V}{3+V}c^2\right)$$

$$= \frac{3.3}{4}\rho\omega^2\left(37.5^2 + \frac{0.7}{3.3}\times 7.5^2\right)$$

$$= 1170\ \rho\omega^2$$

Owing to shrinkage pressure P_c, and the tensile rim loading P_b, from Eq. (12)

$$(\sigma_\theta)_2 = \frac{p_c c^2}{b^2 - c^2}\left(1 + \frac{b^2}{c^2}\right) + 2\frac{p_b b^2}{b^2 - c^2}$$

$$p_c = \frac{7.5^2}{37.5^2 - 7.5^2}\left(1 + \frac{37.5^2}{7.5^2}\right) + 2\frac{44\times 37.5^2}{37.5^2 - 7.5^2}$$

$$= 1.08P_c + 91.7$$

Hence, the combined stress at 7.5 cm radius is

$$\sigma_\theta = 1170\rho\omega^2 + 1.08P_c + 91.7$$

This should be equal to 1164 kgf/cm². Hence,

$$1.08\ P_c = 1164 - 1170\rho\omega^2 - 91.7$$

$$= 1164 - 1170\times(100\pi)^2\times\frac{0.0081}{981} - 91.7$$

$$= 1164 - 953.5 - 91.7 = 118.8$$

Hence,

$$P_c = 110\ \text{kgf/cm}^2$$

The corresponding shrink–fit allowance is obtained from Eq. (8.27), i.e.

$$110 = \frac{E\Delta}{2\times 7.5^2}\times\frac{7.5^2(37.5^2 - 7.5^2)}{37.5^2}$$

$$= 0.064\ E\Delta$$

or $$\Delta = \frac{110\times 10^{-6}}{0.064\times 2.18} = 0.0008\ \text{cm}$$

DISKS OF VARIABLE THICKNESS

Assuming that the stresses do not vary over the thickness of the disk, the method of analysis developed in the previous section for thin disks of constant thickness can be extended also to disks of variable thickness. Let h be the thickness of the disk, varying with radius r. The equation of equilibrium can be obtained by referring to Fig. 9.

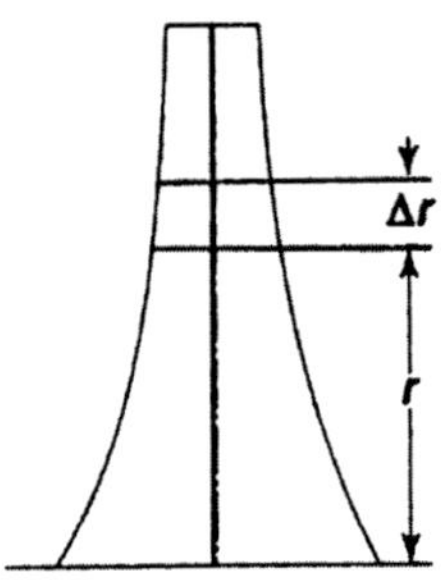

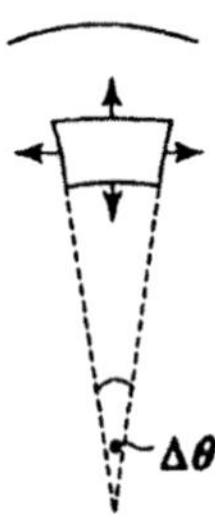

Fig. 9

For equilibrium in the radial direction

$$\left(h\sigma_r + \frac{\partial(h\sigma_r)}{\partial r}\Delta r\right)(r+\Delta r)\Delta\theta + \rho\left(r+\frac{\Delta r}{2}\right)\Delta\theta\left(h+\frac{\Delta h}{2}\right)\omega^2\left(r+\frac{\Delta r}{2}\right) -$$

$$-h\sigma_r\, r\Delta\theta - 2\sigma_\theta\left(h+\frac{\Delta h}{2}\right)\Delta r\sin\frac{\Delta\theta}{2} = 0$$

Simplifying and going to the limit

$$r\frac{d}{dr}(h\sigma_r) + h\sigma_r + \rho\omega^2 r^2 h - \sigma_\theta h = 0$$

or $$\frac{d}{dr}(r\, h\sigma_r) - \sigma_\theta h + \rho\omega^2 r^2 h = 0 \quad (72)$$

Putting $y = rh\sigma_r$ (73a)

$$h\sigma_\theta = \frac{dy}{dr} + h\rho\omega^2 r^2 \quad (73b)$$

The strain components remain as in Eq. (58), i.e.

$$\varepsilon_r = \frac{du_r}{dr} \quad \text{and} \quad \varepsilon_\theta = \frac{u_r}{r}$$

Hence, $\varepsilon_r = \dfrac{d}{dr}(r\varepsilon_\theta)$

and from Hooke's law

$$\frac{1}{E}(\sigma_r - v\sigma_\theta) = \frac{d}{dr}(r\sigma_\theta - vr\sigma_r)$$

Substituting for σ_r and σ_θ from Eqs (73a) and (73b)

$$r^2\frac{r^2 y}{dr^2} + r\frac{dy}{dr} - y + (3+v)\rho\omega^2 hr^3 - \frac{r}{h}\frac{dh}{dr}\left(r\frac{dy}{dr} - vy\right) = 0 \quad (74)$$

In the particular case where the thickness varies according to the equation

$$y = Cr^n$$

in which C is a constant and n any number, Eq. (74) can easily be integrated. The general solution has the form

$$y = mr^{n+2} + Ar^\alpha + Br^\beta$$

in which $m = -\dfrac{(3+v)\rho\omega^2 c}{(vn + 3n + 8)}$

and α and β are the roots of the quadratic equation

$$x^2 - nx + rn - 1 = 0$$

A and B are constants which are determined from the boundary conditions.

Example 11:

Determine the shape for a disk with uniform stress, i.e. $\sigma_r = \sigma_\theta$

From Hooke's law

$$\varepsilon_r = \frac{1}{E}(\sigma_r - v\sigma_\theta), \qquad \varepsilon_\theta = \frac{1}{E}(\sigma_\theta - v\sigma_r)$$

Solution:

if $\sigma_r = \sigma_\theta$ then $\varepsilon_r = \varepsilon_\theta$. From strain-displacement relations

$$\varepsilon_r = \frac{du_r}{dr} \quad \text{and} \quad \varepsilon_\theta = \frac{u_r}{r}$$

we get $\varepsilon_r = \dfrac{d}{dr}(r\varepsilon_\theta) = \dfrac{d}{dr}(r\varepsilon_r)$

Since $\varepsilon_r = \varepsilon_\theta$ the above equation gives

$$\frac{d\varepsilon_r}{dr} = 0$$

i.e. ε_r = constant

Hence, from Hooke's law, σ_r and σ_θ are not only equal but also constant throughout the disk. Let $\sigma_r = \sigma_\theta = \sigma$. Equilibrium Eq. (73) gives

$$h\sigma = \frac{d}{dr}(rh\sigma) + h\rho\omega^2 r^2$$

$$= h\sigma + r\sigma\frac{dh}{dr} + \rho\omega^2 hr^2$$

or
$$\frac{1}{h}\frac{dh}{dr} = -\frac{1}{\sigma}\rho\omega^2 r$$

which upon integration gives

$$\log h = -\frac{\rho\omega^2}{2\sigma}r^2 + C_1$$

or
$$h = \exp\left[-\frac{\rho\omega^2}{2\sigma}r^2 + C_1\right] = C\exp\left(-\rho\omega^2\frac{r^2}{2\sigma}\right)$$

ROTATING SHAFTS AND CYLINDERS

We assumed that the disk was thin and that it was in a state of plane stress with $\sigma_z = 0$. It is also possible to treat the problem as a plane strain problem as in the case of a uniformly torating long circular shaft or a cylinder. Let the z-axis be the axis of rotation. The equation of equilibrium is the same as in Eq. (57):

$$\frac{d}{dr}(r\sigma_r) - \sigma_\theta + \rho\omega^2 r^2 = 0 \tag{76}$$

The strain components are, as before,

$$\varepsilon_r = \frac{du_r}{dr}, \qquad \varepsilon_\theta = \frac{u_r}{r}, \qquad \varepsilon_z = \frac{\partial u_z}{\partial z} = 0 \tag{77}$$

From Hooke's law

$$\varepsilon_r = \frac{1}{E}[\sigma_r - v(\sigma_\theta + \sigma_z)]$$

$$\varepsilon_\theta = \frac{1}{E}[\sigma_\theta - v(\sigma_r + \sigma_z)]$$

$$\varepsilon_z = \frac{1}{E}[\sigma_z - v(\sigma_r + \sigma_\theta)]$$

Since $\varepsilon_z = 0$ (plane strain),

$$\sigma_z = v\,(\sigma_r + \sigma_\theta)$$

and hence, substituting in equations for ε_r and ε_θ

$$\varepsilon_r = \frac{1+v}{E}[(1-v)\sigma_r - v\sigma_\theta]$$

$$\varepsilon_\theta = \frac{1+\nu}{E}[(1-\nu)\sigma_\theta - \nu\sigma_r]$$

From strain-displacement relations given in Eq. (77)

$$\varepsilon_r = \frac{d}{dr}(r\varepsilon_\theta)$$

and using the above expressions for ε_r and ε_θ, we get

$$(1-\nu)\sigma_r + \nu\sigma_\theta = \frac{d}{dr}[(1-\nu)r\sigma_\theta - \nu\, r\, \sigma_r] \tag{78}$$

With $r\sigma_r = y$, Eq. (76) gives for σ_θ

$$\sigma_\theta = \frac{dy}{dr} + \rho\omega^2 r^2$$

Substituting for $\sigma_r = \sigma_\theta$ in Eq. (78)

$$(1-\nu)\frac{y}{r} - \nu\frac{dy}{dr} - \nu\rho\omega^2 r^2 = \frac{d}{dr}\left[(1-\nu)\left(r\frac{dy}{dr} + \rho\omega^2 r^3\right) - \nu y\right]$$

or $$r^2\frac{d^2y}{dr^2} + r\frac{dy}{dr} - y + \frac{3-2\nu}{1-\nu}\rho\omega^2 r^3 = 0$$

The solution for this differential equation is

$$y = Cr + C_1\frac{1}{r} - \frac{(3-2\nu)}{8(1-\nu)}\rho\omega^2 r^3$$

Hence, $$\sigma_r = C + C_1\frac{1}{r^2} - \frac{(3-2\nu)}{8(1-\nu)}\rho\omega^2 r^2 \tag{79a}$$

$$\sigma_\theta = C - C_1\frac{1}{r^2} - \frac{(1+2\nu)}{8(1-\nu)}\rho\omega^2 r^2 \tag{79b}$$

and $$\sigma_z = \nu\left[2C - \frac{1}{2(1-\nu)}\rho\omega^2 r^2\right] \tag{79.c}$$

(i) For a hollow shaft or a long cylinder, $s_r = 0$ at $r = a$ and $r = b$ which are the inner and outer radii. From these

$$C = K(a^2 + b^2) \qquad \text{and} \qquad C_1 = -Ka^2b^2$$

where $$K = \frac{(3-2\nu)}{8(1-\nu)}\rho\omega^2$$

Hence, from Eqs (79a–c)

$$\sigma_r = \frac{(3-2\nu)}{8(1-\nu)}\left[(a^2+b^2) - \frac{a^2b^2}{r^2} - r^2\right]\rho\omega^2 \tag{80}$$

$$\sigma_\theta = \frac{(3-2v)}{8(1-v)}\left[(a^2+b^2)+\frac{a^2b^2}{r^2}-\frac{1+2v}{3-2v}r^2\right]\rho\omega^2 \tag{81}$$

$$\sigma_z = \frac{v}{4(1-v)}\left[(a^2+b^2)(3-2v)-2r^2\right]\rho\omega^2 \tag{82}$$

σ_q assumes a maximum value at $r = a$ and its value is

$$(\sigma_\theta)_{\max} = \frac{(3-2v)}{8(1-v)}\left(2b^2+a^2-\frac{1+2v}{3-2v}a^2\right)\rho\omega^2$$

If a^2/b^2 is very small, we find that

$$(\sigma_\theta)_{\max} \approx \frac{(3-2v)}{8(1-v)}b^2\rho\omega^2 \tag{83}$$

(ii) For a long solid shaft, the constant C_1 must be equal to zero, since otherwise the stresses would become infinite at $r = 0$. Using the other boundary condition that $\sigma_r = 0$ when $r = b$, the radius of the shaft, we find that

$$C = \frac{(3-2v)}{8(1-v)}\rho\omega^2 b^2$$

Hence, the stresses are

$$\sigma_r = \frac{(3-2v)}{8(1-v)}(b^2+r^2)\rho\omega^2 \tag{84}$$

$$\sigma_\theta = \frac{(3-2v)}{8(1-v)}\left(b^2-\frac{1+2v}{3-2v}r^2\right)\rho\omega^2 \tag{85}$$

$$\sigma_z = \frac{v}{4(1-v)}\left[b^2(3-2v)-2r^2\right]\rho\omega^2 \tag{86}$$

The value of σ_θ at $r = 0$ is

$$(\sigma_\theta)_{\max} = \frac{(3-2v)}{8(1-v)}b^2\rho\omega^2 \tag{87}$$

Comparing Eq. (87) with Eq. (83), we find that by drilling a small hole along the axis in a solid shaft, the maximum circumferential stress is doubled in its magnitude.

Example 12:

A solid steel propeller shaft, 60 cm in diameter, is rotating at a speed of 300 rpm. If the shaft is constrained at its ends so that it cannot expand or contract longitudinally, calculate the total longitudinal thrust over a cross-

section due to rotational stresses. Poisson's ratio may be taken as 0.3. The weight of steel may be taken as 0.0081 kgf/cm² (0.07938N/cm²).

Solution:

The total axial force is

$$F_z = \int_0^b \sigma_z 2\pi r\, dr$$

and from Eq. (86), substituting for σ_z,

$$F_z = \frac{\nu}{4(1-\nu)}\left[b^2(3-2\nu)\pi b^2 - \pi b^4\right]\rho\omega^2$$

$$= \frac{\nu}{2}\pi b^4 \rho\omega^2$$

Substituting the numerical values

$$F_z = \frac{0.3}{2}\times\pi\times 30^4\times\frac{0.0081}{981}\times\frac{300^2}{60^2}\times\pi 4^2$$

= 3120 kgf (31576aN) Thinness force

SUMMARY OF RESULTS FOR USE IN PROBLEMS

(i) For a true of internal radius a and external radius b subjected to an internal pressure P_a and an external pressure P_b, the radial and circumferential stresses are given by (according to plane stress theory)

$$\sigma_r = \frac{p_a a^2 - p_b b^2}{b^2 - a^2} - \frac{a^2 b^2}{r^2}\frac{p_a - p_b}{b^2 - a^2}$$

$$\sigma_\theta = \frac{p_a a^2 - p_b b^2}{b^2 - a^2} + \frac{a^2 b^2}{r^2}\frac{p_a - p_b}{b^2 - a^2}$$

$$\sigma_z = 0$$

The stress $\sigma_r < 0$ for all values of p_a and p_b, whereas σ_θ can be greater or less than zero depending on the values of p_a and p_b. σ_θ is greater than zero if

$$p_a > \frac{p_b}{2}\left(\frac{b^2}{a^2}+1\right)$$

The maximum and minimum stresses are

$$(\sigma_r)_{max} = \sigma_r \text{ (at } r = b) = -p_b$$

$$(\sigma_r)_{min} = \sigma_r \text{ (at } r = a) = -p_a$$

$$(\sigma_\theta)_{\max} = \sigma_\theta(\text{at } r = a) = \frac{p_a(a^2+b^2) - 2p_b b^2}{b^2 - a^2}$$

$$(\sigma_\theta)_{\max} = \sigma_\theta(\text{at } r = b) = \frac{2p_a a^2 - p_b(a^2+b^2)}{b^2 - a^2}$$

If $\quad p_a = \frac{1}{2} p_b \left(\frac{b^2}{a^2} + 1 \right)$

then, $\quad (\sigma_\theta)_{\min} = 0$

$$(u_r)_{r=a} = \frac{a}{E}\left[p_a \left(\frac{b^2+a^2}{b^2-a^2} + v \right) - 2p_b \frac{b^2}{b^2-a^2} \right]$$

$$(u_r)_{r=b} = \frac{b}{E}\left[2p_a \frac{a^2}{b^2-a^2} - p_b \left(\frac{b^2+a^2}{b^2-a^2} - v \right) \right]$$

(ii) Built-up cylinders: When the cylinders are of equal length, the contact pressure p_c due to difference Δ between the outer radius of the inner tube and the inner radius of the outer tube is given by

$$p_c = \frac{\Delta / c}{\left[\frac{1}{E_1}\left(\frac{c^2+a^2}{c^2-a^2} - v_1 \right) + \frac{1}{E_2}\left(\frac{b^2+c^2}{b^2-c^2} - v_2 \right) \right]}$$

where E_1, V_1 a and c refer to the inner tube's modulus, poisson's ratio, inner radius and outer radius respectively. E_2, V_2 c and b are the corresponding values for the outer tube.

If $\quad E_1 = E_2$ and $V_1 = V_2$, then

$$p_c = \frac{\Delta E}{2c^3} \frac{(b^2-c^2)(c^2-a^2)}{b^2-a^2}$$

(iii) For a sphere subjected to an internal pressure p_a and an external pressure p_b, the radial and circumferential stresses are given by

$$\sigma_r = \frac{1}{b^3-a^3}\left[-b^3 p_b + a^3 p_a + \frac{a^3 b^3}{r^3} - (p_b - p_a) \right]$$

$$\sigma_\theta = \sigma_\phi = \frac{1}{b^3-a^3}\left[-b^3 p_b + a^3 p_a - \frac{a^3 b^3}{2r^3} - (p_b - p_a) \right]$$

(iv) For a thin solid disk of radius b rotating with an angular velocity ω, the stresses are given by

$$\sigma_r = \frac{3+v}{8} \rho \omega^2 (b^2 - r^2)$$

$$\sigma_\theta = \frac{3+\nu}{8}\rho\omega^2 b^2 - \frac{1+3\nu}{8}\rho\omega^2 r^2$$

These stresses attain their maximum values at the centre $r = 0$, where

$$\sigma_\theta = \sigma_r = \frac{3+\nu}{8}\rho\omega^2 b^2$$

The radial outward displacement at $r = b$ is

$$(u_r)_{r=b} = \frac{1-\nu}{4E}\rho\omega^2 b^3$$

(v) For a thin disk with a hole of radius a, rotating with an angular velocity ω, the stresses are

$$\sigma_r = \frac{3+\nu}{8}\rho\omega^2\left(b^2 + a^2 - \frac{a^2 b^2}{r^2} - r^2\right)$$

$$\sigma_\theta = \frac{3+\nu}{8}\rho\omega^2\left(b^2 + a^2 + \frac{a^2 b^2}{r^2} - \frac{1+3\nu}{3+\nu}r^2\right)$$

$$(\sigma_r)_{\max} = \sigma_r\left(\text{at } r = \sqrt{ab}\right) = \frac{3+\nu}{8}\rho\omega^2(b-a)^2$$

$$(\sigma_\theta)_{\max} = \sigma_\theta(\text{at } r = a) = \frac{3+\nu}{4}\rho\omega^2\left(b^2 + \frac{1+\nu}{3+\nu}a^2\right)$$

and $(\sigma_\theta)_{\max} > (\sigma_r)_{\max}$

The radial displacements are

$$(u_r)_{r=a} = \frac{3+\nu}{4E}\rho\omega^2 a\left(b^2 + \frac{1-\nu}{3+\nu}a^2\right)$$

$$(u_r)_{r=b} = \frac{3+\nu}{4E}\rho\omega^2 b\left(a^2 + \frac{1-\nu}{3+\nu}b^2\right)$$

THERMAL STRESSES

It is well known that changes in temperature cause bodies to expand or contract. The increase in the length of a uniform bar of length L, when its temperature is raised from T_0 to T, is

$$\Delta L = \alpha L(T - T_0)$$

where α is the coefficient of thermal expansion. If the bar is prevented from completely expanding in the axial direction, then average compressive stress induced is

$$\sigma = E\frac{\Delta L}{L}$$

where E is the modulus of elasticity. Thus, for complete restraint, the thermal stress is

$$\sigma = -\alpha E (T - T_0)$$

where the negative sign indicates the compressive nature of the stress. If the expansion is prevented only partially, then the stress induced is

$$\sigma = -k\alpha E (T - T_0)$$

where k represents a restraint coefficient. It is assumed in the above analysis that E and α are independent of temperature. In general, in an elastic continuum, the temperature change is not uniform throughout. It is a function of time and the space coordinates (x, y, x) i.e.

$$T = T(t, x, y, x)$$

The body under consideration may be restrained from expansion or movement in some regions and external tractions may be applied to other regions. The determination of stresses under such situations may be quite complex. In this chapter, we shall restrict ourselves to the analysis of the following problem;

(i) Thin circular disks with symmetrical temperature variation;

(ii) Long circular cylinders - hollow and solid;

(iii) Spheres with purely radial temperature variation - hollow and solid;

(iv) Straight beams of arbitrary cross-section;

(v) Curved beams.

Before these specific problems are analysed, we shall develop the general thermoelastic stress-strain relations and discuss two important general result.

THERMOELASTIC STRESS-STRAIN RELATIONS

Consider a body to be made up of a large number of small cubical elements. If the temperatures of all these elements are uniformly raised and if the boundary of the body is unconstrained, then all the cubical elements will expand uniformly and all will fit together to form a continuous body.

If, however, the temperature rise is not uniform, each element will tend to expand by a different amount and if these elements have to fit together to form a continuous body, then distortions of the elements and consequently stresses should occur in the body.

The total strains at each point of a body are thus made up of two parts. The first part is a uniform expansion proportional to the temperature rise T. For any elementary cubical element of an isotropic body, this expansion is the same in all directions and in this manner only normal strains and no shearing strains occur. If the coefficient of linear thermal expansion is α, this normal strain in any direction is equal to αT. The second part of the strains at each point is due to the stress components. The total strains at each point can, therefore, be written as

$$\begin{aligned} \varepsilon_x &= \frac{1}{E}\left[\sigma_x - v\left(\sigma_y + \sigma_z\right)\right] + \alpha T \\ \varepsilon_y &= \frac{1}{E}\left[\sigma_y - v\left(\sigma_x + \sigma_z\right)\right] + \alpha T \\ \varepsilon_z &= \frac{1}{E}\left[\sigma_z - v\left(\sigma_x + \sigma_y\right)\right] + \alpha T \end{aligned} \tag{1a}$$

$$\gamma_{xy} = \frac{1}{G}\tau_{xy}, \quad \gamma_{yz} = \frac{1}{G}\tau_{yz}, \quad \gamma_{zx} = \frac{1}{G}\tau_{zx} \tag{1b}$$

The stresses can be expressed explicitly in terms of strains by solving Eq. (1a). These are

$$\begin{aligned} \sigma_x &= \lambda e + 2\mu\varepsilon_x - (3\lambda + 2\mu)\,\alpha T \\ \sigma_y &= \lambda e + 2\mu\varepsilon_y - (3\lambda + 2\mu)\,\alpha T \\ \sigma_z &= \lambda e + 2\mu\varepsilon_z - (3\lambda + 2\mu)\,\alpha T \end{aligned} \tag{2a}$$

$$\tau_{xy} = \mu\gamma_{xy}, \quad \tau_{yz} = \mu\gamma_{yz}, \quad \tau_{zx} = \mu\gamma_{zx} \tag{2b}$$

The Lame constants λ and μ (= G) are given by

$$\lambda = \frac{vE}{(1+v)(1-2v)}, \quad \mu = G = \frac{E}{2(1+v)} \tag{3}$$

EQUATIONS OF EQUILIBRIUM

The equations of equilibrium are the same as those of isothermal elasticity since they are based on purely mechanical considerations. In rectangular coordinates these are given by Eq. (65). These are repeated for convenience.

$$\begin{aligned} \frac{\partial \sigma_x}{\partial x} + \frac{\partial \tau_{xy}}{\partial y} + \frac{\partial \tau_{zx}}{\partial z} + \gamma_x &= 0 \\ \frac{\partial \tau_{xy}}{\partial x} + \frac{\partial \sigma_y}{\partial y} + \frac{\partial \tau_{yz}}{\partial z} + \gamma_y &= 0 \\ \frac{\partial \tau_{xz}}{\partial x} + \frac{\partial \tau_{yz}}{\partial y} + \frac{\partial \sigma_z}{\partial z} + \gamma_z &= 0 \end{aligned} \tag{4}$$

where γ_x, γ_y and γ_z are body force components.

STRAIN-DISPLACEMENT RELATIONS

Only geometrical considerations are involved in deriving strain-displacement relations. Hence, the equations are the same as in isothermal elasticity. In rectangular coordinates, these are given by Eqs (18) and (19). To repeat, these are

$$\varepsilon_x = \frac{\partial u_x}{\partial x}, \quad \varepsilon_y = \frac{\partial u_y}{\partial y}, \quad \varepsilon_z = \frac{\partial u_z}{\partial z} \tag{5a}$$

$$\gamma_{xy} = \frac{\partial u_x}{\partial y} + \frac{\partial u_y}{\partial x}, \quad \gamma_{yz} = \frac{\partial u_y}{\partial z} + \frac{\partial u_z}{\partial y}, \quad \gamma_{zx} = \frac{\partial u_z}{\partial x} + \frac{\partial u_x}{\partial z} \tag{5b}$$

SOME GENERAL RESULTS

When the temperature distribution is known, the problem of thermoelasticity consists in determining the following 15 functions:

6 stress components $\sigma_x, \sigma_y, \sigma_z, \tau_{xy}, \tau_{yz}, \tau_{zx}$

6 stress components $\varepsilon_x, \varepsilon_y, \varepsilon_z, \gamma_{xy}, \gamma_{yz}, \gamma_{zx}$

3 displacement components u_x, u_y, u_z

so as to satisfy the following 15 equations throughout the body

3 equilibrium equations, Eq. (4)

3 stress-strain relations, Eq. (1)

6 strain-displacement relations, Eq. (5)

and the prescribed boundary conditions. In most problems, the boundary conditions belong to one of the following two cases:

(i) **Traction Boundary Conditions:** In this case, the stress components determined must agree with the prescribed surface traction at the boundary.

(ii) **Displacement Boundary Conditions:** Here, the displacement components determined should agree with the prescribed displacements at the boundary.

In some cases, the prescribed boundary conditions may be a combination of the above two, i.e. on a part of the boundary, the surface tractions are prescribed and on the remaining part, displacements are prescribed.

(i) The method of arriving at a solution depends in general on the specific nature of the problem. It is shown in books on thermoelasticity that if the temperature distribution in a body is a linear function of the rectangular Cartesian space coordinates, i.e. if

$$T(x, y, z, t) = a(t) + b(t)x + c(t)y + d(t)z \tag{6}$$

where t represents time, then all the stress components are identically zero throughout the body, provided that all external restraints, body forces and displacement discontinuities are absent. Converselly, under those provisions, this is the only temperature distribution for which all stress components are identically zero. These results are obtained immediately by considering the stress compatibility relations.

It, therefore, follows from the above statement and from the linearity of the boundary-value problem as formulated through Eqs (1), (4) and (5) that a linear function may be added to or subtracted from a given temperature distribution without affecting the resulting stress distribution. However, the strains and displacements are elated, as is obvious.

(ii) We shall now show that if a body is subjected to a uniform temperature rise $T = T_0(t)$ and if the boundary of the body is prevented from having any displacements, then the solution of the corresponding thermoelastic problem is

$$u_x = 0, \quad u_y = 0, \quad u_z = 0$$

$$\varepsilon_x = \varepsilon_y = \varepsilon_z = 0, \quad \gamma_{xy} = \gamma_{yz} = \gamma_{zx} = 0$$

$$\tau_{xy} = \tau_{yz} = \tau_{zx} = 0, \quad \sigma_x = \sigma_y = \sigma_z = -\frac{E\alpha}{1-3v}T_0$$

To shown this, we shall apply the principle of superposition. We shall first allow from expansion of the body due to temperature rise T_0 with no restraint whatsoever. Since all cubical elements of the body expand freely, no stresses develop and all elements expand in an identical manner. This has been discussed in Sec. 2. Consequently,

$$\sigma_x = \sigma_y = \sigma_z = 0, \qquad \tau_{xy} = \tau_{yz} = \tau_{zx} = 0$$

$$\gamma_{xy} = \gamma_{yz} = \gamma_{zx} = 0 \qquad \varepsilon_x = \varepsilon_y = \varepsilon_z = aT_0$$

Therefore, $u_x = \alpha T_0 x, \quad u_y = \alpha T_0 y, \quad u_z = \alpha T_0 z$ (7a)

Now we apply boundary tractions to prevent this displacement. If the body is subject to a hydrostatic state of stress, then all elements of the body will experience the same state of stress $(-P)$. With this state of stress, i.e. $\sigma_x = \sigma_y = \sigma_z = -P$ and $\tau_{xy} = \tau_{yz} = \tau_{zx} = 0$, the equations of equilibrium are identically satisfied.

Corresponding to this state of stress, the strain components are

$$\varepsilon_x = \varepsilon_y = \varepsilon_z = \frac{1}{E}[-p - v(-p-p)]$$

$$= \frac{1}{E}(1-2v)p$$

$$\gamma_{xy} = \gamma_{yz} = \gamma_{zx} = 0$$

Therefore, $u_x = \frac{1}{E}(1-2v)px: \quad u_y = -\frac{1}{E}(1-2v)py,$

$$u_z = -\frac{1}{E}(1-2v)pz$$

To get the original problem, the above values of u_x, u_y and u_z [Eq. (7a)] corresponding to free thermal expansion, should give zero displacements. Hence,

$$\alpha T_0 = x - \frac{1}{E}(1-2v)\ px = 0$$

or, $p = \frac{E\alpha}{1-2v} T_0$

Hence, $\sigma_x = \sigma_y = \sigma_z = -p = -\frac{E\alpha}{1-2v} T$ (8)

as stated earlier.

THIN CIRCULAR DISK: TEMPERATURE SYMMETRICAL ABOUT CENTRE

Consider a thin disk subjected to a temperature distribution which varies only with r and is independent of θ. It is assumed that it does not vary over the thickness and consequently, it is taken that the stresses and displacements also do not vary over the thickness. The stresses σ_r and σ_θ therefore, satisfy the equilibrium equation

$$\frac{d\sigma_r}{dr} + \frac{\sigma_r - \sigma_\theta}{r} = 0 \tag{9}$$

Body forces are ignored. Also, because of symmetry, $\tau_{r\theta} = 0$. With $\sigma_z = 0$, the stress-strain relations given by Eq. (1a) take form, in polar coordinates,

$$\varepsilon_r = \frac{1}{E}(\sigma_r - v\sigma_\theta) + \alpha T \tag{10}$$

$$\varepsilon_\theta = \frac{1}{E}(\sigma_\theta - v\sigma_r) + \alpha T$$

Solving the above equation for σ_r and σ_θ we find

$$\sigma_r = \frac{E}{1-v^2}[\varepsilon_r + v\varepsilon_\theta - (1+v)\alpha T]$$

$$\sigma_\theta = \frac{E}{1-\nu^2}[\varepsilon_\theta + \nu\varepsilon_r - (1+\nu)\alpha T]$$

Substituting these in the equation of equilibrium

$$r\frac{d}{dr}(\varepsilon_r + \nu\varepsilon_\theta) + (1+\nu)(\varepsilon_r - \varepsilon_\theta) = (1+\nu)\alpha r\frac{dT}{dr} \tag{12}$$

The strain-displacement relation for a symmetrically strained body, from Eqs (3) and (4), are

$$\varepsilon_r = \frac{du_r}{dr}, \qquad \varepsilon_\theta = \frac{u_r}{r} \tag{13}$$

Substituting in Eq. (12)

$$\frac{d^2u_r}{dr^2} + \frac{1}{r}\frac{du_r}{dr} - \frac{u_r}{r^2} = (1+\nu)\alpha\frac{dT}{dr}$$

This may be written as

$$\frac{d}{dr}\left[\frac{1}{r}\frac{d}{dr}(ru_r)\right] = (1+\nu)\alpha\frac{dT}{dr}$$

Integration of the above equation yields,

$$u_r = (1+\nu)\alpha\frac{1}{r}\int_a^r Tr\,dr + C_1 r + \frac{C_2}{r} \tag{14}$$

It can be observed that the above expression becomes identical to Eq. (8) if T is put equal to zero.

The lower limit a in the integral above depends on the disk. For a disk with a hole, a is the inner radius and for a solid disk, a is zero.

The stress components are determined by substituting the value of u_r in Eq. (13) and using the results in Eq. (11). The result are

$$\sigma_r = -\alpha E\frac{1}{r^2}\int_a^r Tr\,dr + \frac{E}{1-\nu^2}\left[C_1(1+\nu) - C_2(1-\nu)\frac{1}{r^2}\right] \tag{15}$$

$$\sigma_\theta = -\alpha E\frac{1}{r^2}\int_a^r Tr\,dr - \alpha ET + \frac{E}{1-\nu^2}\left[C_1(1+\nu) + C_2(1-\nu)\frac{1}{r^2}\right] \tag{16}$$

The constants C_1 and C_2 are determined by the boundary conditions. We shall now consider two specific cases. It should be observed that a linear variation of temperature with r will also induce stresses.

This does not contradict the statement made that the stresses in a body are zero if the temperature distribution is linear with respect to a Cartesian frame of reference and if the body is free from external restraints and body forces. A linear radial variation will not give a linear variation with respect to the x, y and z-axes. In fact

$$T = kr = k\sqrt{x^2 + y^2 + z^2}$$

(i) Solid Disk of Radius b : In the case of a solid circular disk, $a = 0$ and in Eq. (14) it is observed from L' Hospital's rule that

$$\lim_{r\to 0}\frac{1}{r}\int_0^r Tr\, dr = 0$$

Hence, the constant C_2 should be equal to zero, as otherwise u_r would become infinite at $r = 0$. The remaining constant C_1 is determined from the condition that $s_r = 0$ at $r = b$, the outer radius of the disk. From Eq. (15), therefore, we get

$$C_1(1+\nu)\frac{\alpha}{b^2}\int_0^b Tr\, dr$$

Substituting this, the stresses are

$$\sigma_r = \alpha E\left(\frac{1}{b^2}\int_0^b Tr\, dr - \frac{1}{r^2}\int_0^r Tr\, dr\right) \tag{17}$$

$$\sigma_\theta = \alpha E\left(-T + \frac{1}{b^2}\int_0^b Tr\, dr + \frac{1}{r^2}\int_0^r Tr\, dr\right) \tag{18}$$

From L' Hospital's rule

$$\lim_{r\to 0}\frac{1}{r^2}\int_0^r Tr\, dr = \frac{1}{2}T_0$$

where T_0 is the temperature at the centre of disk. Hence, at $r = 0$

$$\sigma_r(0) = \sigma_\theta(0) = \alpha E\left(\frac{1}{b^2}\int_0^r Tr\, dr - \frac{1}{2}T_0\right) \tag{19}$$

(ii) Disk with a Hole of Radius a: For a disk with a hole and traction free surfaces, $\sigma_r = 0$ at $r = a$ and $r = b$. Substituting these in Eq. (15)

$$C_1(1+\nu) + C_2(1-\nu)\frac{1}{a^2} = 0$$

$$-\alpha E\frac{1}{b^2}\int_a^b Tr\, dr + \frac{E}{1-\nu^2}\left[C_1(1+\nu) - C_2(1-\nu)\frac{1}{b^2}\right] = 0$$

Solving we get

$$C_1 = \alpha(1-\nu)\frac{1}{b^2 - a^2}\int_a^b Tr\, dr; \quad C_2 = \alpha(1-\nu)\frac{a^2}{b^2 - a^2}\int_a^b Tr\, dr$$

Substituting in Eqs (15) and (16)

$$\sigma_r = \frac{\alpha E}{r^2}\left[\frac{r^2 - a^2}{b^2 - a^2}\int_a^b Tr\,dr - \int_a^r Tr\,dr\right] \tag{20}$$

$$\sigma_\theta = \frac{\alpha E}{r^2}\left[\frac{r^2 + a^2}{b^2 - a^2}\int_a^b Tr\,dr + \int_a^r Tr\,dr - Tr^2\right] \tag{21}$$

and from Eq. (14)

$$u_2 = \frac{\alpha}{r}\left[(1+\nu)\int_a^r Tr\,dr + \frac{(1+\nu)r^2 + (1+\nu)a^2}{b^2 - a^2}\int_a^b Tr\,dr\right] \tag{22}$$

If the temperature T is constant, then all the stress components are zero and the radial displacement $u_r = \alpha r T$.

LONG CIRCULAR CYLINDER

We shall now consider the nature of the thermal stresses induced in a long circular cylinder when the temperature is symmetrical about the axis and does not vary along the axis. If the z-axis is the axis of the cylinder and r the radius, then T is a function of r alone and is independent of z. Since the cylinder is long, sections far from the ends can be considered to be in a state of plane strain and we can analyse this problem with u_z, the axial displacement, assumed to be zero.

Once again, owing to symmetry, all the shear stress components are zero and there are now three normal stress components σ_r, σ_θ and σ_z. The stress-strain relations are

$$\varepsilon_r = \frac{1}{E}\left[\sigma_r - \nu(\sigma_\theta + \sigma_z)\right] + \alpha T$$

$$\varepsilon_\theta = \frac{1}{E}\left[\sigma_\theta - \nu(\sigma_r + \sigma_z)\right] + \alpha T$$

$$\varepsilon_z = \frac{1}{E}\left[\sigma_z - \nu(\sigma_r + \sigma_\theta)\right] + \alpha T$$

Since $u_z = 0$, we have $\varepsilon_z = 0$. Hence, from the last equation we get

$$\sigma_z = \nu\,(\sigma_r + \sigma_\theta) - ET\alpha \tag{23}$$

Substituting this in the expressions for ε_z and ε_θ

$$\varepsilon_r = \frac{1-\nu^2}{E}\left(\sigma_r - \frac{\nu}{1-\nu}\sigma_\theta\right) + (1+\nu)\alpha T$$

(24)

$$\varepsilon_\theta = \frac{1-\nu^2}{E}\left(\sigma_\theta - \frac{\nu}{1-\nu}\sigma_r\right) + (1+\nu)\alpha T$$

Let $$\frac{E}{1-\nu^2} = E_1, \quad \frac{V}{1-\nu} = V_1, \quad (1+\nu)\alpha = \alpha_1 \tag{25}$$

Then, Eq. (24) can be written as

$$\varepsilon_r = \frac{1}{E_1}(\sigma_r - \nu_1\sigma_\theta) + \alpha_1 T$$

$$\varepsilon_q = \frac{1}{E_1}(\sigma_\theta - \nu_1\sigma_r) + \alpha_1 T \tag{26}$$

Comparing the above expressions with Eq. (10), it is immediately observed that the expressions for ε_r and ε_θ in the plane strain case is similar to those in the plane stress case if we use E_1, V_1 and α_1, given by Eq. (25), in place of E, V and α respectively. Since the equation of equilibrium is the same as in the plane stress case, further analysis is identical to that in the plane stress case. The expressions for u_r, σ_r and σ_z can, therefore, be written from Eqs (14) and (16) as

$$u_r = \frac{1+\nu}{1-\nu}\alpha\frac{1}{r}\int_a^r Tr\,dr + C_1 r + \frac{C_2}{r} \tag{27}$$

$$\sigma_r = \frac{\alpha E}{1-\nu}\frac{1}{r^2}\int_a^r Tr\,dr + \frac{E}{1+\nu}\left(\frac{C_1}{1-2\nu} + \frac{C_2}{r^2}\right) \tag{28}$$

$$\sigma_\theta = \frac{E}{1-\nu}\frac{1}{r^2}\int_a^r Tr\,dr - \frac{\alpha ET}{1-\nu} + \frac{E}{1+\nu}\left(\frac{C_1}{1-2\nu} + \frac{C_2}{r^2}\right) \tag{29}$$

and from Eq. (22)

$$\sigma_z = \frac{\alpha ET}{1-\nu} + \frac{2\nu EC_1}{(1+\nu)(1-2\nu)} \tag{30}$$

When $T = 0$, the equations become identical to Eqs (8.22a-c). Normal force given by Eq. (30) is necessary to keep $u_z = 0$ throughout. The constants C_1 and C_2 are determined from the boundary conditions. We shall now consider two particular cases.

(i) Solid Cylinder of Radius *b*: As before, from L'Hospital's rule

$$\lim_{r\to 0}\frac{1}{r}\int_0^r Tr\,dr = 0$$

Hence, from Eq. (27) we observe that C_2 should be equal to zero, as otherwise u_r would be infinite at $r = 0$. Since $\sigma_r = 0$ at $r = b$, we get from Eq. (28) with $C_2 = 0$ and $a = 0$ in the lower limit of the integration

$$\frac{C_1}{(1+\nu)(1-2\nu)} = \frac{\alpha}{1-\nu}\frac{1}{b^2}\int_0^b Tr\,dr$$

or
$$C_1 = \frac{(1+\nu)(1-2\nu)}{(1-\nu)}\frac{\alpha}{b^2}\int_0^b Tr\,dr$$

Comparing this with the value of C_1 obtained for the plane stress case, we observe that C_1 for the plane strain case can be obtained from C_1 for the plane stress case, merely by changing E, V and α to E_1, V_1 and α_1, and then converting these according to Eq. (25). The values of σ_r and σ_θ are, accordingly,

$$\sigma_r = \frac{\alpha E}{(1-\nu)}\left(\frac{1}{b^2}\int_0^b Tr\,dr - \frac{1}{r^2}\int_0^r Tr\,dr\right) \tag{31}$$

$$\sigma_\theta = \frac{\alpha E}{(1-\nu)}\left(-T + \frac{1}{b^2}\int_0^b Tr\,dr + \frac{1}{r^2}\int_0^r Tr\,dr\right) \tag{32}$$

and at $r = 0$

$$\sigma_r(0) = \sigma_\theta = \frac{\alpha E}{(1-\nu)}\left(\frac{1}{b^2}\int_0^b Tr\,dr - + \frac{1}{2}T_0\right) \tag{33}$$

where T_0 is the temperature at $r = 0$. Further, from Eq. (23)

$$\sigma_z = \frac{\alpha E}{(1-\nu)}\left(\frac{2\nu}{b^2}\int_0^b Tr\,dr - T\right) \tag{34}$$

The radial displacement is given by

$$u_1 = \frac{1+\nu}{1-\nu}\alpha\left[(1-2\nu)\frac{r}{b^2}\int_0^b Tr\,dr + \frac{1}{r}\int_0^b Tr\,dr\right] \tag{35}$$

Note: In obtaining Eq. (34), we have assumed a plane strain condition with $\varepsilon_z = 0$. Consequently, a stress distribution σ_z as given by Eq. (30) was necessary to maintain $u_z = 0$.

If, however, the ends of the cylinder are free, then the resultant force in z direction should be equal to zero. This condition can be achieved by superposing a uniform stress distribution $\sigma_z' = C_3$ so that the resultant force is zero.

For the solid cylinder, the resultant of σ_z from Eq. (30) is

$$\int_0^b 2\pi r\sigma_z\,dr = -\frac{2\pi\alpha E}{(1-\nu)}\int_0^b Tr\,dr + \frac{2VEC_1}{(1+\nu)(1-2\nu)}\pi b^2$$

The resultant of the superimposed uniform stress $\sigma_z' = C_3$ is $\pi b^2 C_3$. The value of C_3 to make the total force zero in z direction is, therefore, given by

$$C_3 \pi b^2 = \frac{2\pi\alpha E}{(1-\nu)} \int_0^b Tr\,dr - \frac{2\nu E C_1}{(1+\nu)(1-2\nu)} \pi b^2$$

The resultant σ_z distribution is given by

$$\sigma_z = \frac{\alpha E}{(1-\nu)} \left(\frac{2}{b^2} \int_0^b Tr\,dr - T \right)$$

The u_z displacement is then given by Eq. (35) plus $(-\nu C_3 r/E)$.

(ii) Hollow Cylinder with Inner Radius *a*: We shall write the solutions for this from the plane stress case results, Eqs (20) – (22), by putting E_1, V_1 σ_1 in place of E, V σ and then converting these according to Eq. (25). Accordingly, we get

$$\sigma_r = \frac{\alpha E}{(1-\nu)} \frac{1}{r^2} \left[\frac{r^2 - a^2}{b^2 - c^2} \int_a^b Tr\,dr - \int_a^b Tr\,dr \right] \tag{36}$$

$$\sigma_\theta = \frac{\alpha E}{(1-\nu)} \frac{1}{r^2} \left[\frac{r^2 + a^2}{b^2 - a^2} \int_a^b Tr\,dr + \int_a^b Tr\,dr - Tr^2 \right] \tag{37}$$

and

$$u_r = \frac{(1+\nu)\alpha}{(1-\nu)r} \left[\int_a^r Tr\,dr + \frac{(1-2\nu)r^2 + a^2}{(b^2 + a^2)} \int_a^b Tr\,dr \right] \tag{38}$$

Also

$$\sigma_z = \frac{\alpha E}{(1-\nu)} \left(\frac{2}{(b^2 + a^2)} \int_a^b Tr\,dr - T \right) \tag{39}$$

Example 1

The inner surface of a hollow tube is at temperature T_i and the outer surface at zero temperature.

Assuming steady-state conditions, calculate the stresses. What are the values of σ_θ and σ_z near the inner and outer surfaces?

Under steady heat flow conditions, the temperature at any distance *r* from the centre is given by the expression

$$T = \frac{T_i}{\log(b/a)} \log(b/r)$$

Substituting this in Eqs (36) - (38)

$$\sigma_r = \frac{\alpha E T_i}{2(1-\nu)\log(b/a)} \left[-\log\frac{b}{r} - \frac{a^2}{b^2 - a^2}\left(1 - \frac{b^2}{r^2}\right)\log\frac{b}{a} \right]$$

$$\sigma_\theta = \frac{\alpha E T_i}{2(1-\nu)\log(b/a)} \left[1 - \log\frac{b}{r} - \frac{a^2}{b^2 - a^2}\left(1 + \frac{b^2}{r^2}\right)\log\frac{b}{a} \right]$$

$$\sigma_z = \frac{\alpha E T_i}{2(1-\nu)\log(b/a)}\left[1 - 2\log\frac{b}{r} - \frac{2a^2}{b^2 - a^2}\log\frac{b}{a}\right]$$

$\sigma_r = 0$ at $r = 0$ and $r = b$. The stress components σ_θ and σ_z attain their maximum positive and negative values at $r = a$ and $r = b$. These values are

$$(\sigma_\theta)_{r=a} = (\sigma_z)_{r=a} = \frac{\alpha E T_i}{2(1-\nu)\log\dfrac{b}{a}}\left(1 - \frac{2b^2}{b^2 - a^2}\log\frac{b}{a}\right)$$

$$(\sigma_\theta)_{r=b} = (\sigma_z)_{r=b} = \frac{\alpha E T_i}{2(1-\nu)\log\dfrac{b}{a}}\left(1 - \frac{2a^2}{b^2 - a^2}\log\frac{b}{a}\right)$$

If T_i is positive, the radial stress is compressive at all points, whereas σ_θ and σ_z are compressive at the inner surface and tensile at the outer surface. These tensile stresses cause cracks in brittle materials such as stone, brick and concrete.

THE PROBLEM OF A SPHERE

We shall now consider the problem of a sphere subjected to purely radial temperature variation, i.e. T is a function of r alone. Because of symmetry, the shear stresses are all zero and the normal stresses are such that $\sigma_\theta = \sigma_\phi$. The equation of equilibrium in the radial direction is, from Eq. (37),

$$\frac{1}{r^2}\frac{d}{dr}(r^2\sigma_r) - \frac{2}{r}\sigma_\phi = 0$$

or

$$\frac{d}{dr}(r^2\sigma_r) - 2r\sigma_\phi = 0$$

The stress-strain relations are

$$\varepsilon_r \frac{1}{E}(\sigma_r - 2\nu\sigma_\phi) + \alpha T$$

$$\varepsilon_\phi = \varepsilon_\theta = \frac{1}{E}\left[\sigma_\phi - \nu(\sigma_r + \sigma_\phi)\right] + \alpha T$$

Solving the above equations for σ_θ and σ_ϕ

$$\sigma_r = \frac{E}{(1+\nu)(1-2\nu)}\left[(1-\nu)\varepsilon_r + 2\nu\varepsilon_\phi - (1+\nu)\alpha T\right] \tag{40}$$

$$\sigma_\phi = \frac{E}{(1+\nu)(1-2\nu)}\left[\varepsilon_\phi + \nu\varepsilon_r - (1+\nu)\alpha T\right] \tag{41}$$

From strain-displacement relations, we have

$$\varepsilon_r = \frac{du_r}{dr} \quad \text{and} \quad \varepsilon_\phi = \varepsilon_\theta = \frac{u_r}{r} \tag{42}$$

Substituting these in the expressions for σ_r and σ_θ and then substituting these in the equilibrium equation, we get

$$\frac{d^2u_r}{dr^2} + \frac{2}{r}\frac{du_r}{dr} - \frac{2u_r}{r^2} = \frac{1+\nu}{1-\nu}\alpha\frac{dT}{dr}$$

This can also be written as

$$\frac{d}{dr}\left[\frac{1}{r^2}\frac{d}{dr}(r^2u_r)\right] = \frac{1+\nu}{1-\nu}\alpha\frac{dT}{dr}$$

The solution is

$$u_r = \frac{1+\nu}{1-\nu}\alpha\frac{1}{r^2}\int_a^r Tr^2\,dr + C_1r + \frac{C_2}{r^2} \tag{43}$$

where C_1 and C_2 are constants to be determined from boundary conditions. The lower limit of the integral in the above equation is zero if the sphere is solid or is equal to a, the inner radius, if the sphere is hollow. From the expression for u_r, the strain components ε_r and ε_ϕ can be determined from Eq. (42) and substituted in Eq. (40). The results are

$$\sigma_r = -\frac{2\alpha E}{1-\nu}\frac{1}{r^3}\int_a^r Tr^2\,dr + \frac{EC_1}{1-2\nu} - \frac{2EC_2}{1+\nu}\frac{1}{r^3} \tag{44}$$

$$\sigma_\theta = \sigma_\phi = \frac{\alpha E}{1-\nu}\frac{1}{r^3}\int_a^r Tr^2\,dr + \frac{EC_1}{1-2\nu} - \frac{EC_2}{1+\nu}\frac{1}{r^3} - \frac{ET}{1-\nu} \tag{45}$$

we shall consider two specific cases.

(i) Solid Sphere : In this case, the lower limit a in the integrals may be taken as zero. In Eq. (43), the limit

$$\lim_{r\to 0}\left(-\frac{1}{r^2}\int_0^r Tr^2\,dr\right) = 0$$

according to L' Hospital's rule. Consequently, the constant C_2 should be equal to zero, as otherwise, the displacement u_r would become infinite at $r = 0$. The remaining constant C_1 is determined from the condition that $\sigma_r = 0$ at $r = a$ and $r = b$. Hence, from Eq. (44),

$$-\frac{2\alpha E}{(1-\nu)}\frac{1}{b^3}\int_a^r Tr^2\,dr + \frac{EC_1}{1-2\nu} = 0$$

or $$C_1 = \frac{2(1-2\nu)}{(1-\nu)}\int_0^b Tr^2\,dr$$

Substituting this in Eqs (44) and (45)

$$\sigma_r = -\frac{2\alpha E}{1-v}\left(\frac{1}{b^3}\int_0^b Tr^2\,dr - \frac{1}{r^3}\int_0^r Tr^2\,dr\right) \tag{46}$$

$$\sigma_\theta = \sigma_\phi = \frac{\alpha E}{1-v}\left(\frac{2}{b^3}\int_0^b Tr^2\,dr + \frac{1}{r^3}\int_0^r Tr^2\,dr - T\right) \tag{47}$$

(ii) Hollow Sphere : Let a be the radius of the inner cavity and b the outer radius of the sphere. The boundary conditions are $\sigma_r = 0$ at $r = a$ and $r = b$. Hence, form Eq. (44).

$$\frac{EC_1}{1-2v} - \frac{2EC_2}{(1+v)}\cdot\frac{1}{a^3} = 0$$

$$-\frac{2\alpha E}{(1-v)}\frac{1}{b^3}\int_a^b Tr^2\,dr + \frac{EC_1}{1-2v} - \frac{2EC_2}{1+v}\cdot\frac{1}{b^3} = 0$$

The above equations can be solved for C_1 and C_2 and substituted in Eqs (44) and (45). The result is

$$\sigma_r = -\frac{2\alpha E}{1-v}\left[\frac{r^3-a^3}{(b^3-a^3)r^3}\int_a^b Tr^2\,dr - \frac{1}{r^3}\int_a^r Tr^2\,dr\right] \tag{48}$$

$$\sigma_\phi = \frac{2\alpha E}{1-v}\left[\frac{2r^3+a^3}{2(b^3-a^3)r^3}\int_a^b Tr^2\,dr + \frac{1}{2r^2}\int_a^r Tr^2\,dr - \frac{1}{2}T\right] \tag{49}$$

Therefore, the stress components can be calculated if the distribution of temperature is known.

Example 2:

Let the inner surface of a hollow sphere be at temperature T_i and the outer surface at temperature zero. Let the system be in a steady heat flow condition. The temperature distribution is then given by

$$T = \frac{T_i a}{b-a}\left(\frac{b}{r} - 1\right)$$

Determine the stress distribution.

Solution:

Substituting the above expression for T in Eqs (48) and (49), we get

$$\sigma_r = -\frac{\alpha ET_i}{1-v}\frac{ab}{b^3-a^3}\left[a+b-\frac{1}{r}(b^2+ab+a^2)+\frac{a^2b^2}{r^3}\right]$$

$$\sigma_\theta = \sigma_\phi = \frac{\alpha ET_i}{1-v}\frac{ab}{b^3-a^3}\left[a+b-\frac{1}{2r}(b^2+ab+a^2)-\frac{a^2b^2}{2r^3}\right]$$

As can be seen $\sigma_r = 0$ at $r = a$ and $r = b$, according to the boundary conditions. Differentiating the expression for σ_r with respect to r and equating the resulting expression to zero, it is observed that σ_r is a maximum or a minimum when

$$r^2 = \frac{3a^2b^2}{b^2 + ab + a^2}$$

The expression for σ_ϕ shows that its value increases with r for T_i positive, and

$$(\sigma_r)_{r=a} = -\frac{\alpha ET_i}{2(1-\nu)} \frac{b(b-a)(a+2b)}{b^3 - a^3}$$

$$(\sigma_r)_{r=b} = -\frac{\alpha ET_i}{2(1-\nu)} \frac{a(b-a)(2a+b)}{b^3 - a^3}.$$

BEAM WITH RECTANGULAR SECTION

A somewhat more accurate result can be obtained for a curved beam with a rectangular cross-section and temperature independent of θ. This is obtained by superposing the result for a thin circular disk subjected to radial thermal loading with the result for the bending of a curved beam subjected to pure bending moment. If a sectoral element is isolated from a disk,, the ends of the element will be found (Examples 3 and 4) to be subjected to zero resultant circumferential force and some moment F_m, i.e.

$$\int_A \sigma_\theta \, dA = F_\theta = 0$$

$$\int_A \sigma_\theta \, y \, dA = F_m$$

where F_m is the moment about the median line.

If, on this curved beam, we apply an equal and opposite moment F_m, as shown in Fig. 4(c), then we get a free curved beam subjected to thermal loading only.

Example 3:

Show that the resultant circumferential force across any radial section of a hollow disk subjected to thermal loading is zero.

Solution:

From Eq. (21), the value of the circumferential stress σ_θ is

$$\sigma_\theta = \frac{\alpha E}{r^2}\left[\frac{r^2 + a^2}{b^2 - a^2}\int_a^b Tr\,dr + \int_a^r Tr'\,dr' - Tr^2\right]$$

Let the disk be of unit thickness perpendicular to the plane of the paper.

The resultant circumferential force across any section is

$$F_\theta = \int_a^b \sigma_\theta \, dr = \frac{\alpha E}{b^2 - a^2}\left[\int_a^b dr \int_a^b Tr \, dr + \int_a^b \frac{a^2}{r^2} dr \int_\alpha^r Tr \, dr\right]$$

$$+\alpha E\left[\frac{1}{r^2} dr \int_a^r Tr' \, dr' - \int_a^b T \, dr'\right]$$

Let $\quad \int_a^b Tr \, dr = \beta$

Then, $\quad F_\theta = \frac{\alpha E}{b^2 - a^2}\left[\beta(b-a) - \beta a^2\left(\frac{1}{b} - \frac{1}{a}\right)\right]$

$$+\alpha E\left[\left(-\frac{1}{r}\int_a^r Tr' \, dr'\right)\Bigg|_a^b + \int_a^b T \, dr - \int_a^b T \, dr\right]$$

In the above expression, we have made use of the formula

$$\frac{d}{d\alpha}\int_{U(\alpha)}^{V(\alpha)} F(\alpha, x) dx = \int_{U(\alpha)}^{V(\alpha)} \frac{dF(\alpha, x)}{d\alpha} dx + F(V, \alpha)\frac{dV}{d\alpha} - F(U, \alpha)\frac{dU}{d\alpha}$$

Substituting the limits, it is observed that $F_\theta = 0$

There is no resultant circumferential force across any section.

Example 4:

Determine the bending moment due to the circumferential stress across a section of a thin hollow disk subjected to radial thermal variation.

Solution:

If ρ_0 is the radius of the median line and σ_θ the circumferential stress on a fibre at r from the centre of curvature (Fig. 4b), then the moment about the median line is

$$F_m = \int_a^b \sigma_\theta (r - \rho_0) \, dr$$

$$= \int_a^b \sigma_\theta \, r \, dr - \rho_0 \int_a^b \sigma_\theta \, dr$$

The second integral on the right-hand side is zero, from Example 3. Using Eq. (21), the moment becomes

$$F_m = \frac{\alpha E}{b^2 - a^2}\left[\int_a^b r \, dr \int_a^b Tr \, dr + \int_a^b \frac{a^2}{r} dr \int_a^b Tr \, dr\right]$$

$$+\alpha E\left[\int_a^b \frac{1}{r} dr \int_a^r Tr'\, dr' - \int_a^b Tr\, dr\right]$$

Putting $\int_a^b Tr\, dr = \beta$, the above expression becomes

$$F_m = \frac{\alpha E}{b^2 - a^2}\left[\frac{\beta}{2}(b^2 - a^2) + a^2 \beta \log\frac{b}{a}\right]$$

$$+\alpha E\left[\log r \int_a^r Tr'\, dr'\right]_a^b - \alpha E\int_a^b (\log r)\, Tr\, dr - \alpha E\beta$$

$$= \frac{\alpha E\beta}{2(b^2 - a^2)}\left[b^2 + a^2\left(2\log\frac{b}{a} - 1\right)\right] + \alpha E\left[\beta(\log b - 1) - \int_a^b (\log r) Tr\, dr\right]$$

NORMAL STRESSES IN STRAIGHT BEAMS DUE TO THERMAL LOADING

In this section, we shall develop an elementary formula for normal stresses in free beams subjected to thermal loadings. We shall make use of the Bernoulli-Euler assumption mentioned in Chapter 6. According to this assumption, sections which are plane and perpendicular to the axis before loading remain so after loading and the effect of lateral contraction (due to Poisson effect) may be neglected. The beam is assumed to be statically determinate and free of external loads. The temperature variation is arbitrary and the cross-section of the beam is also arbitrary.

Let the y and z-axes lie in the plane of the section and let the z-axis be the axis of the beam. x, y and z-axes form a set of centroidal axes. The analysis is similar to the one used in Chapter 6 for the bending of beams. If the beam is prevented from bending and if warping is not allowed, then the displacement of any section in the axial direction due to temperature rise will be a function of the axial coordinate x. Let this be $f_0(x)$. If now, the beam is allowed to undergo bending with the plane section remaining plane, then the displacement in s direction of any point (y, z) in a plane will be a linear function of the coordinates y and z. This is equivalent to saying that the cross-section rotates about an axis. The section that was plane before bending will, therefore, remain plane after bending and axial displacement. Hence, the total axial displacement, according to the Euler-Bernoulli hypothesis, will be

$$u_x = f_0(x) + yf_1(x) + zf_2(x)$$

where f_1 and f_2 are functions of x alone. The axial strain ε_x is, therefore,

$$\varepsilon_x = \frac{\partial u_x}{\partial x} = f_0'(x) + yf_1'(x) + zf_2'(x) \tag{50}$$

The strain represented by the last two terms on the right-hand side is similar to the one expressed in Chapter 6. We can also assume that the section rotates about an axis, such as BB in Fig. 1(b), and write the strain as $\varepsilon_x = f_0'(x) + ky'$, where y' is the perpendicular distance of a point from BB, which is inclined at β to the y-axis. This is what was done in Chapter 6. The unknowns k and β are now replaced by $f_1'(x)$ and $f_2'(x)$. From Hooke's law, since ε_y and ε_z are assumed to be zero.

$$\sigma_x = E(\varepsilon_x - \alpha T)$$

Substituting for ε_x

$$\sigma_x = E[f_0'(x) + yf_1'(x) + zf_2'(x) \square \alpha T]$$

Since a free beam without external loading is considered, the conditions to be satisfied at any section are

$$\iint \sigma_x \, dA = 0; \quad \iint \sigma_x \, y \, dA = M_z = 0, \quad \iint \sigma_x \, z \, dA = M_y = 0$$

i.e. the resultant force over the section is zero and the moments about the y and z axes should individually vanish. Substituting the expression for σ_z the above conditions become

$$\begin{aligned} f_0' \iint dA + f_1' \iint y \, dA + f_2' \iint z \, dA &= \iint \alpha T \, dA \\ f_0' \iint y \, dA + f_1' \iint y^2 \, dA + f_2' \iint yz \, dA &= \iint \alpha T y \, dA \\ f_0' \iint z \, dA + f_1' \iint yz \, dA + f_2' \iint z^2 \, dA &= \iint \alpha T z \, dA \end{aligned} \tag{52}$$

The integrations extend over the entire cross-section. The expression

$$\iint y \, dA = \iint z \, dA = 0$$

because of the selection of the centroidal axes. Further,

$$\iint dA = A, \quad \iint y^2 dA = I_z, \quad \iint z^2 dA = I_y, \quad \iint yz \, dA = I_{yz}$$

Substituting these, Eq. (52) can be written as

$$\begin{aligned} Af_0'(x) &= \iint \alpha T \, dA \\ f_1' I_z + f_2' I_{yz} &= \iint \alpha T y \, dA \\ f_1' I_{yz} + f_2' I_y &= \iint \alpha T z \, dA \end{aligned} \tag{53}$$

Let $\quad E \iint \alpha T \, dA = p_t, \; E \iint \alpha T y \, dA = -M_{zt}, \; E \iint \alpha T z \, dA = M_{yt}$

A minus sign is used in the second expression in order to make the final result similar to the result. The solutions for f_0', f_1' and f_2' are then given by

$$f_0' = \frac{p_t}{EA}, \quad f_1' = \frac{-I_y M_{zt} + I_{yz} M_{yt}}{E(I_y I_z - I^2_{yz})}, \quad f_2' \frac{I_z M_{yt} + I_{yz} M_{zt}}{E(I_y I_z - I^2_{yz})} \tag{54}$$

Substituting these, the axial stress σ_x is, from Eq. (51),

$$\sigma_x = -\alpha ET + \frac{p_t}{A} - \frac{(I_y M_{zt} + I_{yz} M_{yt})}{(I_y I_z - I^2_{yz})} y + \frac{(I_z M_{yt} + I_{yz} M_{zt})}{(I_y I_z - I^2_{yz})}$$

or

$$\sigma_x = -\alpha ET + \frac{p_t}{A} + \frac{M_{zt}(yI_y - zI_{yz}) + M_{yt}(yI_{yz} - zI_z)}{I^2_{yz} - I_y I_z} \tag{55}$$

Equation (55) bears a very close resemblance to Eq. (14) since the analyses in both cases have proceeded on similar lines.

If the axes chosen happen to be the principal axes of the section, then $I_{yz} = 0$ and Eq. (55) reduces to

$$\sigma_x = -\alpha ET + \frac{p_t}{A} - \frac{M_{zt}}{I_z} y + \frac{M_{yt}}{I_y} z \tag{56}$$

STRESSES IN CURVED BEAMS DUE TO THERMAL LOADING

An elementary analysis of the stresses developed in curved beams may be developed on the same basic assumptions as in the case of straight beams. Consider a free curved beam of arbitrary constant cross-section, the centre line of which is an arc of a circle. It is assumed that this arc lies in one of the principal planes of the beam.

Let the temperature vary as a function of r and θ, i.e. $T(r, \theta)$. We shall follow the notations.

In the isothermal case, the radius of curvature of the neutral surface is given by r_0 [Eq. (33)] such that

$$\iint \frac{y\,dA}{r_0 - y} = 0 \tag{57}$$

The origin 0 lies on the neutral axis and y is measured towards the centre of curvature. A view of the deformed element is given in Fig. 3 Let the elementary length of an undeformed element enclose an angle $\Delta\theta$.

Because of thermal loading, the element deforms and it is assumed that sections which were plane before, remain plane after deformation. A fibre at a distance y from the chosen origin has a length $(r_0 - y)\,\Delta\theta$ before deformation. After deformation, the length of the same fibre becomes

$$\left[r_0' - y - \int_0^y \alpha T\, dy \right] (\Delta\theta + \delta\Delta\theta) \tag{58}$$

The third term in the first bracket above represents the thermal expansion in y direction. The change in the length of the fibre is therefore

$$\left[r_0' - y - \int_0^y \alpha T\, dy\right](\Delta\theta + \delta\Delta\theta) - (r_0 - y)\,\Delta\theta$$

$$= \left[r_0' - r_0 - \int_0^y \alpha T\, dy\right]\Delta\theta + \left[r_0' - y - \int_0^y \alpha T\, dy\right](\delta\,\Delta\theta)$$

Hence, the strain is

$$\varepsilon_\theta = \frac{1}{(r_0 - y)}\left[\left(r_0' - r_0 - \int_0^y \alpha T\, dy\right) + \frac{\delta\Delta\theta}{\Delta\theta}\left(r_0' - y - \int_0^y \alpha T\, dy\right)\right]$$

We observe that

$$\int_0^y \alpha T\, dy << y$$

Hence,

$$\varepsilon_\theta = \frac{1}{(r_0 - y)}\left[r_0' - r_0 - \int_0^y \alpha T\, dy\right] + \frac{\delta\Delta\theta}{\Delta\theta}(r_0' - y) \qquad (59)$$

From Hooke's law, taking only σ_θ into account

$$\sigma_\theta = E(\varepsilon_\theta - \alpha T)$$

Therefore,

$$\sigma_\theta = E\left\{\frac{1}{r_0 - y}\left[\left(r_0' - r_0 - \int_0^y \alpha T\, dy\right) + \frac{\delta\,\Delta\theta}{\Delta\theta}(r_0' - y)\right] - \alpha T\right\}$$

The two unknowns r_0' and $\dfrac{\delta\,\Delta\theta}{\Delta\theta}$ are determined from the boundary conditions of the beam. Since the beam is free of external loadings, we should have

$$\iint \sigma_\theta\, dA = 0; \qquad \iint \sigma_\theta\, y\, dA = 0\,.$$

ENERGY METHODS

The analysis of stress and strain at a point. Except for the condition that the material we considered was a continuum, the shape or size of the body as a whole was not considered. The stresses and strains at a point were related through the material or the constitutive equations. Here too, the material properties rather than the behaviour of the body as such was not considered. Chapter 4, on the theory of failure, also discussed the critical conditions to impend failure at a point. In this chapter, we shall consider the entire body or structural member or machine element, along with the forces acting on it. Hooke's law will relate the force acting on the body to the displacement. When the body deforms under the action of the externally applied forces, the work done by these forces is stored as strain energy inside the body, which can be recovered when the latter is elastic in nature. It is assumed that the forces are applied gradually.

The strain energy methods are extremely important for the solution of many problems in the mechanics of solids and in structural analysis. Many of the theorems developed in this chapter can be used with great advantage to solve displacement problems and statically indeterminate structures and frameworks.

HOOKE'S LAW AND THE PRINCIPLE OF SUPERPOSITION

The rectangular stress components at a point can be related to the rectangular strain components at the same point through a set of linear equations that were designated as the generalised Hooke's Law. In this chapter, however, we shall state Hooke's law as applicable to the elastic body as a whole, i.e. relate the complete system of forces acting on the body to the deformation of the body as a whole. The law asserts that 'deflections are proportional to the forces which produce them; This is a very general assertion without any restriction as to the shape or size of the loaded body.

A force F_1 is applied at point 1, and in consequence, point 2 undergoes a displacement or a deflection, which according to Hooke's law, is proportionate to F_1. This deflection of point 2 may take place in a direction which is quite different from that F_1. If D_2 is the actual deflection, we have

$$D_1 = K_{21}F_1$$

where K_{21} is some proportionality constant.

When F_1 is increased, D_2 also increases proportionately. Let d_2 component of D_2 in a specified direction. If θ is the angle between D_2 and d_2

$$d_2 = D_2 \cos\ \theta = k_{21} \cos\ \theta F_1$$

If we keep θ constant, i.e. if we fix our attention on the deflection in a specified direction, then

$$d_2 = a_{21}F_1$$

where a_{21} is a constant. Therefore, one can consider the displacement of point 2 in any specified direction and apply Hooke's Law. Let us consider the vertical component of the deflection of point 2. If d_2 is the vertical component, then Hooke's law asserts that

$$d_2 = a_{21}F_1$$

where a_{21} is a constant called the 'influence coefficient' for vertical deflection at point 2 due to a force applied in the specified direction (that of F_1) at point 1. If F_1 is a unit force, then a_{21} is the actual value of the vertical deflection at 2.

If force equal and opposite to F_1 is applied at 1, then a deflection equal and opposite to the earlier deflection takes place. If several forces, all having the direction of F_1, are applied simultaneously at 1, the resultant vertical deflection which they produce at 2 will be the resultant of the deflections which they would have produced if applied separately. This is the principle of superposition.

Consider a force F_3 acting alone at point 3, and let d_2' be the vertical component of the deflection of 2. Then, according to Hooke's Law, as stated by Eq. (1)

$$d_2' = a_{23}^{\bullet}F_3$$

where a_{23} is the influence coefficient for vertical deflection at point 2 due to a force applied in the specified direction (that of F_3) at point 3. The question that we now examine is whether the principle of superposition holds true to two or more forces, such as F_1 and F_3, which act in different directions and at different points.

Let F_1 be applied first, and then F_3. The vertical deflection at 2 is

$$d_2 = a_{21}F_1 + a_{23}'F_3 \qquad (3)$$

where a_{23}' may be different from a_{23}. This difference, if it exists, is due to the presence of F_1 when F_3 is applied. Now apply $-\ F_1$. Then

$$d_2'' = \mathrm{a}_{21}F_1 + a_{23}'F_3 - a_{21}'F1$$

a_{21}' may be different from a_{21} since F_3 is acting when $-\ F_1$is applied. Only F_3 is acting now. If we apply $-F_3$, the deflection finally becomes

$$d_2' = a_{21}F_1 + a_{23}'F_3 - a_{21}'F1 - a_{23}F_3 \qquad (4)$$

Since the elastic body is not subjected to any force now, the final deflection given by Eq. (4) must be zero. Hence,

$$a_{21}F_1 + a_{23}'F_3 - a_{21}'F_1 - a_{23}F_3 = 0$$

i.e. $$(a_{21} - a_{21}')F_1 = (a_{23} - a_{23}')F_3$$

or $$\frac{a_{21} - a_{21}'}{F_3} = \frac{a_{23} - a_{23}'}{F_3} \qquad (5)$$

The difference $a_{21} - a_{21}'$, if it exists, must be due to the action of F_3. Hence, the left-hand side is a function of F_3 alone. Similarly, if the difference

$a_{21} - a'_{21}$ exists, it must be due to the action of F_1 and, therefore, the right-hand side must be a function F_1 alone. Consequently, Eq. (5) becomes

$$\frac{a_{21} - a'_{21}}{F_3} = \frac{a_{23} - a'_{23}}{F_3} = k \quad (6)$$

where k is a constant independent of F_1 and F_2. Hence

$$a'_{23} = a_{23} - kF_1$$

Substituting this in Eq. (3)

$$d_2 = a_{21}F_1 + a_{23}F_3 - kF_1F_3$$

The last term on the right-hand side in the above equation is non-linear, which is contradictory to Hooke's law, unless k vanishes. Hence, $k = 0$, and

$$a_{23} = a'_{23} \quad \text{and} \quad a_{21} = a'_{21}$$

The principle of superposition is, therefore, valid for two different forces acting at two different points. This can be extended by induction to include a third or any number of other forces. This means that the deflection at 2 due to any number of forces, Including force F_2 at 2 is

$$d_2 = a_{21}F_1 + a_{22}F_2 + a_{23}F_3 + \ldots \quad (7)$$

CORRESPONDING FORCE AND DISPLACEMENT OR WORK-ABSORBING COMPONENT OF DISPLACEMENT

Consider an elastic body which is in equilibrium under the action of forces $F_1, F_2, F_3, \ldots$ The forces of reaction at the points of support will also be considered as applied forces. The displacement d_1 in a specified direction at point 1 is given by Eq. (7).

If the actual displacement is D_1 and takes place in a direction as shown in Fig. (2), then the component of this displacement in the direction of force F_1 is called the corresponding displacement at point 1. This corresponding displacement is denoted by δ_1.

At every loaded point, a corresponding displacement can be identified. If the points of support *a, b,* and *c* do not yield, then at these points the corresponding displacements are zero. One can apply Hooke's law to these corresponding displacements and obtain from E. (7)

$$\delta_1 = a_{11}F_1 + a_{12}F_2 + a_{13}F_3 + \ldots$$
$$\delta_2 = a_{21}F_1 + a_{22}F_2 + a_{23}F_3 + \ldots \text{ etc.} \quad (8)$$

where $a_{11}, a_{12}, a_{13}, \ldots$, are the influence coefficients of the kind discussed earlier. The corresponding displacement is also called the work-absorbing component of the displacement.

WORK DONE BY FORCES AND ELASTIC STRAIN ENERGY STORED

Equations (8) show that the displacements δ_1, δ_2, . . . etc., depend on all the forces F_1, F_2, . . ., etc. If we slowly increase the forces F_1, F_2, . . ., etc. from zero to their full magnitudes, the deflections also increase similarly. For example, when the forces F_1, F_2, . . ., etc. are one-half of their full magnitudes, the deflections are

$$\frac{1}{2}\delta_1 = a_{11}\left(\frac{1}{2}F_1\right) + a_{12}\left(\frac{1}{2}F_2\right) + \ldots,$$

$$\frac{1}{2}\delta_2 = a_{21}\left(\frac{1}{2}F_1\right) + a_{22}\left(\frac{1}{2}F_2\right) + \ldots, \text{ etc.,}$$

i.e. the deflections reached are also equal to half their full magnitudes. Similarly, when F_1, F_2, . . ., etc. reach two-thirds of their full magnitudes, the deflections reached are also equal to two-thirds of their full magnitudes. Assuming that the forces are increased in constant proportion and the increase is gradual, the work done by F_1 at its point of application will be

$$W_1 = \frac{1}{2}F_1\delta_1$$

$$= \frac{1}{2}F_1(a_{11}F_1 + a_{12}F_2 + a_{13}F_3 + \ldots) \tag{9}$$

Similar expressions hold good for other forces also. The total work done by external forces is, therefore, given by

$$W_1 + W_2 + W_3 + \ldots = \frac{1}{2}(F_1\delta_1 + F_2\delta_2 + F_3\delta_3 + \ldots)$$

If the supports are rigid, then no work is done by the support reactions. When the forces are gradually reduced to zero, keeping their ratios constant, negative work will be done and the total work will be recovered. This shows that the work done is stored as potential energy and its magnitude should be independent of the order in which the forces are applied. If it were not so, it would be possible to store or extract energy by merely changing the order of loading and unloading. This would be contradictory to the principle of conservation of energy.

The potential energy that is stored as a consequence of the deformation of any elastic body is termed elastic strain energy. If F_1, F_2, F_3 are the forces in a particular configuration and δ_1, δ_2, δ_3 are the corresponding displacements then the elastic strain energy stored is

$$U = \frac{1}{2}(F_1\delta_1 + F_2\delta_2 + F_3\delta_3 + \ldots) \tag{10}$$

It must be noted that though this expression has been obtained on the assumption that the forces F_1, F_2, F_3 are increased in constant proportion, the conservation of energy principle and the superposition principle dictate that this expression for U must hold without restriction on the manner or order of the application of these forces.

RECIPROCAL RELATION

It is easy to show that the influence coefficient a_{12} in Eq. (8) is equal to the influence coefficient a_{21}. In general, $a_{ij} = a_{ij}$. To show this, consider a force F_1 applied at point 1, and let δ_1 be the corresponding displacement. The energy stored is

$$U = \frac{1}{2}F_1\delta_1 \quad = \frac{1}{2}a_{11}F_1^2$$

since $\quad \delta_1 = a_{11}F_1$

Next, apply force F_2 at point 2. The corresponding deflection at point 2 is $a_{22}F_2$ and that at point 1 is $a_{22}F_2$. During this displacement, force F_1 is fully acting and hence, the additional energy stored is

$$U_2 = \frac{1}{2}F_2(a_{22}F_2) + F_1(a_{12}F_2)$$

The total elastic energy stored is therefore

$$U = U_1 + U_2 = \frac{1}{2}a_{11}F_1^2 + \frac{1}{2}a_{22}F_2^2 + a_{12}F_1F_2$$

Now, if F_2 is applied before F_1, the elastic energy stored is

$$U' = \frac{1}{2}a_{22}F_2^2 + \frac{1}{2}a_{11}F_1^2 + a_{21}F_1F_2$$

Since the elastic energy stored is independent of the order of application of F_1 and F_2, U and U' must be equal. Consequently,

$$a_{12} = a_{21} \tag{11a}$$

or in general

$$a_{ij} = a_{ij} \tag{11b}$$

The result expressed in Eq. (11b) has great importance in the mechanics of solids, as shown in the next section.

One can obtain an expression for the elastic strain energy in terms of the applied forces, using the above reciprocal relationship. From Eq. (10)

$$U = \frac{1}{2}(F_1\delta_1 + F_2\delta_2 + \ldots + F_n\delta_n)$$

$$= \frac{1}{2}F_1(a_{11}F_1 + a_{12}F_2 + \ldots + a_{1n}F_n)$$

$$+ \ldots \frac{1}{2}F_n(a_{n1}F_1 + a_{n2}F_2 + \ldots + a_{nn}F_n)$$

$$U = \frac{1}{2}(a_{11}F_1^2 + a_{22}F_2^2 + \ldots + a_{nn}F_n^2)$$

$$+ \frac{1}{2}(a_{12}F_1F_2 + a_{13}F_1F_3 + \ldots + a_{1n}F_1F_n + \ldots)$$

$$= \frac{1}{2}\Sigma(a_{11}F_1^2) + \Sigma(a_{12}F_1F_2) \tag{12}$$

MAXWELL-BETTI-RAYLEIGH RECIPROCAL THEOREM

Consider two systems of forces F_1, F_2, . . ., and F_1, F_2, . . ., both systems having the same points of application and the same directions. Let δ_1, δ_2, . . ., be the corresponding displacements caused by F_1, F_2, . . ., and δ_1, δ_2, . . ., the corresponding displacements caused by $F_1', F_2' \ldots,$ Then, making use of the reciprocal relation given by Eq. (11) we have

$$F_1'\delta_1 + F_2'\delta_2 + \ldots + F_n'\delta_n$$

$$= F_1'(a_{11}F_1 + a_{12}F_2 + \ldots + a_{1n}F_n)$$

$$+F_1'(a_{11}F_1 + a_{12}F_2 + \ldots + a_{1n}F_n)$$

$$+F_2'(a_{21}F_1 + a_{22}F_2 + \ldots + a_{2n}F_n)$$

$$+ \ldots + F_n'(a_{n1}F_1 + a_{n2}F_2 + \ldots + a_{nn}F_n)$$

$$= a_{11}F_1\ F_1' + a_{22}F_2\ F_2' + a_{nn}F_n\ F_n'$$

$$+a_{12}(F_1'F_2 + F_2'F_1) + a_{13}(F_1'F_3 + F_3'F_1)$$

$$+ \ldots + a_{1n}(F_1'F_n + F_n'F_1) \tag{13}$$

The symmetry of the expressions between the primed and unprimed quantities in the above expression shows that it is equal to

$$F_1\delta_1' + F_2\delta_2' + \ldots + F_n\delta_n'$$

i.e. $$F_1\delta_1' + F_2\delta_2' + \ldots = F_1\delta_1' + F_2\delta_2' + \ldots \tag{14}$$

In words:

'The forces of the first system (F_1, F_2, . . ., etc.) acting through the corresponding displacements produced by any second system (F_1, F_2, . . ., etc.) do the same amount of work as that done by the second system of forces acting through the corresponding displacements produced by the first system of forces'.

This is the reciprocal theorem of Maxwell, Betti and Rayleigh.

GENERALISED FORCES AND DISPLACEMENTS

In the above discussions, F_1 F_2, . . . , etc. represented concentrated forces and δ_1, δ_2, . . . , etc. the corresponding linear displacements. It is possible to extend the term 'force' to include not only a concentrated force but also a bending moment or a torque. Similarly, the term 'displacement' may mean linear or angular displacement. Consider, for example, the elastic body, subjected to a concentrated force F_1 at point 1 and a couple $F_2 = M$ at point 2. δ_1 will now stand for the corresponding linear displacement of point 1 and δ_2 for the corresponding angular rotation of point 2. If F_1 is a unit force acting alone, then a_{11}, the influence coefficient, gives the linear displacement of point 1 corresponding linear displacement of point 1 caused by a unit couple F_2 applied at point 2. a_{21} gives the corresponding angular rotation of point 2 caused by a unit concentrated force F_1 at point 1.

The reciprocal relation $a_{12} = a_{21}$ can also be interpreted appropriately. For example, making reference to Fig. 3, the above relation reveals that the linear displacement at point 1 in the direction of F_1 caused by a unit couple acting alone at point 2, is equal to the angular rotation at point 2 in the direction of the moment F_2 caused by a unit load acting alone at point 1. This fact will be demonstrated in the next few examples.

With the above generalised definitions for forces and displacements, the work done when the forces are gradually increased from zero to their full magnitudes is given by

$$W = \frac{1}{2}(F_1\delta_1 + F_2\delta_2 + . . . + F_n\delta_n)$$

The reciprocal theorem of Maxwell, Betti and Rayleigh can also be given wider meaning with these extended definitions.

Example 1:

Determine the change in volume of an elastic body subjected to two equal and opposite force, as shown. The distance between the points of application is h and the elastic constants for the material are E and V, This is a very general problem, the solution of which is apparently difficult. However; we can get a solution very easily by applying the reciprocal theorem. Let the elastic body be subjected to a hydrostatic pressure of value σ. Every volume element will be in a state of hydrostatic (isotropic) stress. Consequently, the unit contraction in any direction.

$$\varepsilon = \frac{\sigma}{E} - 2v\frac{\sigma}{E} = (1-2v)\frac{\sigma}{E}$$

Solution:

The two points of application A and B, therefore, move towards each other by a distance

$$\Delta h = h(1-2v)\frac{\sigma}{E}$$

Now we have two systems of forces:

System 1	Force	P
	Volume change	ΔV
System 2	Force	σ
	Distance change	Δh

From the reciprocal theorem

$$P\,\Delta h = \sigma\,\Delta V$$

or

$$\Delta V = \frac{P}{\sigma}\Delta h$$

$$= \frac{Ph}{E}(1-2v)$$

If v is equal to 0.5, the change in volume is zero.

BEGG'S DEFORMETER

In this section, we shall demonstrate the application of the reciprocal theorem to a problem in experimental mechanics. Figure 5.8 shows a structural member subjected to a force P at point E. It is required to determine the forces of reaction at point B.

The reaction forces are V, H and M and these make the displacements (vertical, horizontal and angular) at B equal to zero. A theoretical analysis is quite difficult for an odd structure like the one shown. The reactions at the other supports also are such that the displacement at these supports are zero. To determine V at B we proceed as follows.

A known vertical displacement δ'_2 is imposed at B, keeping A, C, D fixed and preventing angular rotation and horizontal displacement at B. The corresponding displacement at E (i.e. displacement in the direction of P) is measured. Let this be δ_1. During the vertical displacement of B, the forces V', M' and H' that are induced at B are not measured. The two systems involved in the reciprocal theorem are as follows:

System 1 Specified

Forces V, H, M at B (unknown) and other reactive forces at A, C, D (also unknown), P at E (known)

Corresponding displacements 0, 0, 0 at B; 0, 0, 0 at A, C and D; δ_1 (unknown) at E.

System 2 Experimental

Forces V', H', M' at B (unknown) and other reactive forces at A, C, D (all unknown); 0 at E (i.e. point E not loaded)

Corresponding displacements $\delta_{2'}$, 0, 0 at B; 0, 0, 0 at A, C and D; $\delta_{1'}$ at E

Applying the reciprocal theorem.

$$(V \cdot \delta_2) + (H \cdot 0) + (M \cdot 0) + 0 + (P \cdot \delta_1)$$
$$= (V' \cdot 0) + (H' \cdot 0) + (M' \cdot 0) + 0 + (P \cdot \delta_1)$$

i.e.
$$V = -P\frac{\delta_1^{'}}{\delta_2^{'}} \tag{5.15}$$

Since $\delta_2^{'}$ is the known displacement imposed at B and $\delta_1^{'}$ is the corresponding displacement at E that is experimentally measured, the value of V can be determined. It is necessary to remember that the corresponding displacement $\delta_1^{'}$ at E is positive when it is in the direction of P.

To determine H at B, we proceed as above. A known horizontal displacement $\delta_2^{'}$ is imposed t B, with all other displacements being kept zero. The corresponding displacement $\delta_1^{'}$ at E is measured. The result is

$$M = -P\frac{\delta_1^{'}}{\theta'}$$

FIRST THEOREM OF CASTIGLIANO

From Eq. (12), the expression for the elastic strain energy is

$$U = \frac{1}{2}\left(a_{11}F_1^2 + a_{22}F_2^2 + \ldots + a_{nn}F_n^2\right)$$
$$+ (a_{12}F_1F_2 + a_{13}F_1F_3 + \ldots + a_{1n}F_1F_n) \ldots$$

In the above expression, F_1, F_2, etc. are the generalised forces, i.e. concentrated loads, moments or torques. a_{11}, a_{12}, . . . , etc. are the corresponding influence coefficients. The rate at which U increases with F_1 is given by $\frac{\partial U}{\partial F_1}$.

From the above expression for U,

$$\frac{\partial U}{\partial F_1} = a_{11}F_1 + a_{12}F_2 + a_{13}F_3 + \ldots + a_{1n}F_n$$

This is nothing but the corresponding displacement at F_1. Hence, if δ_1 stands for the generalised displacement (linear or angular) corresponding to the generalised force F_1, then

$$\frac{\partial U}{\partial F_1} = \delta_1 \tag{16}$$

In exactly the same way, one can show that

$$\frac{\partial U}{\partial F_2} = \delta_2, \quad \frac{\partial U}{\partial F_3} = \delta_3, \ldots, \text{ etc.}$$

That is to say, 'the partial differential coefficient of the strain energy function with respect to F_r gives the displacement corresponding with F_r'. This is Castigliano's first theorem. In the form derived in Eq. (16), the theorem is applicable to only linearly elastic bodies, i.e. bodies satisfying Hooke's Law.

This theorem is extremely useful in determining the displacements of structures as well as in the solutions of many statically indeterminate structures. Several examples will illustrate these subsequently. We can given an alternative proof for this theorem as follows:

Consider an elastic system in equilibrium under the force $F_1, F_2, \ldots F_n$, etc. Some of these are concentrated loads and some are couples and torques. Let the strain energy stored be U. Now increase one of the forces, say F_n, by ΔF_n and as a result the strain energy increases to $U + \Delta U$, where

$$\Delta U = \frac{\Delta U}{\Delta F_n} \Delta F_n$$

Now we calculate the strain energy in a different manner. Let the elastic system be free of all forces. Let ΔF_n be applied first. The energy stored is

$$\frac{1}{2} \Delta F_n \, \Delta \delta_n$$

where $\Delta \delta_n$ is the elementary displacement corresponding to ΔF_n. This is a quantity of the second order which can be neglected since ΔF_n will be made to tend to zero in the limit. Next, we put all the other forces, F_1 $F_2, \ldots,$ etc. These forces by themselves do an amount of work equal to U. But while these displacements are taking place, the elementary force ΔF_n is acting all the time with full magnitude at the point n which is undergoing a displacement δ_n. Hence, this elementary force does work equal to $\Delta F_n \, \delta_n$. The total energy stored is therefore

$$U + \Delta F_n \, \delta_n + \frac{1}{2} \Delta F_n \, \Delta \delta_n$$

Equating this to the previous expression, we get

$$U + \frac{\Delta U}{\Delta F_n} \Delta F_n = U + \Delta F_n \, \delta_n + \frac{1}{2} \Delta F_n \, \Delta \delta_n$$

In the limit, when $\Delta F_n \to 0$

$$\frac{\Delta U}{\Delta F_n} = \delta_n$$

it is important to note that δ_n is a linear displacement if F_n is a concentrated load, or an angular displacement if F_n is a couple or a torque. Further, we must express the strain energy in terms of the forces (including moments and couples) since it is the partial derivative with respect to a particular force that gives the corresponding displacement. In the next section, expressions for strain energies in terms of forces will be obtained.

EXPRESSIONS FOR STRAIN ENERGY

In this section we shall develop expressions for strain energy when an elastic member is subjected to axial force, shear force, bending moment and torsion. An elastic member subjected to several forces. Consider a section of the member at C. In general, this section will be subjected to three forces F_x, F_y and F_z and three moments M_x, M_y and M_z.

The force F_x is the axial force and forces F_y and F_z are the shear forces across the section. Moment M_x is the torque T and moments M_y and M_z are the bending moments about the y and z axes respectively. Let Δs be an elementary length of the member; then when Δs is very small, we can assume that these forces and moments remain constant over Δs. At the left-hand section of this elementary member, the forces and moments have opposite signs.

During the deformation caused by the axial force F_x alone, the remaining forces and moments do no work. Similarly, during the twist caused by the torque $T = M_x$ no work is assumed to be done (since the deformations are extremely small) by the other forces and moments.

Consequently, the work done by each of these forces and moments can be determined individually and added together to determine the total elastic strain energy stored by Δs while it undergoes deformation. We shall make use of the formulas available from elementary strength of materials.

(i) Elastic energy due to axial force: If δ_x is the axial extension due to F_x, then

$$\Delta U = \frac{1}{2} F_n \delta_n$$

$$= \frac{1}{2} F_x \cdot \frac{F_x}{AE} \Delta s$$

using Hooke's law.

$$\therefore \qquad \Delta U = \frac{F_x^2}{2AE} \Delta s \tag{17}$$

A is the cross-sectional area and E is Young's modulus.

(ii) Elastic energy due to shear force: The shear force F_y (or F_z) is distributed across the section in a complicated manner depending on the shape of the cross-section.

If we assume that the shear force is distributed uniformly across the section (which is not strictly correct), the shear displacement will be $\Delta s\ \Delta\gamma$ and the work done by F_y will be

$$\Delta U = \frac{1}{2} F_y \Delta s\ \Delta\gamma$$

From Hooke's law,

$$\Delta\gamma = \frac{F_y}{AG}$$

where A is the cross-sectional area and G is the shear modulus. Substituting this

$$\Delta U = \frac{1}{2} F_y\ \Delta s \frac{F_y}{AG}$$

or

$$\Delta U = \frac{F_y^2}{2AG} \Delta s$$

It will be shown that the strain energy due to shear deformation is extremely small, which is often ignored. Hence, the error caused in assuming uniform distribution of the shear force across the section will be very small. However, to take into account the different cross-sections and non-uniform distribution, a factor k is introduced. With this

$$\Delta U = \frac{kF_y^2}{2AG} \Delta s \tag{18}$$

A similar expression is obtained for the shear force F_z.

(iii) Elastic energy due to bending moment: If $\Delta\phi$ is the angle of rotation due to the moment M_z (or M_y), the work done is

$$\Delta U = \frac{1}{2} M_z \, \Delta\phi$$

From the elementary flexure formula, we have

$$\frac{M_z}{I_z} = \frac{E}{R}$$

or $$\frac{1}{R} = \frac{M_z}{EI_z}$$

where R is the radius of curvature

and I_z is the area moment of inertia about the z axis. Hence

$$\Delta\phi = \frac{\Delta s}{R} = \frac{M_z}{EI_z} \Delta s$$

Substituting this

$$\Delta U = \frac{M_z^2}{2EI_z} \Delta s \tag{19}$$

A similar expression can be obtained for the moment M_y.

(iv) Elastic energy due to torque: Because of the torque T, the elementary member rotates through an angle $\Delta\theta$ according to the formula for a circular section

$$\frac{T}{I_p} = \frac{G\Delta\theta}{\Delta s}$$

i.e. $$\Delta\theta = \frac{T}{GI_p} \Delta s$$

I_p is the polar moment of inertia. The work done due to this twist is,

$$\Delta U = \frac{1}{2} T \Delta\theta$$

$$= \frac{T^2}{2GI_p} \Delta s \tag{20}$$

Equations (17) - (20) give important expressions for the strain energy stored in the elementary length Δs of the elastic member. The elastic energy for the entire member is therefore

(i) Due to axial force
$$U_1 = \int_0^s \frac{F_x^2}{2AE} ds \qquad (21)$$

(ii) Due to shear force
$$U_2 = \int_0^s \frac{K_y F_y^2}{2AG} ds \qquad (22)$$

$$U_3 = \int_0^s \frac{K_z F_z^2}{2AG} ds \qquad (23)$$

(iii) Due to bending moment
$$U_4 = \int_0^s \frac{M_y^2}{2EI_y} ds \qquad (24)$$

$$U_5 = \int_0^s \frac{M_z^2}{2EI_z} ds \qquad (25)$$

(iv) Due to torque
$$U_6 = \int_0^s \frac{T^2}{2GI_p} ds \qquad (26)$$

Example 2:

Determine the deflection at end A of the cantilever beam

The bending moment at any section x is

$$M = Px$$

Solution:

The elastic energy due to bending moment is, therefore, from Eq. (24)

$$U_1 = \int_0^L \frac{(Px)^2 dx}{2EI} = \frac{P^2 L^3}{6EI}$$

The elastic energy due to shear from Eq. (22) is (putting $k_1 = 1$)

$$U_2 = \int_0^L \frac{P^2}{2AG} dx = \frac{P^2 L}{2AG}$$

One can now show that U_2 is small as compared to U_1. It the beam is of a rectangular section

$$A = bd, \quad I = \frac{1}{12} bd^3$$

and $\quad 2G \approx E$

Substituting these

$$\frac{U_2}{U_1} = \frac{P^2 L}{2bdG} \cdot \frac{6bd^3}{12P^2 L^3} \cdot 2G$$

$$= \frac{d^2}{2L^2}$$

For a member to be designated as beam, the length must be fairly large compared to the cross-sectional dimension. Hence, $L > d$ and the above ratio is extremely small. Consequently, one can neglect shear energy as compared to bending energy. With

$$U = \frac{P^2 L^3}{6EI}$$

we get

$$\frac{\partial U}{\partial P} = \frac{PL^3}{3EI} = \delta_A$$

which agrees with the solution from elementary strength of materials.

FICTITIOUS LOAD METHOD

Castigliano's first theorem described above helps us to determine the displacement at point corresponding to the force acting there. Situations arise where it may be desirable to determine the displacement (either linear or angular) at a point where there is no force (concentrated load or a couple) acting. In such situations, we assume a small fictitious or dummy load to be acting at the point where the displacement is required. Castigliano's theorem is then applied, and in the final result, the fictitious load is put equal to zero. The following example will describe the technique.

Example 3:

Determine the slope at end A of the cantilever in Fig. 18 which is subjected to load P.

To determine the slope by Castigliano's method we have to determine U and take its partial derivative with respect to the corresponding force, i.e. a moment. But no moment is acting at A. So, we assume a fictitious moment M to be acting at A and determine the slope caused by P and M. Since the magnitude of M is actually zero, in the final result, M is equated to zero.

Solution:

The energy due to P and M is,

$$U = \int_{1}^{L} \frac{(Px + M)^2}{2EI} dx$$

$$= \frac{P^2 L^3}{6EI} + \frac{M^2 L}{2EI} + \frac{MPL^2}{2EI}$$

$$\theta = \frac{\partial U}{\partial T} = \frac{ML}{EI} + \frac{PL^2}{2EI}$$

This gives the slope when M and P are both acting. If M is zero, the slope due to P alone is

$$\theta = \frac{PL^2}{2EI}$$

If on the other hand, P is zero and M alone is acting the slope is

$$\theta = \frac{ML}{EI}$$

Example 4:

For the member shown in Fig. 16, Example 8, determine the ratio of L to r if the horizontal and vertical deflections of the loaded end A are equal. P is the only force acting.

Solution:

In addition to the vertical for P at A, apply a horizontal fictitious force F to the right. The bending moment at section θ of the semi-circular part is

$$M_1 = Pr\,(1 - \cos\theta) - Fr\sin\theta)$$

At any section x in the vertical part, the moment is

$$M_2 = 2Pr + Fx$$

Hence,

$$U = \frac{1}{2EI}\int_0^{\pi}[\Pr(1-\cos\theta) - Fr\sin\theta]^2 r\, d\theta \frac{1}{2EI}\int_0^{L}(2\Pr + Fx)^2 dx$$

$$\therefore \quad \frac{\partial U}{\partial F} = -\frac{r^2}{EI}\int_0^{\pi}[\Pr(1-\cos\theta) - Fr\sin\theta]\sin\theta d\theta$$

$$+\frac{1}{EI}\int_0^{L}(2\Pr + Fx)x dx$$

and

$$\left.\frac{\partial U}{\partial F}\right|_{F=0} = \delta_h = -\frac{r^2}{EI}\int_0^{\pi}[\Pr(1-\cos\theta)\sin\theta]d\theta + \frac{1}{EI}\int_0^{L}2\Pr\, x dx$$

$$= -\frac{2pr^3}{EI} + \frac{prL^2}{EI} = \frac{pr}{EI}(-2r^2 + L^2)$$

From Example 8

$$\delta_v = \frac{pr^2}{EI}\left(\frac{3}{2}\pi r + 4L\right)$$

Equating δ_v to δ_h

$$\frac{Pr^2}{EI}\left(\frac{3}{2}\pi r + 4L\right) = \frac{pr}{EI}(-2r^2 + L^2)$$

or $$+L^2 - 4Lr - r^2\left(\frac{3\pi}{2} + 2\right) = 0$$

Dividing by r^2 and putting $\frac{L}{r} = \rho$

$$\rho^2 - 4\rho - \left(\frac{3\pi}{2} + 2\right) = 0$$

Solving, $$\rho = \frac{4 \pm \sqrt{[16 + 4(3\pi/2 + 2)]}}{2}$$

or $$\rho = 2 + \sqrt{6 + \frac{3}{2}\pi}$$

SUPERPOSITION OF ELASTIC ENERGIES

When an elastic body is subjected to several forces, one cannot obtain the total elastic energy by adding the energies caused by individual forces. In other words, the sum of individual energies is not equal to the total energy of the system. The reason for this is simple. Consider an elastic body subjected to two forces F_1 and F_2. When F_1 is applied first, let the energy stored be U_1.

When F_2 is applied next (with F_1 continuing to act), the additional energy stored is equal to U_2 due to F_2 alone, plus the work done by F_1 during the displacement caused by F_2. Hence, the total energy stored when both F_1 and F_2 are acting is equal to $(U_1 + U_2 + U_3)$, where U_1 is the work energy caused by F_1 alone, U_2 is the work energy caused by F_2 alone, and U_3 is the energy due to the work done by F_1 during the displacement caused by F_2. Another way of observing this is to note that the strain energy functions are not linear functions. Hence, individual energies cannot be added to get the total energy. As a specific example, consider the cantilever shown in Fig. 16. Example 10. Let P and M be actual forces acting on the cantilever, i.e. M is not a fictitious force as was assumed in that example. The elastic energy stored due to P and M is given by (a), i.e.

$$U = \frac{P^2L^3}{6EI} + \frac{M^2L}{2EI} + \frac{MPL^2}{2EI}$$

The energy due to P alone is

$$U_1 = \frac{1}{2EI}\int_0^L (px)^2 dx = \frac{P^2L^3}{6EI}$$

Similarly, the energy due to M alone is

$$U_2 = \frac{1}{2EI}\int_0^L M^2 dx = \frac{M^2L}{2EI}$$

Obviously, $U_1 + U_2$ is not equal to U. However, if P is applied first and then M, the total energy is given by

$U_1 + U_2$ + work done by P during the displacement caused by M.

The deflection at the end of the cantilever (where P is acting with full magnitude) caused by M is

$$\delta_A^* = \frac{ML^2}{2EI}$$

During this deflection, the work done by P is

$$U_3 = P\left(\frac{ML^2}{2EI}\right)$$

If this additional energy is added to $U_1 + U_2$, then one gets the previous expression for U. It is immaterial whether P is applied first or M is applied. The order of loading is immaterial. Thus, one should be careful in applying the superposition principle to the energies. However, the individual energies caused by axial force, bending moment and torsion can be added since the force causing one kind of deformation will not do any work during a different kind of deformation caused by another force. For example, an axial force causing linear deformation will not do work during an angular deformation we have been assuming throughout our discussions. Similarly, a bending moment will not do any work during axial or linear displacement caused by an axial force.

STATICALLY INDETERMINATE STRUCTURE

Many statically indeterminate structural problems can be conveniently solved, using Castigliano's theorem. The technique is to determine the forces and moments to produce the required displacement. Example 7 was one such problem. The following example will further illustrate this method.

Example 5(a):

A rectangular frame with all four sides of equal cross section is subjected to forces P, as shown in Fig. 19. Determine the moment at section C and also the increase in the distance between the two points of application of force P.

Solution:

The symmetry conditions indicate that the top and bottom members deform in such a manner that the tangents at the point of loading remain horizontal. Also, there is no shear force across the sections $C - C$. Hence, one can consider only a quarter part of the frame, as shown in (b). Considering only the bending energy and neglecting the energies due to direct tension and shear force, we get

$$U' = \int_0^b \frac{M_0^2}{2EI} dx + \int_0^a \frac{(M_0 - P/2\, x)^2}{2EI} dx$$

$$= \frac{1}{2EI}\left(M_0^2 b + M_0^2 a - M_0 P \frac{a^2}{2} + \frac{1}{12} P^2 a^3 \right)$$

Because of symmetry, the change in slope at section C is zero. Hence

$$\frac{\partial U'}{\partial M_0} = \frac{1}{2EI}\left[2M_0(a+b) - \frac{1}{2} Pa^2 \right]$$

Equating this to zero,

$$M_0 = \frac{Pa^2}{4(a+b)}$$

To determine the increase in distance between the two load points, we determine the partial derivative of $4U'$ with respect to P (assuming that the bottom loaded point is held fixed).

$$U = 4U' = \frac{4}{2EI}\left[\frac{P^2 a^4}{16(a+b)^2}(a+b) - \frac{P^2 a^4}{8(a+b)} + \frac{P^2 a^3}{12} \right]$$

$$\therefore \quad \frac{\partial U}{\partial P} = \frac{Pa^3}{12EI} \frac{(a+4b)}{(a+b)}$$

Example 5(b):

A thin circular ring of radius r is subjected to two diametrically opposite loads P in its own plane. Obtain an expression for the bending moment at any section. Also, determine the change in the vertical diameter.

Because of symmetry, during deformation there is no change in the slopes at A and B. So, one can consider only a quarter of the ring for *calculation. The value of M_0 is such as to cause no change in slope at B. Section at A can be considered as built-in.*

Solution:

Moment at $\theta = M = \frac{P}{2} r(1-\cos\theta) - M_0$

$$U = \frac{1}{2EI} \int_0^{\pi/2} \left[\frac{P}{2} r(1-\cos\theta) - M_0 \right]^2 r d\theta$$

Since there is change in slope at B

$$\frac{\partial U}{\partial M_0} = -\frac{r}{2EI} \int_0^{\pi/2} \left[2\frac{P}{2} r(1-\cos\theta) - M_0 \right] d\theta = 0$$

i.e. $$\int_0^{\pi/2} \left[\frac{P}{2} r(1-\cos\theta) - M_0 \right] d\theta = 0$$

i.e. $$\frac{P}{2} r\left(\frac{\pi}{2} - 1\right) - M_0 \frac{\pi}{2} = 0$$

or $$M_0 = \frac{\Pr}{2}\left(1 - \frac{2}{\pi}\right)$$

$\therefore$ $$M \text{ at } \theta = \frac{P}{2} r(1-\cos\theta) - \frac{P}{2} r\left(1 - \frac{2}{\pi}\right) = \frac{\Pr}{2}\left(\frac{2}{\pi} - \cos\theta\right)$$

To determine the increase in the diameter alone the loads, one has to determine the elastic energy and take the differential.

$$U^* = \int_0^{\pi/2} \frac{1}{2EI} \left[\frac{\Pr}{2}\left(\frac{2}{\pi} - \cos\theta\right) \right]^2 r\, d\theta$$

The differential of this with respect to ($P/2$) will give the vertical deflection of the end B with reference to A. Observe that in order to determine the deflection at B, one ha to take the differential with respect to the particular load that is acting at that point, which is ($P/2$). Putting ($P/2$) = Q.

$$U^* = \frac{1}{2EI} \int_0^{\pi/2} \left[Qr\left(\frac{2}{\pi} - \cos\theta\right) \right]^2 r\, d\theta$$

$$= \frac{Q^2 r^3}{EI} \int_0^{\pi/2} \left(\frac{2}{\pi} - \cos\theta\right)^2 d\theta$$

$\therefore$ $$\frac{\partial U^*}{\partial Q} = \frac{Qr^3}{EI} \int_0^{\pi/2} \left(\frac{4}{\pi^2} + \cos^2\theta - \frac{4}{\pi}\cos\theta \right) d\theta$$

$$= \frac{Qr^3}{EI}\left(\frac{4}{\pi^2}\frac{\pi}{2}+\frac{\pi}{4}-\frac{4}{\pi}\right)$$

$$= \frac{Qr^3}{EI}\left(\frac{\pi}{4}-\frac{2}{\pi}\right)=\frac{\mathrm{Pr}^3}{2EI}\left(\frac{\pi}{4}-\frac{2}{\pi}\right)$$

As this gives only the increase in the radius, the increase in the diameter is twice this quantity, i.e.

$$\delta_v = \frac{\mathrm{Pr}^3}{EI}\left(\frac{\pi}{4}-\frac{2}{\pi}\right)$$

THEOREM OF VIRTUAL WORK

Consider an elastic system subjected to a number of forces (including moments) F_1, F_2, . . . , etc. Let δ_1, δ_2, . . . , etc. be the corresponding displacements. Remember that these are the work absorbing components (linear and angular displacements) in the corresponding directions of the forces.

Let one of the displacements δ_1 be increased by a small quantity $\Delta\delta_1$. During this additional displacement, all other displacements where forces are acting are held fixed, which means that additional forces may be necessary to maintain such a condition. Further, the small displacement $\Delta\delta_1$ that is imposed must be consistent with the constraints acting. For example, if point I is constrained in such a manner that it can move only in a particular direction, then $\Delta\delta_1$ must be consistent with such a constraint. A hypothetical virtual displacement, the forces F_1, F_2, . . . , etc. (except F_1) do no work at all because their points of application do not move (at least in the work-absorbing direction). The only force doing work is F_1 by an amount $F_1\,\Delta\delta_1$ plus a fraction of $\Delta F_1\,\Delta\delta_1$, caused by the change in F_1. This additional work is stored as strain energy ΔU. Hence

$$\Delta U = F_1\,\Delta\delta_1 + k\,\Delta F_1\,\Delta\delta_1$$

or
$$\frac{\Delta U}{\Delta\delta_1} = F_1 + k\,\Delta F_1$$

and
$$\underset{\Delta\delta_1\to 0}{Lt}\frac{\Delta U}{\Delta\delta_1} = \frac{\partial U}{\partial\delta_1} = F_1 \tag{27}$$

This is the theorem of virtual work. Note that in this case, the strain energy must be expressed in terms of δ_1, δ_2, . . . , etc. whereas in the application of Castigliano's theorem U had to be expressed in terms of F_1, F_2, . . . , etc.

It is important to observe that in obtaining the above equation, we have not assumed that the material is linearly elastic, i.e. that it obeys Hooke's law.

The theorem is applicable to any elastic body, linear or nonlinear, whereas Castigliano's first theorem, as derived in Eq. (16), is strictly applicable to linear elastic or Hookean materials.

Example 6:

There elastic members AD, BD, and CD are connected by smooth pins. All the members have the same cross-sectional areas and are of the same material. BD is 100 cm long and members AD and CD are each 200 cm long. What is the deflection of D under load W?

Solution:

Under the action of load W, it is possible for D to move vertically and horizontally. If δ_1 and δ_2 are the vertical and horizontal displacements, then according to the principle of virtual work.

$$\frac{\partial U}{\partial \delta_1} = W, \frac{\partial U}{\partial \delta_2} = 0$$

where U is the total strain energy of the system.

Because of δ_1, BD will not undergo any changes in length but AD will extend by $\delta_1 \cos\theta$ and CD will contract by the same amount, From Fig. (a),

$$\cos\theta = \sqrt{\frac{3}{2}}$$

Because of δ_2 and BD will extend by δ_2 and AD and CD each will extend by 1/2 δ_2. Hence, the total extension of each member is

AD extends by $\frac{1}{2}\left(\sqrt{3}\delta_1 + \delta_2\right)$ cm

BD extends by δ_2 cm

CD extends by $\frac{1}{2}\left(-\sqrt{3}\delta_1 + \delta_2\right)$ cm

To calculate the strain energy, one needs to know the force-deformation equation for the non-Hookean members. This aspect will be taken up in Sec. 17, and Example 17. For the present example, assuming Hooke's law, the forces in the members are (with δ as corresponding extensions)

$$\text{in } AD: \frac{aE\delta}{L} = aE\frac{1}{2}\left(\sqrt{3}\delta_1 + \delta_2\right)\frac{1}{200}$$

$$\text{in } BD: \frac{aE\delta}{L} = aE\delta_2\frac{1}{100}$$

$$\text{in } CD: \frac{aE\delta}{L} = aE\frac{1}{2}\left(-\sqrt{3}\delta_1 + \delta_2\right)\frac{1}{200}$$

The total elastic strain energy taking only axial forces into account is

$$U = \Sigma \frac{p^2 L}{2aE} = \frac{aE}{2}\left[\frac{1}{800}\left(\sqrt{3}\,\delta_1 + \delta_2\right)^2 + \frac{1}{100}\delta_2^2 + \frac{1}{800}\left(-\sqrt{3}\,\delta_1 + \delta_2\right)^2\right]$$

$$= aE\left(\frac{3}{800}\delta_1^2 + \frac{1}{160}\delta_2^2\right)$$

$$\therefore \qquad W = \frac{\partial U}{\partial \delta_1} = \frac{3aE}{400}\delta_1$$

and $$0 = \frac{\partial U}{\partial \delta_2} = \frac{aE}{80}\delta_2$$

Hence, δ_2 is zero, which means that D moves only vertically under W and the value of this vertical deflection δ_1 is

$$\delta_1 = \frac{400}{3aE}W$$

KIRCHHOFF'S THEOREM

In this section, we shall prove an important theorem dealing with the uniqueness of solution. First, we observe that the applied forces taken as a whole work on the body upon which they act. This means that some of the products $F_n\,\delta_n$ etc. may be negative but the sum of these products taken as a whole is positive. When the body is elastic, this work is stored as elastic strain energy. This amounts to the statement that U is an essentially positive quantity. If this were not so, it would have been possible to extract energy by applying an appropriate system of forces. Hence, every portion of the body must store positive energy or no energy at all. Accordingly, U will vanish only when every part of the body is undeformed. On the basis of this and the superposition principle, we can prove Kirchhoff's uniqueness theorem, which states the following:

An elastic body for which displacements are specified at some points and forces at others, will have a unique equilibrium configuration.

Let the specified displacements be $\delta_1, \delta_2, \ldots, \delta_r$ and the specified forces be $F_s, F_t, \ldots, F_n$. It is necessary to observe that it is not possible to prescribe simultaneously both force and displacement for one and the same point. Consequently, at those points where displacements are prescribed, the corresponding forces are $F_1^{'}, F_2^{'}, \ldots, F_r^{'}$ and at those points where forces are prescribed, the corresponding displacement are $\delta_s^{'}, \delta_t^{'}, \ldots, \delta_n^{'}$. Let this be the equilibrium configuration. If this system is not unique, then there should be another equilibrium configuration in which the forces corresponding to the displacements $\delta_1', \delta_2', \ldots, \delta_r'$ have the values $F_1^*, F_2^*, \ldots, F_r^*$ and the displacements corresponding to the forces F_s, F_t, . . . , F_n have the values $\delta_s^*, \delta_t^*, \ldots, \delta_n^*$. We therefore have two distinct systems.

First System Forces $F_1', F_2', \ldots, F_r', \; F_s, \; F_t, \ldots, \; F_n$

Corresponding displacements $\delta_1, \delta_2, \ldots, \delta_r, \; \delta_s', \; \delta_t', \ldots, \; \delta_n'$

Second System Forces $F_1^*, \; F_2^*, \ldots, F_r^* \; F_s, \; F_t, \ldots, \; F_n$

Corresponding displacements $\delta_1, \; \delta_2, \ldots, \; \delta_r \quad \delta_s^*, \; \delta_t^*, \ldots, \delta_n^*$

We have assumed that these are possible equilibrium configurations. Hence, by the principle of superposition the difference between these two systems must also be an equilibrium configuration. Subtracting the second system from the first, we get the third equilibrium configuration as

Forces $(F_1' - F_1^*), (F_2' - F_2^*), \ldots (F_r' - F_r^*); \quad 0, \ldots 0$

Corresponding displacements $0, 0, \ldots, 0 \qquad (\delta_s' - \delta_s^*), (\delta_t' - \delta_t^*), \ldots, (\delta_n' - \delta_n^*)$

The strain energy corresponding to the third system is $U = 0$. Consequently the body remains completely undeformed. This means that the first and second systems are identical, i.e. there is a unique equilibrium configuration.

SECOND THEOREM OF CASTIGLIANO OR MENABREA'S THEOREM

This theorem is of great importance in the solution of redundant structures or frames. Let a framework consist of m number of members and j number of joints. Then, if

$$M > 3j - 6$$

the frame is termed a redundant frame. The reason is as follows. For each joint, we can write three force equilibrium equations (in a general three-dimensional case), thus giving a total of $3j$ number of equations. However, all these equation are not independent, since all the external forces by themselves are in equilibrium and, therefore, satisfy the three force equilibrium equations and the three moment equilibrium equations.

Hence, the number of independent equations are $3j - 6$ and if the number of members exceed $3j - 6$, the frame is redundant. The number

$$N = m - 2j + 6$$

is termed the order of redundancy of the framework. If the skeleton diagram lies wholly in one plane, the framework is termed a plane frame. For a plane framework, the degree of redundancy is given by the number

$$N = m\ 2j + 3$$

Castigliano's second theorem (also known as Menabrea's theorem) can be stated as follows:

The forces developed in a redundant framework are such that the total elastic strain energy is a minimum.

Thus, if F_1, F_2 and F_r are the forces in the redundant members of a framework and U is the elastic strain energy, then

$$\frac{\partial U}{\partial F_1} = 0, \frac{\partial U}{\partial F_2} = 0, \ldots, \frac{\partial U}{\partial F_r} = 0$$

This is also called the principle of least work and can be proven as follows:

Let r be the number of redundant members. Remove the latter and replace their actions by their respective forces. Assuming that the values of these redundant forces $F_1, F_2, \ldots, F_r$ are known, the framework will have become statically determinate and the elastic strain energy of the remaining members can be determined. Let U_s be the strain energy of these members. Then by Castigliano's first theorem, the 'increase' in the distance between the joints a and b is given as

$$\delta'_{ab} = -\frac{\partial U_s}{\partial F_i} \tag{28}$$

The negative appears because of the direction of F_i. The reactive force on the redundant members ab being F_i, its length will increase by

$$\delta_{ab} = \frac{F_i\, l_i}{A_i\, E_i} \tag{29}$$

where l_i is the length and A_i is the sectional area of the member. The increase in the distance given by Eq. (28) must be equal to the increase in the length of the member ab, given by Eq. (29). Hence

$$-\frac{\partial U_s}{\partial F_i} = \frac{F_i\, l_i}{A_i\, E_i} \tag{30}$$

The elastic strain energies of the redundant members are

$$U_1 = \frac{F_1^2\, l_1}{2A_1\, E_1}, \quad U_2 = \frac{F_2^2\, l_2}{2A_2\, E_2}, \ldots, \quad U_r = \frac{F_r^2\, l_r}{2A_r\, E_r}$$

Hence, the total elastic energy of all redundant members is

$$U_1 + U_2 + \dots U_r = \frac{F_1^2\, l_1}{2A_1\, E_1} + \frac{F_2^2\, l_2}{2A_2\, E_2} + \dots + \frac{F_r^2\, l_r}{2A_r\, E_r}$$

$$\therefore \qquad \frac{\partial}{\partial F_1}(U_1 + U_2 + \dots U_r) = \frac{F_i\, l_i}{A_i\, E_i}$$

since all terms, other than the *i*th term on the right-hand side, will vanish when differentiated with respect to F_i. Substituting this in Eq. (30)

$$-\frac{\partial U_s}{\partial F_i} = \frac{\partial}{\partial F_i}(U_1 + U_2 + \dots U_r) = 0$$

or
$$\frac{\partial}{\partial F_i}(U_1 + U_2 + \dots U_r + U_s) = 0$$

The sum of the terms inside the parentheses is the total energy of the entire framework including the redundant members. If U is this total energy

$$\frac{\partial U}{\partial F_i} = 0$$

Similarly, by considering the redundant members one-by-one, we get

$$\frac{\partial U}{\partial F_1} = 0, \frac{\partial U}{\partial F_2} = 0, \dots, \frac{\partial U}{\partial F_r} = 0 \qquad (31)$$

This is the principle of least work.

Example 7:

The framework contains a redundant bar. All the members are of the same section and material. Determine the force in the horizontal redundant member.

Solution:

Let T be the tension in the member AB. The forces in the members are

Members	***Length***	***Force***
AB	$2\sqrt{3}\, h$	$+T$
AC, BD	h	$T/\sqrt{3} - P$
AF, BF	$2h$	$-2T/\sqrt{3} + 0$
CF, DF	$\sqrt{3}\, h$	$-T + P\sqrt{3}$
CE, DE	$2h$	$2T/\sqrt{3} - 2P$
FE	h	$-2T/\sqrt{3} + 0$

The total strain energy is

$$U = \frac{h}{2EA}\left[2\sqrt{3}T^2 + 2\left(P^2 + \frac{T^2}{3} - \frac{2PT}{\sqrt{3}}\right) + \frac{16T^2}{3} + \right.$$

$$2\sqrt{3}\left(T^2 + 3P^2 - 2PT\sqrt{3}\right) +$$

$$\left. 16\left(\frac{T^2}{3} + P^2 - \frac{2PT}{\sqrt{3}}\right) + \frac{4T^2}{3}\right]$$

The condition for minimum strain energy or least work is

$$\frac{\partial U}{\partial T} = 0 = \frac{h}{2EA}\left[4\sqrt{3T} + \frac{4T}{3} - \frac{4P}{\sqrt{3}} + \frac{32T}{3} + 4\sqrt{3}T - \right.$$

$$\left. 12P + \frac{32T}{3} - \frac{32}{\sqrt{3}}P + \frac{8T}{3}\right]$$

$$\therefore \qquad T\left(4\sqrt{3} + \frac{4}{3} + \frac{32}{3} + 4\sqrt{3} + \frac{32}{3} + \frac{8}{3}\right) = P\left(\frac{4}{\sqrt{3}} + 12\frac{32}{\sqrt{3}}\right)$$

or
$$T = \frac{9\left(\sqrt{3}+1\right)}{6\sqrt{3}+19}P$$

Example 8:

A cantilever is supported at the free end by an elastic spring of spring constant k.. The cantilever beam is uniformly loaded. The intensity of loading is W.

Solution:

Let R be the unknown reaction at A, i.e. R is the force on the spring. The strain energy in the spring is

$$U_1 = \frac{1}{2}R\delta = \frac{1}{2}R\frac{R}{K} = \frac{R^2}{2K}$$

where δ is the deflection of the spring. The strain energy in the beam is

$$U_2 = \int_0^L \frac{M^2 dx}{2EI}$$

$$\int_0^L \frac{(Rx - wx^2/2)^2 dx}{2EI}$$

$$= \frac{1}{EI}\left(\frac{1}{6}R^2L^3 + \frac{1}{40}w^2L^5 - \frac{1}{8}RwL^4\right)$$

Hence, the total strain energy for the system is

$$U+U_1+U_2=\frac{R^2}{2k}+\frac{1}{EI}\left(\frac{1}{6}R^2L^3+\frac{1}{40}w^2L^5-\frac{1}{8}RwL^4\right)$$

From Castigliano's second theorem

$$\frac{\partial U}{\partial R}=\frac{R}{k}+\frac{1}{EI}\left(\frac{1}{3}RL^3-\frac{1}{8}wL^4\right)=0$$

$$\therefore \quad R=\frac{3kwL^4}{8(3EI+kL^3)}$$

EXERCISES

1. A thick-walled tube has an internal radius of 4 cm and an external radius of 8 cm. It is subjected to an external pressure of 1000 kPa (10.24 kgf/cm^2). If E = 1.2 × 10^8 kPa (1.23 × 10^6 kgf/cm^2) and ν = 0.24, determine the internal pressure according to Mohr's theory of failure, which says that

$$(\sigma)_{max} - n\ (\sigma)_{min} \le \sigma_{\text{tenslie strength}}$$

where n is the ratio of σ-tenslie strength to σ-compressive strength. For the present problem, assume σ-tensile strength = 30000 kPa (307.2 kgf/cm^2) and σ-compressive strength = 120000 kPa (1228.8 kgf/cm^2). [*Ans.* P = 17000 kPa (174 kgf/cm^2)]

2. In the above problem, determine the changes in the radii.

$$\left[\begin{array}{l} Ans.\ \Delta r_1 = 0.01\ \text{m} \\ \quad\quad \Delta r_2 = 0.007\ \text{mm} \end{array}\right]$$

3. In Example 1, if one uses the energy of distortion theory, what will be the external radius of the cylinder? The rest of the data remain the same.

[*Ans.* = 6.05 cm]

4. A thick-walled tube with an internal radius of 10 cm is subjected to an internal pressure of 2000 kgf/cm^2 (196000 kPa). E = 2 × 10^6 kgf/cm^2 (196 × 10^6 kPa) and ν = 0.3. Determine the value of the external radius if the maximum shear stress developed is limited to 3000 kgf/cm^2 (294 × 10^6 kPa). Calculate the change in the internal radius due to the pressure.

$$\left[\begin{array}{l} Ans.\ r_2 = 17.3\ \text{cm} \\ \quad\quad \Delta r_1 = 0.023\ \text{cm} \end{array}\right]$$

5. Determine the pressure P_0 between the concrete tube and the perfectly rigid core. Assume $E_c = 2 \times 10^6$ kgf/cm², $r_c = 0.16$. Take $r_1/r_2 = 0.5$

[*Ans.* $P_0 = 17.4$ kgf/cm²]

6. Determine the dimensions of a two-piece composite tube of optimum dimensions if the internal pressure is 2000 kgf/cm² (196000 kPa), external pressure $P_c = 0$, internal radius $r_1 = 8$ cm and $E = 2 \times 10^6$ kgf/cm² (196 × 10⁶ kPa). The maximum shear stress is to be limited to 1500 kgf/cm² (147 × 10⁶ kPa). Check the strength according to the maximum shear theory.

$$\left[\begin{array}{l} Ans.\ r_2 \approx 14 \text{ cm};\ r_3 = 24 \text{ cm} \\ \Delta \text{ cm} = 0.014 \text{ cm} \\ P_c = 500 \text{ kgf/cm}^2 \\ (49030 \text{ kPa}) \end{array}\right]$$

7. Determine the radial and circumferential stresses due to the internal pressure $P = 2000$ kgf/cm² (196,000 kPa) in a composite tube consisting of an inner copper tube of radii 10 cm and 20 cm and an outer steel tube of external radius 40 cm. $V_{st} = 0.3$, $V_{cu} = 0.34$, $E_{st} = 2 \times 10^6$ kgf/cm² (196 × 10⁶ kPa) and $E_{cu} = 10^6$ kgf/cm² (98 × 10⁶ kPa). Calculate the stresses at the inner and outer radius points of each tube. Determine the contact pressure also.

$$\left[\begin{array}{l} Ans.\ \text{For inner tube} \\ \sigma_r = -2000 \text{ kgf/cm}^2\ (-196000 \text{ kPa}) \\ \sigma_r = -577 \text{ kgf/cm}^2\ (-56546 \text{ kPa}) \\ \sigma_t = 1800 \text{ kgf/cm}^2\ (176400 \text{ kPa}) \\ \sigma_t = 371 \text{ kgf/cm}^2\ (36358 \text{ kPa}) \\ \text{For outer tube:} \\ \sigma_r = -577 \text{ kgf/cm}^2\ (-56546 \text{ kPa}) \\ \sigma_r = 0 \\ \sigma_t = 962 \text{ kgf/cm}^2\ (94276 \text{ kPa}) \\ \sigma_t = 385 \text{ kgf/cm}^2\ (37730 \text{ kPa}) \\ P_c = 577 \text{ kgf/cm}^2\ (56546) \end{array}\right]$$

8. A thick-walled tube is subjected to an external pressure P_2. Its internal and external radii are 10 cm and 15 cm respectively, $N = 0.3$ and $E = 200000$ MPa (2041×10^3 kgf/cm^2). If the maximum shear stress is limited to 200000 kPa (2041 kgf/cm^2), determine the value of $P2$ and also the change in the external radius.

$$\left[\begin{array}{l} Ans.\ P_2 = 111 \text{ MPa } (1133 \text{ kgf/cm}^2 \\ \quad \Delta r_2 = -0.19 \text{ mm} \end{array}\right]$$

9. In problem 7, if the inner tube is made of steel (radii 10 cm and 20 cm) and the outer tube is of copper (outer radius 40 cm), determine the circumferential and radial stresses at the inner and outer radii points of each tube.

$$\left[\begin{array}{l} Ans.\ \text{For inner tube} \\ \quad \sigma_r = -2000 \text{ kgf/cm}^2 \ (-196000 \text{ kPa}) \\ \quad \sigma_r = -248 \text{ kgf/cm}^2 \ (-24304 \text{ kPa}) \\ \quad \sigma_t = 2672 \text{ kgf/cm}^2 \ (262032 \text{ kPa}) \\ \quad \sigma_t = 920 \text{ kgf/cm}^2 \ (90221 \text{ kPa}) \\ \text{For outer tube:} \\ \quad \sigma_r = -248 \text{ kgf/cm}^2 \ (-24304 \text{ kPa}) \\ \quad \sigma_r = 0 \\ \quad \sigma_t = 413 \text{ kgf/cm}^2 \ (40474 \text{ kPa}) \\ \quad \sigma_t = 165 \text{ kgf/cm}^2 \ (16170 \text{ kPa}) \\ \quad P_c = 248 \text{ kgf/cm}^2 \ (24304 \text{ kPa}) \end{array}\right]$$

10. A steel disk of 50 cm outside diameter and 10 cm inside diameter is shrunk on a steel shaft so that the pressure between the shaft and disk at standstill is 364 kgf/cm^2 (3562 kPa). Take $\rho = 0.0081/g$ kgm/cm^3.

    (a) Assuming that the shaft does not change its dimensions because of its own centrifugal force, find the speed at which the disk is just free on the shaft.

    (b) Solve the problem without making assumption (a).

$$\left[\begin{array}{l} Ans.\ \text{(a)}4013 \text{ rpm} \\ \quad \text{(b)}4028 \text{ rpm.} \end{array}\right]$$

11. A composite tube is made of an inner copper tube of radii 10 cm and 20 cm and an outer steel tube of external radius 40 cm. If the temperature of the assembly is raised by 100°C, determine the radial and tangential stresses at the inner and outer radius points of each tube. $\alpha_{cu} = 16.5 \times 10^{-6}$; $\alpha_{st} = 12.5 \times 10^{-6}$; E_{st}, V_{st}, E_{cu} and V_{cu} are as in problem 7.

[*Ans.* For inner tube

$$\sigma_r = 0$$

$$\sigma_r = -173 \text{ kgf/cm}^2 \ (-16954 \text{ kPa})$$

$$\sigma_t = -461 \text{ kgf/cm}^2 \ (-45080 \text{ kPa})$$

$$\sigma = -288 \text{ kgf/cm}^2 \ (-28243 \text{ kPa})$$

For outer tube:

$$\sigma_r = -173 \text{ kgf/cm}^2 \ (-16954 \text{ kPa})$$

$$\sigma_r = 0$$

$$\sigma_t = 288 \text{ kgf/cm}^2 \ (28243 \text{ kPa})$$

$$\sigma_t = 115 \text{ kgf/cm}^2 \ (11270 \text{ kPa})]$$

12. The radial displacement at the outside of a thick cylinder subjected to an internal pressure P_a is

$$\frac{\left(2p_a r_b r_a^2\right)}{E\left(r_b^2 - r_a^2\right)}$$

By Maxwell's reciprocal theorem, find the inward radial displacement at the inside of a thick cylinder subjected to external pressure.

$$\left[\textit{Ans.} \quad u_r = \frac{2p_b ab^2}{E(b^2 - a^2)}\right]$$

13. A thin spherical shell of thickness h and radius R is subjected to an internal pressure P. Determine the mean radial stress, the circumferential stress and the radial displacement.

[*Ans.* $u_r = PR^2(1-v)/(2Eh)$

$$\sigma_\theta = \sigma_\phi = \frac{PR}{2h}$$

$$(\sigma_r) \text{ average } = \frac{1}{2}P]$$

14. An infinite elastic medium with a spherical cavity of radius R is subjected to hydrostatic compression P at the outside. Determine the radial and circumferential stresses at point r. Show that the circumferential stress at the surface of the cavity exceeds the pressure at infinity.

$$\left[\begin{aligned} Ans.\quad \sigma_r &= -p\left(1-\frac{R^2}{r^3}\right) \\ \sigma_\theta = \sigma_\phi &= -P\left[1+\frac{R^3}{2r^3}\right] \\ (\sigma_\theta) \text{ at cavity } &= -\frac{3}{2}P \end{aligned}\right]$$

15. A perfectly rigid spherical body of radius a is surrounded by a thick spherical shell of thickness h. If the shell is subjected to an external pressure P, determine the radial and circumferential stresses at the inner surface of the shell ($b = a + h$).

$$\left[Ans.\quad \sigma_r = \frac{3(1-v)b^3 p}{2(1-2v)a^3+(1+v)b^3}\right]$$

16. Determine for the composite three-piece tube:
    (a) Stresses due to the heavy-force fits with interferences of $D_1 = 0.06$ mm and $D_2 = 0.12$ mm in diameters
    (b) Stresses due to the internal pressure $P = 2400$ kgf/cm^2
    $r_1 = 80$ mm, $r_2 = 100$ mm, $r_3 = 140$ mm,
    $r_4 = 200$ mm,
    $E = 2.2 \times 10^6$ kgf/cm^2.

17. A steel disk of 75 cm diameter is shrunk on a steel shaft of 7.5 cm diameter. The interference on the diameter is 0.0045 cm
    (a) Find the maximum tangential stress in the disk when it is at a standstill.
    (b) Find the rotation speed at which the contact pressure is zero.
    (c) What is the maximum tangential stress at the above speed.

[*Ans.* (*a*) 6620 kgf/cm^2
(*b*) 3460 rpm
(*c*) 13090 kgf/cm^2 (130850 kPa)]

18. A steel shaft of 7.5 cm diameter has an aluminium disk of 25 cm outside diameter shrunk on it. The shrink allowance is 0.001 cm/cm. Calculate the rpm of rotation at which the shrink-fit loosens up. Neglect the expansion of the shaft caused by rotation $V_{al} = 0.3$, E_{st} = 7.3 × 105 kgf/cm^2 (175 × 10^5 kPa); $\gamma = 2.76\ 10^{-3}$ kgf/cm^3.

[*Ans.* 13420 rpm]

19. A disk of thickness t and outside diameter $2b$ is shrunk on to a shaft of diameter $2a$, producing a radial interface pressure p in the non-rotating condition. It is then rotated with an angular velocity ω rad/s. If f is the coefficient of friction between disk and shaft and ω_0 is the value of the angular velocity for which the interface pressure falls to zero, show that

(a) the maximum horsepower is transmitted when $\omega_0 / \sqrt{3}$ and

(b) this maximum horsepower is equal to 0.000366 $a^2 t\, fp\, \omega_0$, where dimension are in inches and pounds.

5

Analysis of Stress

INTRODUCTION

In this book we shall deal with the mechanics of deformable solids. The starting point for discussion can be either the analysis of stress or the analysis of strain. In book on the theory of elasticity, one usually starts with the analysis of strain, which deals with the geometry of deformation without considering the forces that cause the deformation. However, one is more familiar with forces, though the measurement of force is usually done through the measurement of deformations caused by the force. Books on the strength of materials, begin with the analysis of stress. The concept of stress has already been introduced in the elementary strength of materials. When a bar of uniform cross–section, say a circular rod of diameter *d*, is subjected to a tensile force *F* along the axis of the bar, the average stress induced across any transverse section perpendicular to the axis of the bar and away form the region of loading is given by

$$\sigma = \frac{F}{\text{Area}} = \frac{4F}{\pi d^2}$$

It is assumed that the reader is familiar with the elementary flexural stress and torsional stress concepts. In general, a structural member or a machine element will not possess uniform geometry of shape or size, and the loads acting on it will also be complex. For example, an automobile crankshaft or a piston inside on engine cylinder or an aircraft wing are subject to loadings that are both complex as well as dynamic in nature. In such cases, one will have to introduce the concept of the state of stress at a point and its analysis, which will be the subject of discussion in this chapter. However, we shall not deal with forces that vary with time.

It will be assumed that the matter of the body that is being considered is continuously distributed over its volume, so that if we consider a small volume element of the matter surrounding a point and shrink this volume, in the limit we shall not come across a void. In reality, however, all materials are composed of many discrete particles, which are often microscopic, and when an arbitrarily selected volume element is shrunk, in the limit one may end up in a void. But in our analysis, we assume that the matter is continuously distributed. Such a body is called a continuous medium and the mechanics of such a body or bodies is called continuum mechanics.

BODY FORCE, SURFACE FORCE AND STRESS VECTOR

Consider a body B occupying a region of space referred to a rectangular coordinate system *Oxyz,* as shown in Fig. 1. In general, the body will be subjected to two types of forces–body forces and surface forces. The body forces act on each volume element of the body. Examples of this kind of force are the gravitational force, the inertia force and the magnetic force. The surface forces act on the surface or area elements of the body. When the area considered lies on the actual boundary of the body, the surface force distribution is often termed surface traction. In Fig. 1. 1, the surface forces F_1, F_2, F_3 . . . F_r, are concentrated forces, while p is a distributed force. The support reactions R_1, R_2 and R_3 are also surface forces. It is explicitly assumed that under the action .of both body forces and surface forces, the body is in equilibrium.

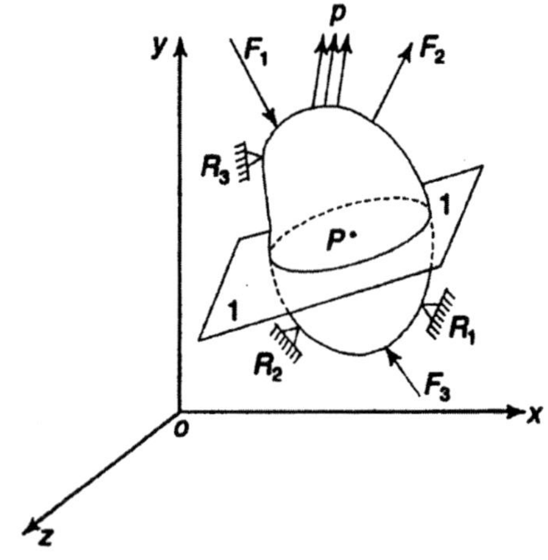

Fig 1

Let p be a point inside the body with coordinates (x, y, z). Let the body be cut into two parts C and D by a plane 1–1 passing through point p, as shown in Fig. 1.2. If we consider the free– body diagrams of C and D, then each part is in equilibrium under the action of the externally applied forces and the internally distributed forces across the interface. In part D, let ΔA be a small area surrounding the point p. In part C, the corresponding area at p' is A'.

These two areas are distinguished by their outward drawn normals $\overset{1}{n}$ and $\overset{1}{n'}$. The action of part C on ΔA at point P can be represented by the force vector $\Delta\overset{1}{T}$ and the action of part D on ΔA' at P' can be represented by the force vector $\Delta\overset{1}{T}$. We assume that as ΔA tends to zero, the ratio $\dfrac{\Delta\overset{1}{T}}{\Delta A}$ tends to a definite limit, and further, the moment of the forces acting on area ΔA αβουτ any point within the area vanishes in the limit. The limiting vector is written as

$$\lim_{\Delta A \to 0} \frac{\Delta \overset{1}{T}}{\Delta A} = \frac{d\overset{1}{T}}{dA} = \overset{1}{T} \tag{1}$$

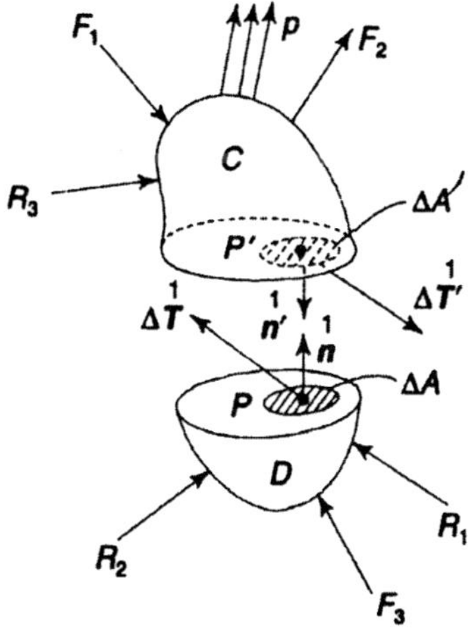

Fig. 2

Similarly, at point P', the action of part D on C as ΔA' tends to zero, can be represented by a vector

$$\lim_{\Delta A' \to 0} \frac{\Delta \overset{1}{T}'}{\Delta A'} = \frac{d\overset{1}{T}'}{dA'} = \overset{1}{T} \tag{2}$$

Vectors $\overset{1}{T}$ and $\overset{1}{T'}$ are called the stress vectors and they represent forces per unit area acting respectively at P and P' on planes with outward drawn normals $\overset{1}{n}$ and $\overset{1}{n'}$.

We further assume that stress vector $\overset{1}{T}$ representing the action of C on D at P is equal in magnitude and opposite in direction to stress vector $\overset{1}{T'}$ representing the action of D on C at corresponding point P'. This assumption is similar to Newton's third law, which is applicable to particles. We thus have

$$\overset{1}{T} = -\overset{1}{T'} \tag{3}$$

If the body in Fig. 1 is cut by a different plane 2–2 with outward drawn normals $\overset{2}{n}$ and $\overset{2}{n'}$ passing through the same point P, then the stress vector representing the action of C_2 on D_2 will be represented by $\overset{2}{T}$ (Fig. (3), i.e.

$$\overset{2}{T} = \lim_{\Delta A \to 0} = \frac{\Delta \overset{2}{T}}{\Delta A}$$

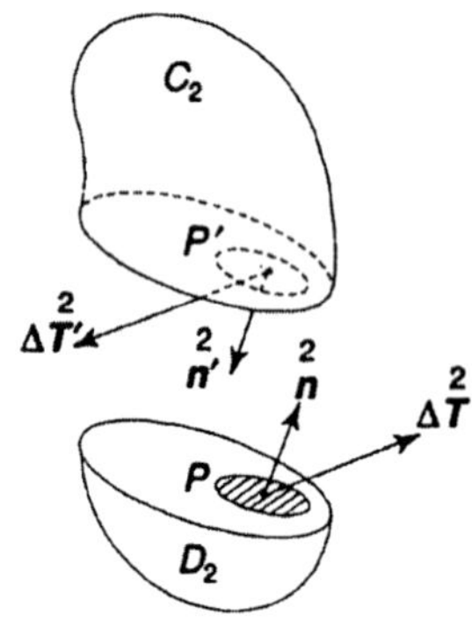

Fig. 3

In general, stress vector $\overset{1}{T}$ acting at point P on a plane with outward drawn normal $\overset{1}{n}$ will be different from stress vector $\overset{2}{T}$ acting at the same point P, but on a plane with outward drawn normal $\overset{2}{n}$. Hence the stress at a point depends not only on the locatio of the point (identified by coordinates x, y, z) but also on the plane passing through the point (identified by direction cosines n_x, n_y, n_z of the outward drawn normal).

THE STATE OF STRESS AT A POINT

Since an infinite number of planes can be drawn through a point, we get an infinite number of stress vectors acting at a given point, each stress vector characterised by the corresponding plane on which it is acting. The totality of all stress vectors acting on every possible plane passing through the point is defined to be the state of stress at the point. It is the knowledge of this state of stress that is of importance to a designer in determining the critical planes and the respective critical stresses. It will be shown in Sec. 1.6 that if the stress vectors acting on three mutually perpendicular planes passing through the point are known, we can determine the stress vector acting on any other arbitrary plane at that point.

NORMAL AND SHEAR STRESS COMPONENTS

Let $\overset{n}{T}$ be the stress vector at point P acting on a plane whose outward drawn normal is n (Fig. 4). This can be resolved into two components, one along the normal n and the other perpendicular to n. The component parallel to n is called the normal stress and is generally denoted by $\overset{n}{\sigma}$. The component perpendicular to n is known as the tangential stress or shear stress component and is denoted by $\overset{n}{\tau}$. We have, therefore, the relation:

$$\left|\overset{n}{T}\right|^2 = \overset{n}{\sigma}^2 + \overset{n}{\tau}^2 \tag{4}$$

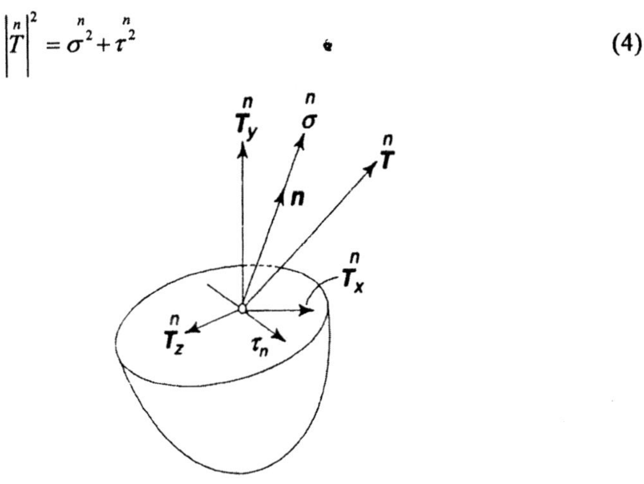

Fig. 4

where $\left|\overset{n}{T}\right|$ is the magnitude of the resultant stress. Stress vector $\overset{n}{T}$ can also be resolved into three components parallel to the x, y, z axes. If these components are denoted by $\overset{n}{T}_x$, $\overset{n}{T}_y$, $\overset{n}{T}_z$, we have

$$\left|\overset{n}{T}\right|^2 = \overset{n}{T}_x^2 + \overset{n}{T}_y^2 + \overset{n}{T}_z^2 \tag{5}$$

RECTANGULAR STRESS COMPONENTS

Let the body B, shown in Fig. 1, be cut by a plane parallel to the yz plane. The normal to this plane is parallel to the x axis and hence, the plane is called the x plane.

The resultant stress vector at p acting on this will be $\overset{x}{T}$. This vector can be resolved into three components parallel to the x, y, z axes. The component parallel to the x axis, being normal to the plane, will be denoted by σ_x (instead of by $\overset{x}{\sigma}$). The components parallel to the y and z axes are shear stress components and are denoted by τ_{xy} and τ_{xz} respectively (Fig. 5).

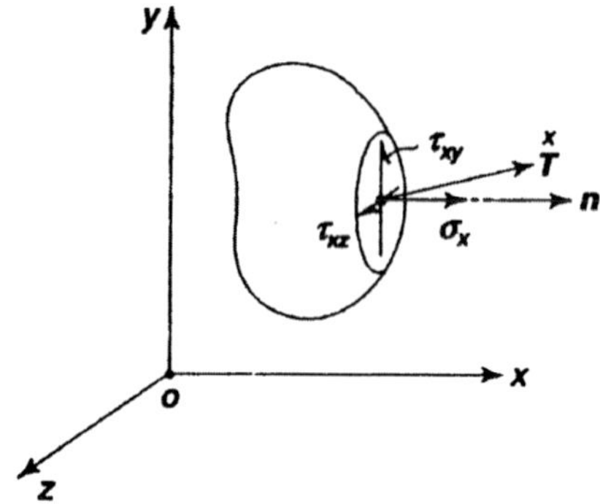

Fig 5

In the above designation, the first subscript x indicates the plane on which the stresses are acting and the second subscript (y or z) indicates the direction of the component.

For example, τ_{xy} is the stress component on the x plane in y direction. Similarly, τ_{xz} is the stress component on the x plane in z direction. To maintain consistency, one should have denoted the normal stress component as τ_{xx}. This would be the stress component on the x plane in the x direction. However, to distinguish between a normal stress and a shear stress, the normal stress is denoted by σ and the shear stress by τ.

At any point P, one can draw three mutually perpendicular planes, the x plane, the y plane and the z plane. Following the notation mentioned above, the normal and shear stress components on these planes are

$$\sigma_x, \tau_{xy}, \tau_{xz} \text{ on } x \text{ plane}$$

$$\sigma_y, \tau_{yx}, \tau_{yz} \text{ on } y \text{ plane}$$

$$\sigma_z, \tau_{zx}, \tau_{zy} \text{ on } z \text{ plane}$$

These components are shown acting on a small rectangular element surrounding the point P in Fig. 6.

One should observe that the three visible faces of the rectangular element have their outward drawn normals along the positive x, y and z axes respectively. Consequently, the positive axes. The three hidden faces have their outward drawn normals in the negative x, y and z axes. The positive stress components on these faces will, therefore, be directed along the negative axes. For example, the bottom face has its outward drawn normal along the negative y axis. Hence, the positive stress components on this face, i.e., σ_y, τ_{yx} and τ_{yz} are directed respectively along the negative y, x and z axes.

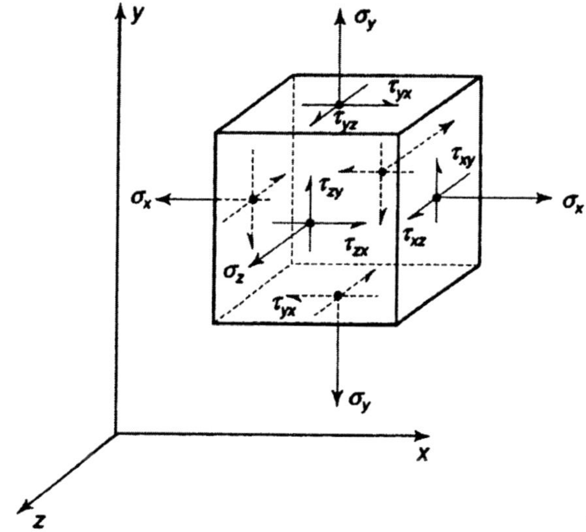

Fig. 6

STRESS COMPONENTS ON AN ARBITRARY PLANE

It was stated that a knowledge of stress components acting on three mutually perpendicular planes passing through a point will enable one to determine the stress components acting on any plane passing through that point. Let the three mutually perpendicular planes be the *x, y* and *z* planes and let the arbitrary plane be identified by its outward drawn normal *n* whose direction cosines are n_x, n_y and n_z. Consider a small tetrahedron at *P* with three of its faces normal to the coordinate axes, and the inclined face having its normal parallel to *n*. Let h be the perpendicular distance from *P* to the inclined face. If the tetrahedron is isolated from the body and a free–body diagram is drawn, then it will be in equilibrium under the action of the surface forces and body forces. The free–body diagram is shown in Fig. 7.

Since the size of the tetrahedron considered is very small and in the limit as we are going to make *h* tend to zero, we shall speak in terms of the average stresses over the faces. Let $\overset{n}{T}_x$, $\overset{n}{T}_y$, $\overset{n}{T}_z$, parallel to the three axes *x, y* and *z*. On the three faces, the rectangular stress components are σ_x, τ_{xy}, τ_{xz}, σ_y, τ_{yz}, τ_{yx}, σ_z, τ_{zx}, and τ_{zy}. If *A* is the area of the inclined face then

Area of *BPC* = projection of area *ABC* on the *yz* plane

$$=An_x$$

Area of *CPA* = projection of area *ABC* on the *xz* plane

$$=An_y$$

Area of APB = projection of area ABC on the xy plane

$=An_z$

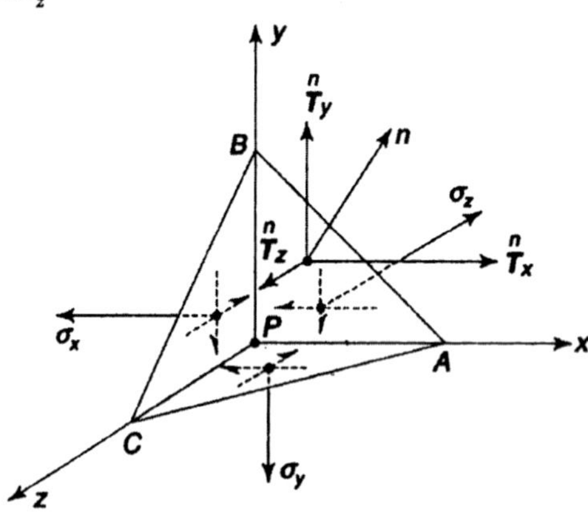

Fig. 7

Let the body force components in x, y and z directions be γ_x, γ_y and γ_z respectively, per unit volume. The volume of the tetrahedron is equal to 1/ 3 Ah where h is the perpendicular distance from p to the inclined face. For equilibrium of the tetrahedron, the sum of the forces in x, y and z directions must individually vanish. Thus, for equilibrium in x direction

$$\overset{n}{T}_x A - \sigma_x A n_x - \tau_{yx} A n_y - \tau_{zx} A n_z + \frac{1}{3} A h y_x = 0$$

Cancelling A,

$$\overset{n}{T}_x = \sigma_x n_x + \tau_{yx} n_y + \tau_{zx} n_z - \frac{1}{3} h \gamma_x \tag{6}$$

Similarly, for equilibrium in y and z directions

$$\overset{n}{T}_y = \tau_{xy} n_x + \sigma_y n_y + \tau_{zy} n_z - \frac{1}{3} h \gamma_y \tag{7}$$

and $$\overset{n}{T}_z = \tau_{xz} n_x + \tau_{yz} n_y + \sigma_z n_z - \frac{1}{3} h \gamma_z \tag{8}$$

In the limit as h tends to zero, the oblique plane ABC will pass through point p, and the average stress components acting on the faces will tend to their respective values at point p acting on their corresponding planes. Consequently, one gets from Eqs (6) – (8)

$$\overset{n}{T}_x = n_x \sigma_x + n_y \tau_{yx} + n_z \tau_{zx}$$

$$\overset{n}{T}_y = n_x \tau_{xy} + n_y \sigma_y + n_z \tau_{zy} \tag{9}$$

$$\overset{n}{T}_z = n_x \tau_{xz} + n_y \tau_{yz} + n_z \sigma_z$$

Equation (9) is known as Cauchy' s stress formula. This equation shows that the nine rectangular stress components at p will enable one to determine the stress components on any arbitrary plane passing through point p. It will be shown in Sec. 1.8 that among these nine rectangular stress components only six are independent. This is because $\tau_{xy} = \tau_{yx}$, $\tau_{zy} = \tau_{yz}$ and $\tau_{zx} = \tau_{xz}$. This is known as the equality of cross shears. In anticipation of this result, one can write Eq. (9) as

$$\overset{n}{T}_i = n_x \tau_{ix} + n_y \tau_{iy} + n_z \tau_{iz} \sum_j n_j \tau_{ij} \tag{10}$$

where i and j can stand for x or y or z, and $\sigma_{xx} = \tau_{xx}$, $\sigma_y = \tau_{yy}$ and $\sigma_z = \tau_{zz}$. If $\overset{n}{T}$ is the resultant stress vector on plane ABC, we have

$$\left|\overset{n}{T}\right|^2 = \overset{n}{T}{}_x^2 + \overset{n}{T}{}_y^2 + \overset{n}{T}{}_z^2$$

If σ_n and τ_n are the normal and shear stress components, we have

$$\left|\overset{n}{T}\right|^2 = \sigma_n^2 + \tau_n^2 \tag{11}$$

Since the normal stress is equal to the projection of $\overset{n}{T}$ along the normal, it is also equal to the sum of the projections of its components $\overset{n}{T}_x, \overset{n}{T}_y$ and $\overset{n}{T}_z$ along n. Hence,

$$\sigma_n = n_x \overset{n}{T}_x + n_y \overset{n}{T}_y + n_z \overset{n}{T}_z \tag{12a}$$

Substituting for $\overset{n}{T}_x, \overset{n}{T}_y$ and $\overset{n}{T}_z$ from Eq. (9)

$$\sigma_n = n_x^2 \sigma_x + n_y^2 \sigma_y + n_z^2 \sigma_z + 2n_x n_y \tau_{xy} + 2n_y n_z \tau_{yz} + 2n_z n_x \tau_{zx} \tag{12b}$$

equation (11) can then be used to obtain the value of τ_n

Example 1:

A rectangular steel bar having a cross–section 2 cm × 3 cm is subjected to a tensile force of 6000 N (612.2 kgf). If the axes are chosen as shown in Fig. 8, determine the normal and shear stresses on a plane whose normal has the following direction cosines:

(*i*) $n_x = n_y = \dfrac{1}{\sqrt{2}}, n_z = 0$

(*ii*) $n_x = 0, n_y = n_x = \frac{1}{\sqrt{2}}$

(*iii*) $n_x = n_y = n_x = \frac{1}{\sqrt{3}}$

Solution:

Area of section = 2 × 3 = 6 cm². The average stress on this plane is 6000/6 = 1000 Ncm². This is the normal stress σ_y. The other stress components are zero.

(*i*) Using Eqs (9), (11) and (12)

$$\overset{n}{T}_x = 0, \quad \overset{n}{T}_y = \frac{1000}{\sqrt{2}}, \quad \overset{n}{T}_z = 0$$

$$\sigma_n = \frac{1000}{2} = 500\text{N/cm}^2$$

$$\tau_n^2 = \left|\overset{n}{T}\right|^2 - \sigma_n^2 = 250{,}000\text{N}^2/\text{cm}^4$$

$$\tau_n = 500\text{N/cm}^2(51\text{kgf/cm}^2)$$

(*ii*) $\overset{n}{T}_x = 0, \quad \overset{n}{T}_y = \frac{1000}{\sqrt{3}}, \quad \overset{n}{T}_z = 0$

σ_n = 500 N/cm², and τ_n = 500 N/cm² (51 kgf/cm²)

(*iii*) $\overset{n}{T}_x = 0, \quad \overset{n}{T}_y = \frac{1000}{\sqrt{3}}, \quad \overset{n}{T}_z = 0$

$$\sigma_n = \frac{1000}{3}\ \text{N/cm}^2$$

$$\tau_n = 817\ \text{N/cm}^2(83.4\ \text{kgf/cm}^2)$$

Example 2:

At a point p in a body, σ_x *= 10,000 N/cm² (1020kgf/cm²),* σ_y *= –5,000 N/cm² (–510 kgf/cm²),* σ_z *= 5,000 N/cm²,* $\tau_{xy} = \tau_{yz} = \tau_{zx}$ *= 10,000 N/cm². Determine the normal and shearing stresses on a plane that is equally inclined to all the three axes.*

Solution:

A plane that is equally inclined to all the three axes will have

$$n_x = n_y = n_z = \frac{1}{\sqrt{3}} \text{ since } n_x^2 + n_y^2 + n_z^2 = 1$$

Form Eq. (12)

$$\sigma_n = \frac{1}{3}[10000 - 5000 - 5000 + 20000 + 20000 + 20000]$$

$$= 20000 \text{ N/cm}^2$$

From Eqs (6) – (8)

$$\overset{n}{T}_x = \frac{1}{\sqrt{3}}(10000 + 10000 + 10000) = 10000\sqrt{3} \text{ N/cm}^2$$

$$\overset{n}{T}_y = \frac{1}{\sqrt{3}}(10000 - 5000 + 10000) = 5000\sqrt{3} \text{ N/cm}^2$$

$$\overset{n}{T}_z = \frac{1}{\sqrt{3}}(10000 - 10000 - 5000) = 5000\sqrt{3} \text{ N/cm}^2$$

$$\therefore \quad \left|\overset{n}{T}\right|^2 = 3[(10^8) + (25\times 10^6) + (25\times 10^6)] \text{ N}^2/\text{cm}^4$$

$$= 450 \times 10^6 \text{ N}^2/\text{cm}^4$$

$$\therefore \quad \tau_n^2 = 450\times 10^6 - 400\times 10^6 = 50\times 10^6 \text{ N}^2/\text{cm}^4$$

or $\quad \tau_n = 7000 \text{ N/cm}^2$ (approximately)

Example 3:

A cantilever beam in the form of a trapezium of uniform thickness loaded by a force p at the end. If it is assumed that the bending stress on any vertical section of the beam is distributed according to the elementary flexure formula, show that the normal stress σ on a section perpendicular to the top edge of the beam at point A is $\frac{\sigma_1}{\cos^2\theta}$, where σ_l is the flexural stress $\frac{Mc}{I}$,

Solution:

At point A, let axes x and y be chosen along and perpendicular to the edge. On the x plane, i.e. the plane perpendicular to edge EF, the resultant stress is along the normal (i.e., x axis). There is no shear stress on this plane since the top edge is a free surface (see Sec. 1.9). But on plane AB at point A there can exist a shear stress. The normal to plane AB makes an angle θ with the x axis. Let the normal and shearing stresses on this plane be σ_1 and τ_1.

We have

$$\sigma_x = \sigma, \qquad \sigma_y = \sigma_z = 0 \qquad \tau_{xy} = \tau_{yz} = \tau_{zx} = 0$$

The direction cosines of the normal to plane AB are

$$n_x = \cos\theta, \qquad n_y = \sin\theta, \qquad n_z = 0$$

The components of the stress vector action on plane AB are

$$\overset{n}{T}_x = \sigma_1 = n_x\,\sigma_x + n_y\,\tau_{yx} + n_z\,\tau_{zy} = \sigma\cos\theta$$

$$\overset{n}{T}_y = n_y\,\tau_{xy} + n_y\,\sigma_y + n_z\,\tau_{zy} = 0$$

$$\overset{n}{T}_z = n_x\,\tau_{xz} + n_y\,\sigma_{yz} + n_z\,\sigma_z = 0$$

Therefore, the normal stress on plane

$$AB = \sigma_n = n_x\overset{n}{T}_x + n_y\overset{n}{T}_y + n_z\overset{n}{T}_z = \sigma\cos^2\theta.$$

Since $\sigma_n = \sigma_1$

$$\sigma = \frac{\sigma_1}{\cos^2\theta} = \frac{Mc}{I\cos^2\theta}$$

Further, the resultant stress on plane AB is

$$\left|\overset{n}{T}\right|^2 = \overset{n}{T}{}_x^2 + \overset{n}{T}_y + \overset{n}{T}{}_z^2 = \sigma^2\cos^2\theta$$

Hence $\tau^2 = \sigma^2\cos^2\theta - \sigma_n^{\,2}$

$= \sigma^2\cos^2\theta - \sigma^2\cos^4\theta$

or $$\tau = \frac{1}{2}\sigma\sin 2\theta$$

DIGRESSION ON IDEAL FLUID

By definition, an ideal fluid cannot sustain any shearing forces and the normal force on any surface is compressive in nature. This can be represented by

$$\overset{n}{T} = -pn, \;\; p \geq 0$$

The rectangular components of $\overset{n}{T}$ along the x, y and z axes. If n_x, n_y and n_z are the direction cosines of n, then

$$\overset{n}{T}_x = -pn_x, \quad \overset{n}{T}_y = -pn_y, \quad \overset{n}{T}_z = -pn_z \tag{13}$$

Since all shear stress components are zero, one has from Eqs (9),

$$\overset{n}{T}_x = n_x\,\sigma_x, \quad \overset{n}{T}_y = n_y\,\sigma_y, \quad \overset{n}{T}_z = n_z\,\sigma_z \tag{14}$$

Comparing Eqs (13) and (14)

$$\sigma_x = \sigma_y = \sigma_z = -p$$

Since plane n was chosen arbitrarily, one concludes that the resultant stress vector on any plane is normal and is equal to $-p$. This is the type of stress that a small sphere would experience when immersed in a liquid. Hence, the state of stress at a point where the resultant stress vector on any plane is normal to the plane and has the same magnitude is known as a hydrostatic or anisotropic state if stress. The word isotropy means 'independent dent of orientation' or 'same in all directions'.

EQUALITY OF CROSS SHEARS

We shall now show that of the nine rectangular stress components σ_x, τ_{xy}, τ_{xz}, σ_y, τ_{yx}, τ_{yz}, σ_z, τ_{zx} and τ_{zy} only six are independent. This is because $\tau_{xy} = \tau_{yx}$, $\tau_{yz} = \tau_{zy}$ and $\tau_{zx} = \tau_{xz}$. These are known as cross–shears. Consider an infinitesimal rectangular parallelepiped surrounding point p. Let the dimensions of the sides be Δx, Δy and Δz (Fig. 8).

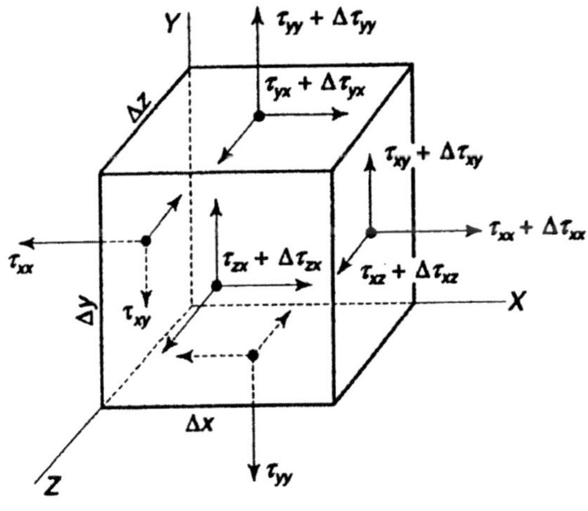

Fig. 8

Since the element considered is small, we shall speak in terms of average stresses over the faces. The stress vector action on the faces are shown in the figure. On the left x plane, the stress vectors are τ_{xx}, τ_{xy} and τ_{xz}. On the right face, the stresses are $\tau_{xx} + \Delta\tau_{xx}$, $\tau_{xy} + \Delta\tau_{xy}$ and $\tau_{xz} + \Delta\tau_{xz}$, These changes are because the right face is at a distance Δx from the left face. To the first order of approximation we have

$$\Delta\tau_{xx} = \frac{\partial\tau_{xx}}{\partial x}\Delta x, \quad \Delta\tau_{xy} = \frac{\partial\tau_{xy}}{\partial x}\Delta x, \quad \Delta\tau_{xz} = \frac{\partial\tau_{xz}}{\partial x}\Delta x$$

Similarly, the stress vectors on the bottom face are τ_{yy}, $+ \Delta\tau_{yy}$, $\tau_{yx} + \Delta\tau_{yx}$ and $\tau_{yz} + \Delta\tau_{yz}$ where

$$\Delta\tau_{yy} = \frac{\partial\tau_{yy}}{\partial y}\Delta y, \quad \Delta\tau_{yx} = \frac{\partial\tau_{yx}}{\partial y}\Delta y, \quad \Delta\tau_{yz} = \frac{\partial\tau_{yz}}{\partial y}\Delta y$$

On the front and rear faces, the components of stress vectors are respectively

$$\tau_{zz}, \ \tau_{zx}, \ \tau_{zy}$$

$$\tau_{zz} + \Delta\tau_{zz}, \ \tau_{zx} + \Delta\tau_{zx} \ \tau_{zy} + \Delta\tau_{zy}$$

where

$$\Delta\tau_{zz} = \frac{\partial\tau_{zz}}{\partial z}\Delta z, \quad \Delta\tau_{zx} = \frac{\partial\tau_{zx}}{\partial z}\Delta z, \quad \Delta\tau_{zy} = \frac{\partial\tau_{zy}}{\partial z}\Delta z$$

For equilibrium, the moments of the forces about the x, y and z aces must vanish individually. Taking moments about the z axis, one gets

$$\tau_{xx}\Delta y\,\Delta z\,\frac{\Delta y}{2} - (\tau_{xx} + \Delta\tau_{xx})\Delta y\,\Delta z\,\frac{\Delta y}{2} +$$

$$(\tau_{xy} + \Delta\tau_{xy})\,\Delta y\,\Delta z\;\;\Delta x - \tau_{yy}\;\Delta x\,\Delta z\;\;\frac{\Delta x}{2} +$$

$$(\tau_{yy} + \Delta\tau_{yy})\,\Delta x\,\Delta z\;\;\frac{\Delta z}{2} - (\tau_{yx} + \Delta\tau_{yx})\Delta x\,\Delta z\;\;\Delta y +$$

$$\tau_{zy}\;\Delta x\,\Delta y\;\;\frac{\Delta x}{2} - \tau_{zx}\;\Delta x\,\Delta y\;\;\frac{\Delta y}{2} - (\tau_{zy} + \Delta\tau_{zy})\Delta x\,\Delta y\,\frac{\Delta x}{2} +$$

$$(\tau_{zx} + \Delta\tau_{zx})\Delta x\,\Delta y\,\frac{\Delta y}{2} = 0$$

Substituting for $\Delta\tau_{xx}, \Delta\tau_{xy}$ etc. and dividing by $\Delta x\,\Delta y\,\Delta z$

$$-\frac{\partial\tau_{xx}}{\partial x}\frac{\Delta y}{2} + \tau_{xy} + \frac{\partial\tau_{xy}}{\partial x}\Delta x + \frac{\partial\tau_{yy}}{\partial y}\frac{\Delta y}{2} -$$

$$\tau_{yx} - \frac{\partial\tau_{yx}}{\partial y}\Delta y - \frac{\partial\tau_{zy}}{\partial z}\frac{\Delta x}{2} + \frac{\partial\tau_{zx}}{\partial z}\frac{\Delta y}{2} = 0$$

In the limit as Δx, Δy and Δz tend to zero, the above equation gives $\tau_{xy} = \tau_{yx}$. Similarly, taking moments about the other two axes, we get $\tau_{yz} = \tau_{zy}$ and $\tau_{zx} = \tau_{xz}$. Thus, the cross shears are equal, and of the nine rectangular components, only six are independent. The six independent rectangular stress components are σ_x, σ_y, σ_z, τ_{xy}, τ_{yz} and τ_{zx}.

A MORE GENERAL THEOREM

The fact that cross shears are equal can be used to prove a more general theorem which states that if n and n' define two planes (not necessarily orthogonal) with corresponding stress vectors $\overset{n}{T}$ and $\overset{n'}{T}$, then the projection of $\overset{n}{T}$ along n' is equal to the projection of $\overset{n'}{T}$ along n, i.e. $\overset{n}{T} \cdot n' = \overset{n'}{T} \cdot n$ (see Fig. 9).

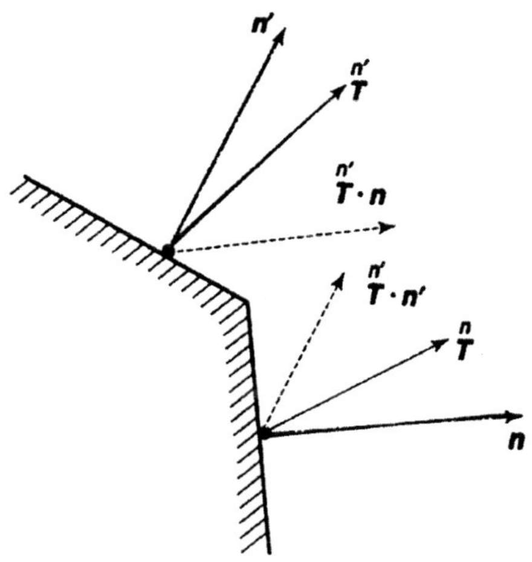

Fig. 9

The proof is straightforward. If n'_x, n'_y, and n'_z, are the direction cosines of n', then

$$\overset{n}{T} \cdot n' = \overset{n}{T}_x n'_x + \overset{n}{T}_y n'_y + \overset{n}{T}_z n'_z$$

From Eq. (9), substituting for $\overset{n}{T}_x, \overset{n}{T}_y$ and $\overset{n}{T}_z$

$$\overset{n}{T} \cdot n' = \sigma_x n_x n'_x + \sigma_y n_y n'_y + \sigma_z n_z n'_z + \tau_{xy} n_x n'_y + \tau_{yx} n_y n'_x +$$
$$\tau_{yz} n_y n'_z + \tau_{zy} n_z n'_y + \tau_{zx} n_z n'_x + \tau_{xz} n_x n'_z$$

Using the result $\tau_{xy} = \tau_{yx}, = \tau_{zy}$ and $\tau_{zx} = \tau_{zx}$

$$\overset{n}{T} \cdot n' = \sigma_x n_x n'_x + \sigma_y n_y n'_y + \sigma_z n_z n'_z + \tau_{xy} (n_x n'_y + n_y n'_x) +$$
$$\tau_{yz} (n_y n'_z + n_z n'_y) + \tau_{zx} (n_z n'_x + n_x n'_z)$$

Similarly,

$$\overset{n}{T} \cdot n' = \sigma_x n_x n'_x + \sigma_y n_y n'_y + \sigma_z n_z n'_z + \tau_{xy} (n_x n'_y + n_y n'_x) +$$
$$\tau_{yz} (n_y n'_z + n_z n'_y) + \tau_{zx} (n_z n'_x + n_x n'_z)$$

Comparing the above two expressions, we observe

$$\overset{n}{T} \cdot n' = \overset{n'}{T} \cdot n'$$

Note: An important fact is that cross shears are equal. This can be used to prove that a shear cannot cross a free boundary. For example, consider a beam of rectangular cross–section as shown in Fig. 10.

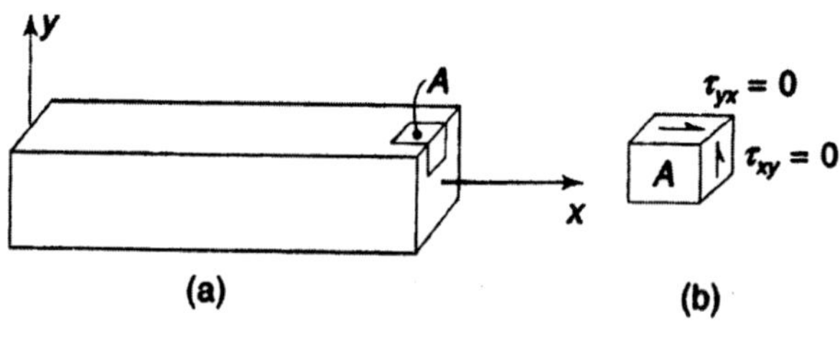

Fig. 10

If the top surface is a free boundary, then at point A, the vertical shear stress component $\tau_{xy} = 0$ because if τ_{xy} were not zero, it would call for a complementary shear τ_{yx} on the top surface. But as the top surface is an unloaded or a free surface, τ_{yx} is zero and hence, τ_{xy} is also zero.

PRINCIPAL STRESSES

We have seen that the normal and shear stress components can be determined on any plane with normal n, using Cauchy's formula given by Eqs (9). From the strength or failure considerations of materials, answers to the following questions are important:

(i) Are there any planes passing through the given point on which the resultant stresses are wholly normal (in other words, the resultant stress vector is along the normal)?

(ii) What is the plant on which the normal stress is a maximum and what is its magnitude?

(iii) What is the plane on which the tangential or shear stress is a maximum and what it is its magnitude?

Answers to these questions are very important in the analysis of stress, and the next few sections will deal with these. Let us assume that there is a plane n with direction cosines n_x, n_y and n_z on which the stress is wholly normal. Let s be the magnitude of this stress vector. Then we have

$$\overset{n}{T} = \sigma n \tag{16}$$

The components of this along the x, y and z axes are

$$\overset{n}{T}_x = \sigma n_x, \quad \overset{n}{T}_y = \sigma n_y, \quad \overset{n}{T}_z = \sigma n_z \tag{17}$$

Also, from Cauchy's formula, i.e. Eqs (9),

$$\overset{n}{T}_x = \sigma_x n_x + \tau_{xy} n_y + \tau_{xz} n_z$$

$$\overset{n}{T}_y = \tau_{xy} n_x + \sigma_y n_y + \tau_{yz} n_z$$

$$\overset{n}{T}_z = \tau_{xz} n_x + \tau_{yz} n_y + \sigma_z n_z$$

Subtracting Eq. (17) from the above set of equations we get

$$\begin{aligned} (\sigma_x - \sigma)\, n_x + \tau_{xy} n_y + \tau_{xz} n_z &= 0 \\ \tau_{xy} n_x + (\sigma_y - \sigma)\, n_y + \tau_{yz} n_z &= 0 \\ \tau_{xz} n_x + \tau_{yz} n_y + (\sigma_z - \sigma)\, n_z &= 0 \end{aligned} \tag{18}$$

We can view the above set of equations as three simultaneous equations involving the unknowns n_x, n_y and n_z. These direction cosines define the plane on which the resultant stress is wholly normal. Equation (18) is a set of homogeneous equations. The trivial solution is $n_x = n_y = n_z = 0$. For the existence of a non–trivial solution, the determinant of the coefficients of n_x, n_y and n_z must be equal to zero, i.e.

$$\begin{vmatrix} (\sigma_x - \sigma) & \tau_{xy} & \tau_{xz} \\ \tau_{xy} & (\sigma_y - \sigma) & \tau_{yz} \\ \tau_{xz} & \tau_{yz} & (\sigma_z - \sigma) \end{vmatrix} = 0 \tag{19}$$

Expanding the above determinant, one gets a cubic equation in σ as

$$\sigma^3 - (\sigma_x + \sigma_y + \sigma_z)\,\sigma^2 + (\sigma_x \sigma_y + \sigma_y \sigma_z + \sigma_z \sigma_x - \tau^2_{xy} - \tau^2_{yz} - \tau^2_{zx})\,\sigma$$

$$- (\sigma_x \sigma_y \sigma_z + 2\tau_{xy} \tau_{yz} \tau_{zx} - \sigma_x - \tau^2_{yz} - \sigma_y \tau^2_{xz} - \sigma_z \tau^2_{xy}) = 0 \tag{20}$$

The three roots of the cubic equation can be designated as σ_1, σ_2 and σ_3. It will be shown subsequently that all these three roots are real. We shall later give a method (Example 4) to solve the above cubic equation. Substituting any one of these three solutions in Eqs (18), we can solve for the corresponding n_x, n_y and n_z. In order to avoid the trivial solution, the condition.

$$n_x^2 + n_y^2 + n_z^2 = 1 \tag{21}$$

is used along with any two equations from the set of Eqs (18). Hence, with each σ there will be an associated plane. These planes on each of which the stress vector is wholly normal are called the principal planes, and the corresponding stresses, the principal stresses. Since the resultant stress is along the normal, the tangential stress component on a principal plane is zero, and consequently, the principal plane is also known as the shearless plane. The normal to a principal plane is called the principal stress axis.

STRESS INVARIANTS

The coefficients of σ^2, σ and the last term in the cubic Eq. (20) can be written as follows:

$$I_1 = \sigma_x + \sigma_y + \sigma_z \quad (22)$$

$$I_2 = \sigma_x\sigma_y + \sigma_y\sigma_z + \sigma_z\sigma_x - \tau^2_{xy} - \tau^2_{yz} - \tau^2_{zx}$$

$$\begin{vmatrix} \sigma_x & \tau_{xy} \\ \tau_{xy} & \sigma_y \end{vmatrix} + \begin{vmatrix} \sigma_y & \tau_{xz} \\ \tau_{yz} & \sigma_z \end{vmatrix} + \begin{vmatrix} \sigma_x & \tau_{xz} \\ \tau_{xz} & \sigma_z \end{vmatrix} \quad (23)$$

$$I_3 = \sigma_x\, \sigma_y\, \sigma_z + 2\tau_{xy}\; \tau_{yz}\; \tau_{zx} - \sigma_x\, \tau^2_{yz} - \sigma_y\, \tau^2_{zx} - \sigma_z\, \tau^2_{xy}$$

$$= \begin{vmatrix} \sigma_x & \tau_{xy} & \tau_{zx} \\ \tau_{xy} & \sigma_y & \tau_{yz} \\ \tau_{zx} & \tau_{yz} & \sigma_z \end{vmatrix} \quad (24)$$

Equation (20) can then be written as

$$\sigma^3 - I_1\sigma^2 + I_2\sigma - I_3 = 0$$

The quantities I_1 I_2 and I_3 are known as the first, second and third invariants of stress respectively. An invariant is one whose value does not change when the frame of reference is changed. In other words if x', y', z', is another frame of reference at the same point and with respect to this frame of reference, the rectangular stress competence are $\sigma_{x'}$, $\sigma_{y'}$, $\sigma_{z'}$, $\tau_{x'y'}$, $\tau_{y'z'}$ and $\tau_{z'x'}$ then the value of I_1, I_2 and I_3, calculate as in Eqs (22) – (24), will show that

$$\sigma_x + \sigma_y + \sigma_z = \sigma'_x + \sigma'_y + \sigma'_z$$

i.e. $\quad I_1 = I'_1$

and similarly, $\quad I_1 = I'_2$ and $I_3 = I'_3$

The reason for this can be explained as follows. The principal stresses at a point depend only on the state of stress at that point and not on the frame of reference describing the rectangular stress components. Hence, if xyz and $z'y'z'$ are two orthogonal frames of reference at the point, then the following cubic equations

$$\sigma^3 \square\, I_1\, \sigma^2 + I_2\, \sigma \,\square\, I_3 = 0$$

and $\quad \sigma^3 \square\, I'_1\, \sigma^2 + I'_2\, \sigma \,\square\, I'_3 = 0$

must give the same solutions for σ. Since the two systems of axes were arbitrary, the coefficients of σ^2 and σ and the constant terms in the two equations must be equal, i.e.

$$I_1 = I'_1, \quad I_2 = I'_2 \quad \text{and} \quad I_3 = I'_3$$

In terms of the principal stresses, the invariants are

$$I_1 = \sigma_1 + \sigma_2 + \sigma_3$$
$$I_2 = \sigma_1\sigma_2 + \sigma_2\sigma_3 + \sigma_3\sigma_1$$
$$I_3 = \sigma_1\sigma_2\sigma_3$$

PRINCIPAL PLANES ARE ORTHOGONAL

The principal planes corresponding to a given state of stress at a point can be shown to be mutually orthogonal. To prove this, we make use of the general theorem. Let n and n' be the two principal planes and σ_1 and σ_2, the corresponding principal stresses. Then the projection of σ_1 in direction n' is equal to the projection of σ_2 in direction n, i.e.

$$\sigma_1 n' \cdot n = \sigma_2 n' \cdot n \tag{1.25}$$

If n_x, n_y and n_z are the direction cosines of n and n'_x, n'_y and n'_z those of n', then expanding Eq. (25)

$$\sigma_1\,(n_x n'_x + n_y n'_y + n_z n'_z) = \sigma_2\,(n_x n'_x + n_y n'_y + n_z n'_z)$$

Since in general, σ_1 and σ_2 are not equal, the only way the above equation can hold is

$$n_x n'_x + n_y n'_y + n_z n'_z = 0$$

i.e. n and n' are perpendicular to each other. Similarly, considering two other planes n' n'' on which the principal stresses σ_1 and σ_3 are acting, and following the same argument as above, one finds that n' and n'' are perpendicular to each other. Similarly, n and n'' are perpendicular to each other. Consequently, the principal planes are mutually perpendicular.

CUBIC EQUATION HAS THREE REAL ROOTS

In Sec. 1.10, it was stated that Eq. (20) has three real roots. The proof is as follows. Dividing Eq. (20) by σ^2,

$$\sigma - I_1 + \frac{I_2}{\sigma} - \frac{I_3}{\sigma^2} = 0$$

For appropriate values of σ, the quantity on the left–hand side will be equal to zero. For other values, the quantity will not be equal to zero and one can write the above function as

$$\sigma - I_1 + \frac{I_2}{\sigma} - \frac{I_3}{\sigma^2} = f(\sigma) \tag{1.26}$$

Since I_1, I_2 and I_3 are finite, $f(\sigma)$ can be made positive for large positive values of σ. Similarly, $f(\sigma)$ can be made negative for large negative values of σ. Hence, if one plots $f(\sigma)$ for different values of σ as shown in Fig. 11, the curve must cut the σ axis at least once as shown by the dotted curve and for this value of σ, $f(\sigma)$ will be equal to zero. Therefore, there is at least one real root.

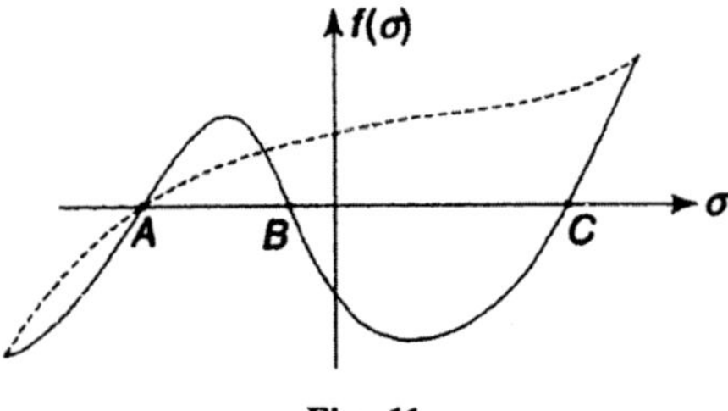

Fig. 11

Let σ_3 be this root and n the associated plane. Since the state of stress at the point can be characterised by the six rectangular components referred to any orthogonal frame of reference, let us choose a particular one $x'y'z'$, where the z' axis is along n and the other two axes, x' and y', are arbitrary. With reference to this system, the stress matrix has the form.

$$\begin{bmatrix} \sigma_{x'} & \sigma_{x'y'} & 0 \\ \tau_{x'y'} & \sigma_{y'} & 0 \\ 0 & 0 & \sigma_3 \end{bmatrix} \tag{27}$$

Figure 12 shows these stress vectors on a rectangular element. The shear stress components $\tau_{x'z'}$ and $\tau_{y'z'}$ are zero since the z' plane is chosen to be the principal plane.

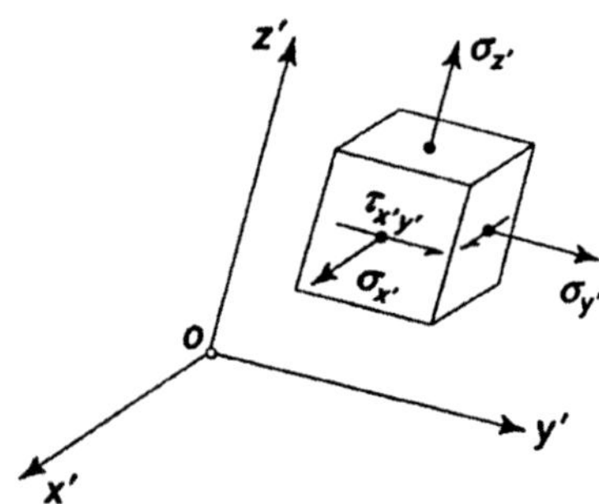

Fig. 12

With reference to this system, Eq. (19) becomes

$$\begin{vmatrix} (\sigma_{x'} - \sigma) & \tau_{x'y'} & 0 \\ \tau_{x'y'} & (\sigma_{y'} - \sigma) & 0 \\ 0 & 0 & (\sigma_3 - \sigma) \end{vmatrix} = 0 \tag{28}$$

Expanding $(\sigma_3 \,\Box\, \sigma)\,[\sigma^2 \,\Box\, (\sigma_{x'} + \sigma_{y'})\,\sigma + \sigma_{x'}\sigma_{y'} \,\Box\, \tau^2_{x'y'}] = 0$

This is a cubic in σ. One of the solutions is $\sigma = \sigma_3$. The two other solutions are obtained by solving the quadratic inside the brackets. The two solutions are

$$\sigma_{1,2} = \frac{\sigma_{x'} + \sigma_{y'}}{2} \pm \left[\left(\frac{\sigma_{x'} - \sigma_{y'}}{2} \right)^2 + \tau^2_{x'y'} \right]^{\frac{1}{2}} \qquad (29)$$

The quantity under the square root $\left(\text{power}\,\frac{1}{2}\right)$ is never negative and hence, σ_1 and σ_2 are real. This means that the curve for $f(\sigma)$ in Fig. 12 will cut the σ axis at three points *A*, *B* and *C* in general. In the next section we shall study a few particular cases.

PARTICULAR CASES

(i) If σ_1, σ_2 and σ_3 are distinct, i.e. σ_1, σ_2 and σ_3 have different values, then the three associated principal axes n_1, n_2 and n_3 are unique and mutually perpendicular. This follows from Eq. (25). Since σ_1, σ_2 and σ_3 are distinct, we get three distinct axes n_1, n_2 and n_3 from Eqs (18), and being mutually perpendicular they are unique.

(ii) If $\sigma_1 = \sigma_2$ and σ_3 is distinct, the axis of n_3 is unique and every direction perpendicular to n_3 is a principal direction associated with $\sigma_1 = \sigma_2$. This is shown in Fig. 13.

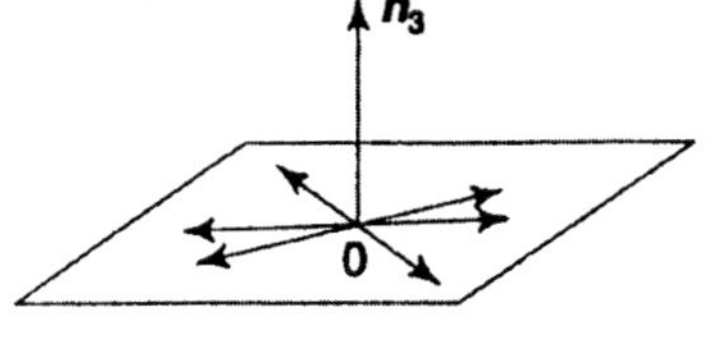

Fig. 13

To prove this, let us choose a frame of reference $Ox'y'z'$ such that the z' axis is along n_3 and the x' and y' axes are arbitrary. From Eq. (29), if $\sigma_1 = \sigma_2$ then the quantity under the radical must be zero. Since this is the sum of two squared quantities, this can happen only if

$$\sigma_{x'} = \sigma_{y'} \quad \text{and} \quad \tau_{x'y'} = 0$$

But we have chosen x' and y' axes arbitrarily, and consequently the above condition must be true for any frame of reference with the z' axis along n_3. Hence, the x' and y' planes are shearless planes, i.e. principal planes. Therefore, every direction perpendicular to n_3 is a principal direction associated with $\sigma_1 = \sigma_2$.

(iii) If $\sigma_1 = \sigma_2 = \sigma_3$, then every direction is a principal direction. This is the hydrostatic or the isotropic state of stress. For proof, we can repeat the argument given in (ii). Choose a coordinate system $Ox'y'z'$ with the z' axis along n_3 corresponding to σ_3. Since $\sigma_1 = \sigma_2$ every direction perpendicular to n_3 is a principal direction. Next, choose the z' axis parallel to n_2 corresponding to σ_2. Then every direction perpendicular to n_2 is a principal direction since $\sigma_1 = \sigma_3$. Similarly, if we choose the z' axis parallel to n_1 corresponding to σ_1, every direction perpendicular to n_1 is also a principal direction. Consequently, every direction is a principal direction.

Choosing $Oxyz$ coinciding with n_1, n_2 and n_3, the stress vector on any arbitrary plane n has value σ, the direction of σ coinciding with n. Hence, every plane is a principal plane. Such a state of stress is equivalent to a hydrostatic state of stress or an isotropic state of stress.

RECAPITULATION

The material discussed in the last few sections is very important and it is worthwhile to put it in the form of definitions and theorems.

Definition

For a given state of stress at point P, if the resultant stress vector $\overset{n}{T}$ on any plane n is along n having a magnitude σ, then σ is a principal stress at P, n is the principal direction associated with σ, the axis of σ is a principal axis and the plane is a principal plane at P.

Theorem

In every state of stress there exist at least three mutually perpendicular principal axes and at most three distinct principal stresses. The principal stresses σ_1, σ_2 and σ_3 are the roots of the cubic equation

$$\sigma^3 \square\, l_1\sigma^2 + l_2\sigma \square\, l_3 = 0$$

where $l_1 l_2$ and l_3 are the first, second and third invariants of stress. The principal directions associated with σ_1, σ_2 and σ_3 are obtained by substituting σ_i ($i = 1, 2, 3$) in the following equations and solving for n_x, n_y and n_z.

$$(\sigma_x \square\, \sigma_i)\, n_x + \tau_{xy}\, n_y + \tau_{xz}\, n_z = 0$$

$$\tau_{xy}\, n_x + (\sigma_y \square\, \sigma_i)\, n_y + \tau_{yz}\, n_z = 0$$

$$n_x^2 + n_y^2 + n_z^2 = 1$$

If σ_1, σ_2 and σ_3 are distinct, then the axes of n_1, n_2 and n_3 are unique and mutually perpendicular. If, say $\sigma_1 = \sigma_2 \neq \sigma_3$, then the axis of n_3 is a unique and every direction perpendicular to n_3 is a principal direction associated with $\sigma_1 = \sigma_2$. If $\sigma_1 = \sigma_2 = \sigma_3$, then every direction is a principal direction.

Standard Method of Solution

Consider the cubic equation $y^3 + py^2 + qy + r = 0$, where p, q and r are constants.

Substitute $\quad y = x - \dfrac{1}{3}p$

This gives $\quad x^3 + ax + b = 0$

where $\quad a = \dfrac{1}{3}(3q - p^2), \qquad b = \dfrac{1}{27}(2p^3 - 9pq + 27r)$

Put
$$\cos\phi = -\frac{b}{2\left(-\dfrac{a^3}{27}\right)^{1/2}}$$

Determine ϕ, the putting $g = 2\sqrt{-a/3}$, the solutions are

$$y_1 = g\cos\frac{\phi}{3} - \frac{p}{3}$$

$$y_2 = g\cos\left(\frac{\phi}{3} + 120°\right) - \frac{p}{3}$$

$$y_3 = g\cos\left(\frac{\phi}{3} + 240°\right) - \frac{p}{3}$$

Example 4:

At a point P, the rectangular stress components are

$$\sigma_x = 1,\ \sigma_y = \square 2,\ \sigma_z = 4,\ \tau_{xy} = 2,\ \tau_{yz} = \square 3,\ \text{and}\ \ \tau_{xz} = 1$$

all in units of kPa. Find the principal stresses and check for invariance. The given stress matrix is

$$\left[\tau_{ij}\right] = \begin{bmatrix} 1 & 2 & 1 \\ 2 & -2 & -3 \\ 1 & -3 & 4 \end{bmatrix}$$

Solution:

From Eqs (22) – (24),

$$I_1' = 1 - 2 + 4 = 3$$

$$I_2 = (-2 - 4) + (-8 - 9) + (4 - 1) = -20$$

$$I_3 = 1(-8 - 9) - 2(8 + 3) + 1(-6 + 2) = -43$$

$\therefore \qquad f(\sigma) = \sigma^3 - 3\sigma^2 - 20\sigma + 43 = 0$

For this cubic, following the standard method,

$$y = \sigma, \; p = -3, \; q = -20, \; r = 43$$

$$a = \frac{1}{3}(-60-9) = -23$$

$$b = \frac{1}{27}(-54-540+1161) = 21$$

$$\cos\phi = -\frac{\left(\frac{21}{2}\right)}{\left(\frac{12167}{27}\right)^{1/2}}$$

$\therefore \qquad \phi = -119^\circ\ 40$

The solutions are

$$\sigma_1 = y_1 = 4.25 + 1 = 5.25 \text{ kPa}$$

$$\sigma_2 = y_2 = -5.2 + 1 = -4.2 \text{ kPa}$$

$$\sigma_3 = y_3 = 0.95 + 1 = 1.95 \text{ kPa}$$

Renaming such that $\sigma_1 \geq \sigma_2 \geq \sigma_3$ we have,

$$\sigma_1 = 5.25 \text{ kPa}, \quad \sigma_2 = 1.95 \text{ kPa}, \quad \sigma_3 = -4.2 \text{ kPa}$$

The stress invariants are

$$l_1 = 5.25 + 1.95 - 4.2 = 3.0$$

$$l_2 = (5.25 \times 1.95) - (1.95 \times 4.2) - (4.2 \times 5.25) = -20$$

$$l_3 = -(5.25 \times 1.95 \times 4.2) = -43$$

These agree with their earlier values.

Example 5:

With respect to the frame of reference Oxyz, the following state of stress exists. Determine the principal stresses and their associated directions. Also, check on the invariances of l_1, l_2, l_3.

$$\left[\tau_{ij}\right] = \begin{bmatrix} 1 & 2 & 1 \\ 2 & 1 & 1 \\ 1 & 1 & 1 \end{bmatrix}$$

Solution:

For this state

$$l_1 = 1 + 1 + 1 = 3$$

$$I_2 = (1-4) + (1-1) + (1-1) = -3$$
$$I_3 = 1(1-1) - 2(2-1) + 1(2-1) = 1$$
$$f(\sigma) = \sigma^3 - I_1\sigma^2 + I_2\sigma - I_3 = 0$$

i.e., $\sigma^3 - 3\sigma^2 - 3\sigma + 1 = 0$

or $(\sigma^3 + 1) - 3\sigma(\sigma + 1) = 0$

i.e., $(\sigma + 1)(\sigma^2 - \sigma + 1) - 3\sigma(\sigma + 1) = 0$

or $(\sigma + 1)(\sigma^2 - 4\sigma + 1) = 0$

Hence, one solution is $\sigma = -1$. The other two solutions are obtained from the solution of the quadratic equation, which are $\sigma = 2 \pm \sqrt{3}$.

$\therefore$ $\sigma_1 = -1,\ \sigma_2 = 2+\sqrt{3},\ \sigma_3 = 2-\sqrt{3}$

Check on the invariance:

With the set of axes chosen along the principal axes, the stress matrix will have the form

$$\left[\tau_{ij}\right] = \begin{bmatrix} -1 & 0 & 0 \\ 0 & 2+\sqrt{3} & 0 \\ 0 & 0 & 2-\sqrt{3} \end{bmatrix}$$

Hence, $I_1 = -1 + 2 + \sqrt{3} + 2 - \sqrt{3} = 3$

$$I_2 = (-2-\sqrt{3}) + (4-3) + (-2+\sqrt{3}) = -3$$

$$I_3 = -1 - (4-3) = -1$$

Directions of principal axes:

(i) For $\sigma_1 = -1$, from Eqs (18) and (21)

$$(1 + 1)n_x + 2n_y + n_z = 0$$
$$2n_x + (1 + 1)n_y + n_z = 0$$
$$n_x + n_y + (1 + 1)n_z = 0$$

together with

$$n_x^2 + n_y^2 + n_z^2 = 1$$

From the second and third equations above, $n_z = 0$. Using this in the third and fourth equations and solving,

$n_x = \pm\left(1/\sqrt{2}\right), n_y = \pm\left(1/\sqrt{2}\right)$.

Hence $\sigma_1 = -1$ is in the direction $\left(+1/\sqrt{2}, -1/\sqrt{2}, 0\right)$.

It should be noted that the plus and minus signs associated with n_x, n_y and n_z represent the same line.

(ii) For $\sigma_2 = 2 + \sqrt{3}$

$$(-1-\sqrt{3})n_x + 2n_y + n_z = 0$$

$$2n_x + (-1-\sqrt{3})n_y + n_z = 0$$

$$n_x + n_y(-1-\sqrt{3})n_z = 0$$

together with

$$n_x^2 + n_y^2 + n_z^2 = 1$$

Solving, we get

$$n_x = n_y = \left(1 + \frac{1}{\sqrt{3}}\right)^{1/2} \quad n_z = \frac{1}{\left(3+\sqrt{3}\right)^{1/2}}$$

(iii) For $\sigma_3 = 2 - \sqrt{3}$

We can solve for n_x, n_y and n_z in a manner similar to the proceeding one or get the solution from the condition that n_1, n_2, and n_3 form a right–angled triad, i.e. $n_3 = n_1 \times n_2$.

The Solution is

$$n_x = n_y = -\frac{1}{2}\left(1 - \frac{1}{\sqrt{3}}\right)^{1/2}, \quad n_z = \frac{1}{\sqrt{2}}\left(1 + \frac{1}{\sqrt{3}}\right)^{1/2}$$

Example 6:

For the given state of stress, determine the principal stresses and their directions.

$$\left[\tau_{ij}\right] = \begin{bmatrix} 0 & 1 & 1 \\ 1 & 0 & 1 \\ 1 & 1 & 0 \end{bmatrix}$$

Solution:

$$I_1 = 0,\ I_2 = -3,\ I_3 = 2$$

$$f(\sigma) = -\sigma^3 + 3\sigma + 2 = 0$$

$$= (-\sigma^3 - 1) + (3\sigma + 3)$$

$$= -(\sigma + 1)(\sigma^2 - \sigma + 1) + 3(\sigma + 1)$$

$$= (\sigma + 1)(\sigma - 2)(\sigma + 1) = 0$$

$$\therefore \quad \sigma_1 = \sigma_2 = -1 \quad \text{and } \sigma_3 = 2$$

Since two of the three principal stresses are equal, and σ_3 is different, the axis of σ_3 is unique and every direction perpendicular to σ_3 is a principal direction associated with $\sigma_1 = \sigma_2$. For $\sigma_3 = 2$

$$-2n_x + n_y + n_z = 0$$

$$n_x + 2n_y + n_z = 0$$

$$n_x + n_y - 2n_z = 0$$

$$n_x^2 + n_y^2 + n_z^2 = 1$$

These give $n_x = n_y = n_z = \dfrac{1}{\sqrt{3}}$

Example 7:

The state of stress at a point is such that

$$\sigma_x = \sigma_y = \sigma_z = \tau_{xy} = \tau_{yz} = \tau_{zx} = \rho$$

Determine the principal stresses and their directions

For the given state,

$$I_1 = 3\rho, \quad I_2 = 0, \quad I_3 = 0$$

Solution:

Therefore the cubic is $\sigma^3 - 3\rho\sigma^2 = 0$; the solutions are $\sigma_1 = 3\rho$, $\sigma_2 = \sigma_3 = 0$. For $\sigma_1 = 3\rho$

$$(\rho - 3\rho)\, n_x + \rho n_y + \rho n_z = 0$$

$$\rho n_x + (\rho - 3\rho)\, n_y + \rho n_z = 0$$

$$\rho n_x + \rho n_y + (\rho - 3\rho)\, n_z = 0$$

or

$$-2n_x + n_y + n_z = 0$$

$$n_x - 2n_y + n_z = 0$$

$$n_x + n_y - 2n_z = 0$$

The above equations give

$$n_x = n_y = n_z$$

With $n_x^2 + n_y^2 + n_z^2 = 1$, one gets $n_x = n_y = n_z = 1/\sqrt{3}$.

Thus, on a plane that is equally inclined to *xyz* axes, there is a tensile stress of magnitue 3ρ. This is the case of a uniaxial tension, the axis of loading making equal angles with the given *xyz* axes. If one denotes this loading axis by z', the other two axes, x' and y', can be chosen arbitrarily, and the planes normal to these, i.e. x' plane and y' plane, are stress free.

THE STATE OF STRESS REFERRED TO PRINCIPAL AXES

In expressing the state of stress at a point by the six rectangular stress components, we can choose the principal axes as the coordinate axes and refer the rectangular stress components accordingly. We then have for the stress matrix

$$\left[\tau_{ij}\right] = \begin{bmatrix} \sigma_1 & 0 & 0 \\ 0 & \sigma_2 & 0 \\ 0 & 0 & \sigma_3 \end{bmatrix} \tag{30}$$

On any plane with normal n, the components of the stress vector are, from Eq. (9),

$$\overset{n}{T}_x = \sigma_1 n_x, \qquad \overset{n}{T}_y = \sigma_2 n_y, \qquad \overset{n}{T}_z = \sigma_3 n_z \tag{31}$$

The resultant stress has a magnitude

$$\left|\overset{n}{T}\right| \sigma_1^2 n_x^2 + \sigma_2^2 n_y^2 + \sigma_2^2 n_z^2 \tag{32}$$

If σ is the normal and τ the shearing stress on this plane, then

$$\sigma = \sigma_1 n_x^2 + \sigma_2 n_y^2 + \sigma_3 n_z^2 \tag{33}$$

and
$$\tau^2 = \left|\overset{n}{T}\right|^2 - \sigma^2 \tag{34}$$

$$= n_x^2 n_y^2 (\sigma_1 - \sigma_2)^2 + n_y^2 n_z^2 (\sigma_2 - \sigma_3)^2 + n_z^2 n_x^2 (\sigma_3 - \sigma_1)^2$$

The stress invariants assume the form

$$\begin{aligned} l_1 &= \sigma_1 + \sigma_2 + \sigma_3 \\ l_2 &= \sigma_1\sigma_2 + \sigma_2\sigma_3 + \sigma_3\sigma_1 \\ l_3 &= \sigma_1\sigma_2\sigma_3 \end{aligned} \tag{35}$$

MOHR'S CIRCLES FOR THE THREE–DIMENSIONAL STATE OF STRESS

We shall now describe a geometrical construction that brings out some important results. At a given point p, let the frame of reference $Pxyz$ be chosen along the principal stress axes. Consider a plane with normal n at point P. Let σ be the normal stress and τ the shearing stress on this plane. Take another set of axes σ and τ. In this plane we can mark a point Q with coordinates (Q, τ) representing the values of the normal and shearing stress on the plane n. For different planes passing through point p, we get different values of σ and τ. Corresponding to each plane n, a point Q can be located with coordinates

(σ, τ). The plane with the σ axis and the τ axis is called the stress plane π. (No numerical value is associated with this symbol). The problem now is to determine the bounds for $Q\ (\sigma, \tau)$ for all possible directions n.

Arrange the principal stresses such that algebraically

$$\sigma_1 \geq \sigma_2 \geq \sigma_3$$

Mark off σ_1, σ_2 and σ_3 along the σ axis and construct three circles with diameters $(\sigma_1 - \sigma_2)$, $(\sigma_2 - \sigma_3)$ and $(\sigma_1 - \sigma_3)$ as shown in Fig. 14.

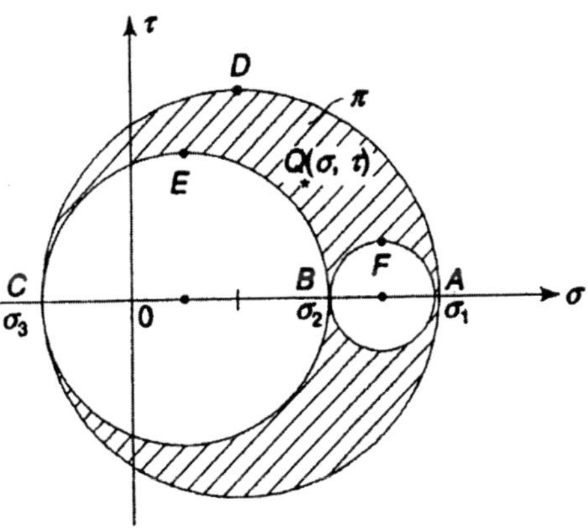

Fig. 14

The point $Q\ (\sigma, \tau)$ for all possible n will lie within the shaded area. This region is called Mohr's stress plane π and the three circles are known as mohr's circles. From Fig. 14, the following points can be observed:

(i) Points A, B and C represent the three principal stresses and the associated shear stresses are zero.

(ii) The maximum shear stress is equal to $1/2\ (\sigma_1 - \sigma_3)$ and the associated normal stress is $1/2\ (\sigma_1 + \sigma_3)$. This is indicated by point D on the outer circle.

(iii) Just as there are three extremum values σ_1, σ_2 and σ_3 for the normal stresses, there are three extremum values for the shear stresses, these being $\dfrac{\sigma_1 - \sigma_3}{2}$, $\dfrac{\sigma_2 - \sigma_3}{2}$ and $\dfrac{\sigma_1 - \sigma_2}{2}$. The planes on which these shear stresses act are called the principal shear planes. While the planes on which the principal normal stresses act are free of shear stresses, the principal shear planes are not free from normal stresses. The normal stresses associated with the principal shears are respectively $\dfrac{\sigma_1 + \sigma_3}{2}$, $\dfrac{\sigma_2 + \sigma_3}{2}$ and $\dfrac{\sigma_1 - \sigma_2}{2}$. These are indicated by points D, E and F in Fig. 14. The principal shear planes are at 45° to the principal normal planes. The principal shears are denoted by τ_1, τ_2 and τ_3 where

$$2\tau_3 = (\sigma_1 - \sigma_2), \quad 2\tau_2 = (\sigma_1 - \sigma_3), \quad 2\tau_1 = (\sigma_2 - \sigma_3) \tag{36}$$

(iv) When $\sigma_1 = \sigma_2 \neq \sigma_3$ or $\sigma_1 \neq \sigma_2 = \sigma_3$, the three circles reduce to only one circle and the shear stress on any plane will not exceed $\frac{1}{2}(\sigma_1 - \sigma_3)$ or $\frac{1}{2}(\sigma_1 - \sigma_2)$ according as $\sigma_1 = \sigma_2$ or $\sigma_2 = \sigma_3$.

(v) When $\sigma_1 = \sigma_2 = \sigma_3$, the three circles collapse to a single point on the σ axis and every plane is a shearless plane.

MOHR'S STRESS PLANE

If was stated in the previous section that when points with coordinates (σ, τ) for all possible planes passing through a point are marked on the $\sigma = \tau$ plane, as in Fig. 14, the points are bounded by the three Mohr's circles. In this, section we shall prove this.

Choose the coordinate frame of reference $Pxyz$ such that the axes are along the principal axes. On any plane with normal n, the resultant stress vector $\overset{n}{T}$ and the normal stress σ are such that from Eqs (32) and (33)

$$\left|\overset{n}{T}\right|^2 = \sigma^2 + \tau^2 = \sigma_1^2 n_x^2 + \sigma_2^2 n_y^2 + \sigma_3^2 n_z^2 \tag{37}$$

$$\sigma = \sigma_1 n_x^2 + \sigma_2 n_y^2 + \sigma_3 n_z^2 \tag{38}$$

and also
$$1 = n_x^2 + n_y^2 + n_z^2 \tag{39}$$

The above three equations can be used to solve for n_x^2, n_y^2 and n_z^2 yielding

$$n_x^2 = \frac{(\sigma - \sigma_2)(\sigma - \sigma_3) + \tau^2}{(\sigma_1 - \sigma_2)(\sigma_1 - \sigma_3)} \tag{40}$$

$$n_y^2 = \frac{(\sigma - \sigma_3)(\sigma - \sigma_1) + \tau^2}{(\sigma_2 - \sigma_3)(\sigma_2 - \sigma_1)} \tag{41}$$

$$n_z^2 = \frac{(\sigma - \sigma_1)(\sigma - \sigma_2) + \tau^2}{(\sigma_3 - \sigma_1)(\sigma_3 - \sigma_2)} \tag{42}$$

Since n_x^2, n_y^2 and n_z^2 are all positive, the right–hand side expressions in the above equations must all be positive. Recall that we have arranged the principal stresses such that $\sigma_1 \geq \sigma_2 \geq \sigma_3$. There are three cases one can consider.

Case (i) $\sigma_1 > \sigma_2 > \sigma_3$

Case (ii) $\sigma_1 = \sigma_2 > \sigma_3$

Case (iii) $\sigma_1 = \sigma_2 = \sigma_3$

We shall consider these cases individually.

Case (i) $\sigma_1 > \sigma_2 > \sigma_3$

For this case, the denominator in Eq. (40) is positive and hence, the numerator must also be positive. In Eq. (41), the denominator being negative, the numerator must also be negative. Similarly, the numerator in Eq. (42) must be positive. Therefore.

$$(\sigma - \sigma_2)(\sigma - \sigma_3) + \tau^2 \geq 0$$

$$(\sigma - \sigma_3)(\sigma - \sigma_1) + \tau^2 \leq 0$$

$$(\sigma - \sigma_1)(\sigma - \sigma_2) + \tau^2 \geq 0$$

The above three inequalities can be rewritten as

$$\tau^2 + \left(\sigma - \frac{\sigma_2 - \sigma_3}{2}\right)^2 \geq \left(\sigma - \frac{\sigma_2 - \sigma_3}{2}\right)^2$$

$$\tau^2 + \left(\sigma - \frac{\sigma_3 + \sigma_1}{2}\right)^2 \leq \left(\frac{\sigma_3 - \sigma_1}{2}\right)^2$$

$$\tau^2 + \left(\sigma - \frac{\sigma_1 + \sigma_2}{2}\right)^2 \geq \left(\frac{\sigma_1 - \sigma_2}{2}\right)^2$$

According to the first of the above equations, the point (σ, τ) must lie on or outside a circle of radius $1/2\ (\sigma_2 - \sigma_3)$ with its centre at $1/2\ (\sigma_2 + \sigma_3)$ along the σ axis (Fig. 14). This is the circle with *BC* as diameter. The second equation indicates that the point (σ, τ) must lie inside or on the circle *ADC* with radius $1/2\ (\sigma_1 - \sigma_2)$ and centre at $1/2\ (\sigma_1 + \sigma_2)$.

Hence, for this case, the point $Q\ (\sigma, \tau)$ should lie inside the shaded area of Fig. 14.

Case (ii) $\sigma_1 = \sigma_2 > \sigma_3$

Following arguments similar to the ones given above, one has for this case from Eqs (40) – (42)

$$\tau^2 + \left(\sigma - \frac{\sigma_2 + \sigma_3}{2}\right)^2 = \left(\frac{\sigma_2 - \sigma_3}{2}\right)^2$$

$$\tau^2 + \left(\sigma - \frac{\sigma_3 + \sigma_1}{2}\right)^2 = \left(\frac{\sigma_3 - \sigma_1}{2}\right)^2$$

$$\tau^2 + \left(\sigma - \frac{\sigma_1 + \sigma_2}{2}\right)^2 \geq \left(\frac{\sigma_1 - \sigma_2}{2}\right)^2$$

From the first two of these equations, since $\sigma_1 = \sigma_2$, point (σ, τ) must lie on the circle with radius $1/2\ (\sigma_1 - \sigma_3)$ with its centre at $1/2\ (\sigma_1 + \sigma_3)$. The last equation indicates that the point must lie outside a circle of zero radius (since $\sigma_1 = \sigma_2$). Hence, in this case, the Mohr's circles will reduce to a circle BC and a point circle B. The point Q lies on the circle BEC.

Case (iii) $\sigma_1 = \sigma_2 = \sigma_3$

This is a trivial case since this is the isotropic or the hydrostatic state of stress. Mohr's circles collapse to a single point on the σ axis.

The graphical determination of the normal and shear stresses on an arbitrary plane, using Mohr's circles.

PLANES OF MAXIMUM SHEAR

The Fig. 14 for the case $\sigma_1 > \sigma_2 > \sigma_3$, the maximum shear stress is $1/2\,(\sigma_1 - \sigma_3) = \tau_2$ and the associated normal stress is $1/2\,(\sigma_1 + \sigma_3)$. Substituting these values in Eqs. (37) – (39), one gets $n_x = \pm\sqrt{1/2}, n_y = 0$ and $n_z = \pm\sqrt{2}$.

This means that the planes (there are two of them) on which the shear stress takes on an extremum value, make angles of 45° and 135° with the σ_1 and σ_2 planes as shown in Fig. 15.

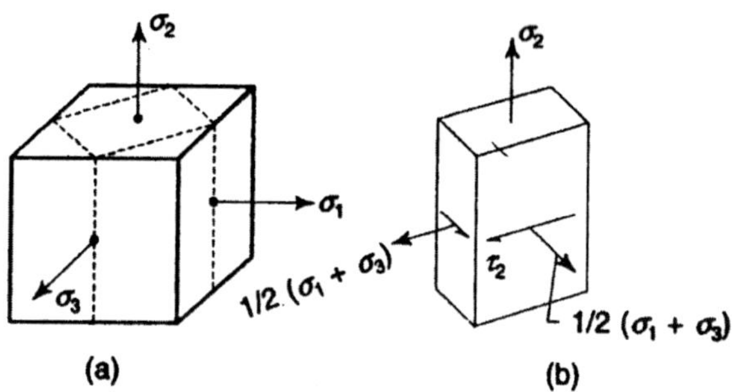

Fig 15

If $\sigma_1 = \sigma_2 > \sigma_3$, then the three Mohr's circles reduce to one circles BC (Fig. 14) and the maximum shear stress will be $1/2\ (\sigma_2 - \sigma_3) = \tau_1$. with the associated normal stress $1/2\ (\sigma_2 + \sigma_3)$. Substituting these values in Eqs (37) – (39), we get $n_x = 0/0, n_y = 0/0$ and $n_z = \pm\sqrt{2}$ i.e. n_x and n_y are indeterminate.

This means that the planes on which τ_1 is acting makes angles of 45° and 135° with the σ_3 axis but remains indeterminate with respect to σ_1 and σ_2 axes. This is so because, since $\sigma_1 = \sigma_2 \neq \sigma_3$, the axis of σ_3 is unique, whereas, every direction perpendicular to σ_3 is a principal direction associated with $\sigma_1 = \sigma_2$. The principal shear plane will, therefore, make a fixed angle with σ_3 axis (45° or 135°)but will have different values depending upon the selection of σ_1 and σ_3 axes.

OCTAHEDRAL STRESSES

Let the frame of reference be again chosen along σ_1, σ_2 and σ_3 axes. A plane that is equally inclined to these three axes is called an octahedral plane. Such a plane will have $n_x = n_y = n_z$. Since $n_x^2 + n_y^2 + n_z^2 = 1$, an octahedral plane will be defined by $n_x = n_y = n_z = \pm 1/\sqrt{3}$. There are eight such planes, as shown in Fig. 16.

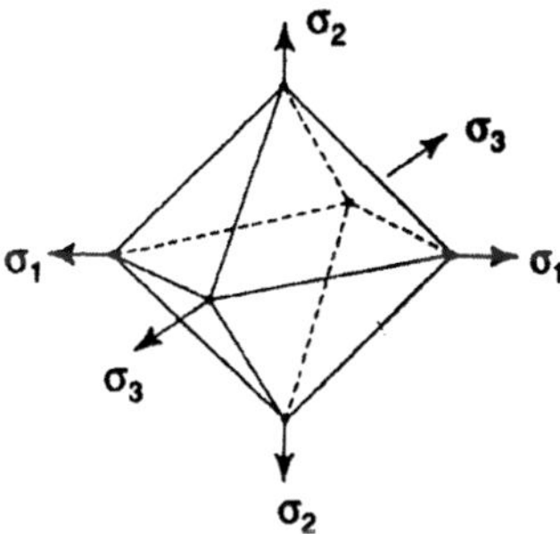

Fig. 16

The normal and shearing stresses on these planes are called the octahedral normal stress and octahedral shearing stress respectively. Substituting $n_x = n_y = n_z = \pm 1/\sqrt{3}$ in Eqs (33) and (34),

$$\sigma_{oct} = \frac{1}{3}(\sigma_1 + \sigma_1 + \sigma_1) = \frac{1}{3} l_1 \tag{43}$$

and
$$\tau_{oct}^2 = \frac{1}{9}[(\sigma_1 - \sigma_1)^2 + (\sigma_2 - \sigma_3)^2 + (\sigma_3 - \sigma_1)^2] \tag{44a}$$

or
$$9\tau_{oct}^2 = 2(\sigma_1 + \sigma_2 + \sigma_3)^2 - 6(\sigma_1\sigma_2 + \sigma_2\sigma_3 + \sigma_3\sigma_1)^2 \tag{44b}$$

or
$$\tau_{oct} = \frac{\sqrt{2}}{3}(l_1^2 - 3l_2)^{1/2} \tag{44c}$$

It is important to remember that the octahedral planes are defined with respect to the principal axes and not with reference to an arbitrary frame of reference. Since σ_{oct} and τ_{oct} have been expressed in terms of the stress invariants, one can express these in terms of σ_x, σ_y, σ_z τ_{xy}, τ_{yz} and τ_{zx} also, Using Eqs (22) and (23),

$$\sigma_{oct} = \frac{1}{3}(\sigma_x + \sigma_y + \sigma_z) \tag{45}$$

$$9\tau_{oct}^2 = (\sigma_x - \sigma_y)^2 + (\sigma_y - \sigma_z)^2 + (\sigma_z - \sigma_x)^2 + 6(\tau_{xy}^2 + \tau_{yz}^2 + \tau_{zx}^2)^2 \tag{46}$$

The octahedral normal stress being equal to 1/3 I_1, it may be interpreted as the mean normal stress at a given point in a body. If in a state of stress, the first invariant ($\sigma_1 + \sigma_2 + \sigma_3$) is zero, then the normal stresses on the octahedral planes will be zero and only the shear stresses will act.

Example 8:

The state of stress at a point is characterised by the components

$$\sigma_x = 100 \text{ MPa}, \ \sigma_y = -40 \text{ MPa}, \ \sigma_z = 80 \text{ MPa},$$

$$\tau_{xy} = \tau_{yz} = \tau_{zx} = 0$$

Determine the extremum values of the shear stresses, their associated normal stresses, the octahedral shear stress and its associated normal stress.

Solution:

The given stress components are the principal stresses, since the shears are zero. Arranging the terms such that $\sigma_1 \geq \sigma_2 \geq \sigma_3$,

$$\sigma_1 = 100 \text{ MPa}, \ \sigma_2 = 80 \text{ MPa}, \ \sigma_3 = -40 \text{ MPa}$$

Hence from Eq. (1.36),

$$\tau_1 = \frac{\sigma_2 - \sigma_3}{2} = \frac{80+40}{2} = 60 \text{ MPa}$$

$$\tau_2 = \frac{\sigma_3 - \sigma_1}{2} = \frac{-40-100}{2} = -70 \text{ MPa}$$

$$\tau_3 = \frac{\sigma_1 - \sigma_2}{2} = \frac{100-80}{2} = 10 \text{ MPa}$$

The associated normal stresses are

$$\sigma_1^{\bullet} = \frac{\sigma_2 + \sigma_3}{2} = \frac{80-40}{2} = 20 \text{ MPa}$$

$$\sigma_2^{\bullet} = \frac{\sigma_3 + \sigma_1}{2} = \frac{-40+100}{2} = 30 \text{ MPa}$$

$$\sigma_3^{\bullet} = \frac{\sigma_1 + \sigma_2}{2} = \frac{100+80}{2} = 90 \text{ MPa}$$

$$\tau_{oct} = \frac{1}{3}\left[(\sigma_1 - \sigma_2)^2 + (\sigma_2 - \sigma_3)^2 + (\sigma_3 - \sigma_1)^2\right]^{1/2} = 61.8 \text{ MPa}$$

$$\sigma_{oct} = \frac{1}{3}(\sigma_1 + \sigma_2 + \sigma_3) = \frac{140}{3} = 46.7 \text{ MPa}$$

THE STATE OF PURE SHEAR

The state of stress at a point can be characterised by the six rectangular stress components referred to a coordinate frame of reference. The magnitudes of these components depend on the choice of the coordinate system. It, for at least one particular choice of the frame of reference, we find that $\sigma_x = \sigma_y = \sigma_z = 0$, then a state of pure shear is said to exist at point *P*. For such a state, with that particular choice of coordinate system, the stress matrix will be

$$\left[\tau_{ij}\right] = \begin{bmatrix} 0 & \tau_{xy} & \tau_{xz} \\ \tau_{xy} & 0 & \tau_{yz} \\ \tau_{xz} & \tau_{yz} & 0 \end{bmatrix}$$

For this coordinate system, $I_1 = \sigma_x + \sigma_y + \sigma_z = 0$. Since I_1 is an invariant, this must be true for any choice of coordinate system selected at *P*. Hence, the necessary condition for a state of pure shear to exist is that $I_1 = 0$.

It was remarked in the previous section that when $I_1 = 0$, an octahedral plane is subjected to pure shear with no normal stress. Hence, for a pure shear stress state, the octahedral plane (remember that this plane is defined with respect to the principal axes and not with respect to an arbitrary set of axes) is free from normal stress.

DECOMPOSITION INTO HYDROSTATIC AND PURE SHEAR STATES

It will be shown in the present section that an arbitrary state of stress can be resolved into a hydrostatic state and a state of pure shear. Let the given state referred to a coordinate system be

$$\left[\tau_{ij}\right] = \begin{bmatrix} \sigma_x & \tau_{xy} & \tau_{xz} \\ \tau_{xy} & \sigma_y & \tau_{yz} \\ \tau_{xz} & \tau_{yz} & \sigma_z \end{bmatrix}$$

Let $P = 1/3\ (\sigma_x + \sigma_y + \sigma_z) = 1/3 I_1$ (47)

The given state can be resolved into two different states, as shown:

$$\begin{bmatrix} \sigma_x & \tau_{xy} & \tau_{xz} \\ \tau_{xy} & \sigma_y & \tau_{yz} \\ \tau_{xz} & \tau_{yz} & \sigma_z \end{bmatrix} = \begin{bmatrix} p & 0 & 0 \\ 0 & p & 0 \\ 0 & 0 & p \end{bmatrix} + \begin{bmatrix} \sigma_x - p & \tau_{xy} & \tau_{xz} \\ \tau_{xy} & \sigma_y - p & \tau_{yz} \\ \tau_{xz} & \tau_{yz} & \sigma_z - p \end{bmatrix} \quad (48)$$

The first state on the right–hand side of the above equation is a hydrostatic state.

The second state is a state of pure shear since the first invariant for this state is

$$I'_1 = (\sigma_x - p) + (\sigma_y - p) + (\sigma_x - p)$$

$$= \sigma_x + \sigma_y + \sigma_z - 3p$$
$$= 0 \text{ from Eq. (47)}$$

It the given state is referred to the principal axes, the decomposition into a hydrostatic state and a pure shear state can once again be done as above, i.e.

$$\begin{bmatrix} \sigma_1 & 0 & 0 \\ 0 & \sigma_2 & 0 \\ 0 & 0 & \sigma_3 \end{bmatrix} = \begin{bmatrix} p & 0 & 0 \\ 0 & p & 0 \\ 0 & 0 & p \end{bmatrix} + \begin{bmatrix} \sigma_1 - p & 0 & 0 \\ 0 & \sigma_2 - p & 0 \\ 0 & 0 & \sigma_3 - p \end{bmatrix} \tag{49}$$

where, as before, $p = 1/3\ (\sigma_1 + \sigma_2 + \sigma_3) = 1/3 l_1$.

The pure shear state of stress is also known as the deviatoric state of stress or simply as stress deviator.

Example 9:

The state of stress characterised by τ_{ij} is given below. Resolve the given state into a hydrostatic state and a pure shear state. Determine the normal and shearing stresses on an octahedral plane. Compare these with the σ_{oct} and τ_{oct} calculated for the hydrostatic and the pure shear states. Are the octahedral planes for the given state, the hydrostatic state and the pure shear state the same or are they different? Explain why.

$$\left[\tau_{ij}\right] = \begin{bmatrix} 10 & 4 & 6 \\ 4 & 2 & 8 \\ 6 & 8 & 6 \end{bmatrix}$$

$$l_1 = 10 + 2 + 6 = 18, \qquad \frac{1}{3} l_1 = 6$$

Solution:

Resolving into hydrostatic and pure shear states, Eq. (47),

$$\left[\tau_{ij}\right] = \begin{bmatrix} 6 & 0 & 0 \\ 0 & 6 & 0 \\ 0 & 0 & 6 \end{bmatrix} + \begin{bmatrix} 4 & 4 & 6 \\ 4 & -4 & 8 \\ 6 & 8 & 0 \end{bmatrix}$$

For the given state, the octahedral normal and shear stresses are:

$$\sigma_{oct} \frac{1}{3} l_1 = 6$$

From Eq. (44)

$$\tau_{oct} = \frac{\sqrt{2}}{3}\left(l_1^2 - 3l_2^2\right)^{1/2}$$

$$= \frac{\sqrt{2}}{3}[18^2 - 3(20 - 16 + 12 - 64 + 60 - 36)]^{1/2}$$

$$= \frac{\sqrt{2}}{3}(396)^{1/2} = 2\sqrt{22}$$

For the hydrostatic state, $\sigma_{oct} = 6$, since every plane is a principal plane with $\sigma = 6$ and consequently, $\tau_{oct} = 0$.

For the pure shear state, $\sigma_{oct} = 0$ since the first invariant of stress for the pure shear state is zero. The value of the second invariant of stress for the pure shear state is

$$I_2 = (-16 - 16 + 0 - 64 + 0 - 36) = -132$$

Hence, the value of τ_{oct} for the pure shear state is

$$\tau_{oct} = \frac{\sqrt{2}}{3}(396)^{1/2} = 2\sqrt{22}$$

Hence, the value of σ_{oct} for the given state is equal to the value of σ_{oct} for the hydrostatic state, and τ_{oct} for the given state is equal to τ_{oct} for the pure shear state.

The octahedral planes for the given state (which are identified after determining the principal stress directions), the hydrostatic state and the pure state are all identical. For the hydrostatic state, every direction is a principal direction, and hence, the principal stress directions for the given state and the pure shear state are identical. Therefore, the octahedral planes corresponding to the given state and the pure shear state are identical.

Example 10:

A cylindrical boiler, 180 cm in diameter, is made of plates, 1.8 cm thick, and is subjected to an internal pressure 1400 kPa. Determine the maximum shearing stress in the plate at point P and the plane on which it acts.

Solution:

From elementary strength of materials, the axial stress in plate is $\frac{pd}{4t}$ where p is the internal pressure, d the diameter and t the thickness. The circumferential or the hoop stress is $\frac{pd}{2t}$. The state of stress acting on an element.

The principal stresses when arranged such that $\sigma_1 \geq \sigma_2 \geq \sigma_3$ are

$$\sigma_1 = \frac{pd}{2t}; \quad \sigma_2 = \frac{pd}{4t}; \quad \sigma_3 = -p$$

The maximum shear stress is therefore,

$$\tau_{max} = \frac{1}{2}(\sigma_1 - \sigma_3) = \frac{1}{2}p\left(\frac{d}{2t} + 1\right)$$

Substituting the values

$$\tau_{max} = \frac{1400}{2}\left(\frac{1.8 \times 100}{2 \times 1.8} + 1\right) = 35{,}700 \text{ kPa}$$

CAUCHY'S STRESS QUADRIC

We shall now describe a geometrical description of the state of stress at a point P. Choose a frame of reference whose axes are along the principal axes. Let σ_1, σ_2 and σ_3 be the principal stresses. Consider a plane with normal n. The normal stress on this plane is from Eq. (33),

$$\sigma = \sigma_1 n_x^2 + \sigma_2 n_y^2 + \sigma_3 n_z^2$$

Along the normal n to the plane, choose a point Q such that

$$PQ = R = 1/\sqrt{\sigma} \tag{50}$$

As different planes n are chosen at P, we get different values for the normal stress σ and correspondingly different PQs. If such Qs are marked for every plane passing through P, then we get a surface S. This surface determines the normal component of stress on every plane passing through P. This surface is known as the stress surface of Cauchy. This has a very interesting property. Let Q be a point on the surface, Fig. 17(a). By the previous definition, the length $PQ = R$ is such that the normal stress on the plane whose normal is along PQ is given by

$$\sigma = \frac{1}{R^2} \tag{51}$$

If m is

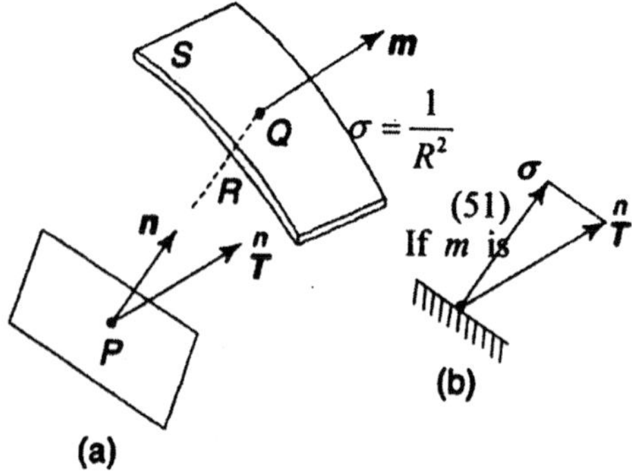

Fig. 17

normal to the tangent plane to the surface S at point Q, then this normal m is parallel to the resultant stress vector $\overset{n}{T}$ at P.

Since the direction of the resultant vector $\overset{n}{T}$ is known, and its component σ along the normal is known, the resultant stress vector $\overset{n}{T}$ can be easily determined, as shown in Fig. 1720(b).

We shall now show that the normal m to the surface S is parallel to $\overset{n}{T}$, the resultant stress vector. Let $Pxyz$ be the principal axes at P (Fig. 18). n is the normal to a particular plane at P. The normal stress on this plane, as before, is

$$\sigma = \sigma_1 n_x^2 + \sigma_2 n_y^2 + \sigma_3 n_z^2$$

If the coordinates of the point Q are (x, y, x) and the length $PQ = R$, then

$$n_x = \frac{x}{R}, \quad n_y = \frac{y}{R}, \quad n_z = \frac{z}{R} \tag{52}$$

Substituting these in the above equation for σ

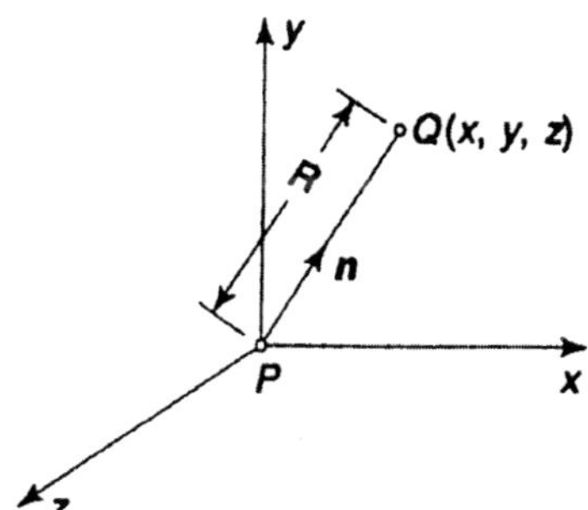

Fig. 18

From Eq. (51), we have $\sigma R^2 = \pm 1$. The plus sign is used when σ is tensile and the minus sign is used when σ is compressive. Hence, the surface S has the equations (a surface of second degree)

When σ is tensile

$$\sigma_1 x^2 + \sigma_2 y^2 + \sigma_3 z^2 = +1$$

When σ is compressive

$$\sigma_1 x^2 + \sigma_2 y^2 + \sigma_3 z^2 = -1$$

We know form calculus that for a surface with equation $F(x, y, z) = 0$, the normal to the tangent plane at a point Q on the surface has direction cosines proportional to $\frac{\partial F}{\partial x}, \frac{\partial F}{\partial y}$ and $\frac{\partial F}{\partial z}$. From Fig. (1.17), m is the normal perpendicular to the tangent plane to S at Q. Hence, if m_x, m_y and m_z are the direction cosines of m, then

$$m_x = \alpha \frac{\partial F}{\partial x}, \qquad m_y = \alpha \frac{\partial F}{\partial y}, \qquad m_z = \alpha \frac{\partial F}{\partial z}$$

From Eq. (53a) or Eq. (53b)

$$m_x = 2\alpha\sigma_1 x, \qquad m_y = 2\alpha\sigma_2 y, \qquad m_z = 2\alpha\sigma_3 z \tag{54}$$

where α is constant of proportionality.

$\overset{n}{T}$ is the resultant stress vector on plane n and its components $\overset{n}{T}_x, \overset{n}{T}_y$ and $\overset{n}{T}_z$ according to Eq. (31), are

$$\overset{n}{T}_x = \sigma_1 n_x, \quad \overset{n}{T}_y = \sigma_2 n_y, \quad \overset{n}{T}_z = \sigma_3 n_z$$

Substituting for n_x, n_y and n_z from Eq. (52)

$$\overset{n}{T}_x = \frac{1}{R}\sigma_1 x, \quad \overset{n}{T}_y = \frac{1}{R}\sigma_2 y, \quad \overset{n}{T}_z = \frac{1}{R}\sigma_3 z$$

or $$\sigma_1 x = R\overset{n}{T}_x, \quad \sigma_2 y = R\overset{n}{T}_y, \quad \sigma_3 z = R\overset{n}{T}_z$$

Substituting these in Eq. (54)

$$m_x = 2\alpha\, R\overset{n}{T}_x, \qquad m_y = 2\alpha\, R\overset{n}{T}_y, \qquad m_z = 2\alpha\, R\overset{n}{T}_z$$

i.e. m_x, m_y and m_z are proportional to $\overset{n}{T}_x$, $\overset{n}{T}_y$ and $\overset{n}{T}_z$.

Hence, m and $\overset{n}{T}$ are parallel.

The stress surface of Cauchy, therefore, has the following properties:

(i) If Q is a point on the stress surface, then $PQ = 1/\sqrt{\sigma}$ where s is the normal stress on a plane whose normal is PQ.

(ii) The normal to the surface at Q is parallel to the resultant stress vector $\overset{n}{T}$ on the plane with normal PQ.

Therefore, the stress surface of Cauchy completely defines the state of stress at P. It would be of interest to know the shape of the stress surface for different states of stress. This aspect will be discussed in Appendix 3.

LAME'S ELLIPSOID

Let $Pxyz$ be a coordinate frame of reference at point P, parallel to the principal axes at P. On a plane passing through P with normal n, the resultant stress vector is $\overset{n}{T}$ and its components, according to Eq. (31), are

$$\overset{n}{T}_x = \sigma_1 n_x, \quad \overset{n}{T}_y = \sigma_2 n_y, \quad \overset{n}{T}_z = \sigma_3 n_z$$

Let PQ be along the resultant stress vector and its length be equal to its magnitude, i.e. $PQ = \overset{n}{T}$. The coordinates (x, y, z) of the point Q are then

$$x = \overset{n}{T}_x, \qquad y = \overset{n}{T}_y, \qquad z = \overset{n}{T}_z$$

Since $n_x^2 + n_y^2 + n_z^2 = 1$, we get from the above two equations.

$$\frac{x^2}{\sigma_1^2} + \frac{y^2}{\sigma_2^2} + \frac{z^2}{\sigma_3^2} = 1 \tag{55}$$

This is the equation of an ellipsoid referred to the principal axes. This ellipsoid is called the stress ellipsoid or Lame's ellipsoid. One of its three semi–axes is the longest, the other the shortest, and the third in–between (Fig. 19). These are the extremum values.

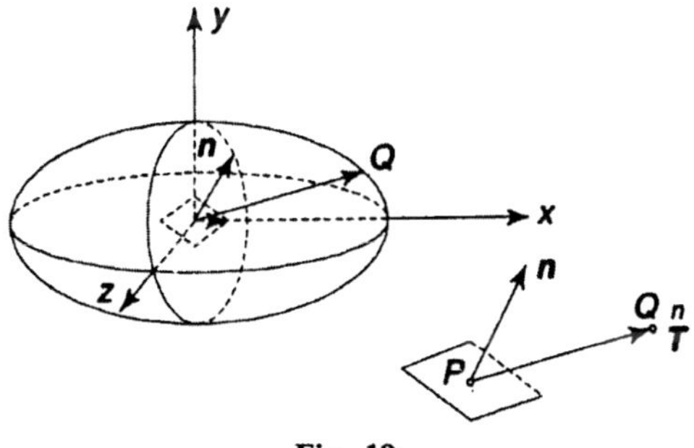

Fig. 19

If two of the principal stresses are equal, for instance $\sigma_1 = \sigma_2$, Lame's ellipsoid is an ellipsoid of revolution and the state of stress at a given point is symmetrical with respect to the third principal axis Pz. If all the principal are equal, $\sigma_1 = \sigma_2 = \sigma_3$, Lame's ellipsoid becomes a sphere.

Each radius vector PQ of the stress ellipsoid represents to a certain scale, the resultant stress on one of the planes through the centre of the ellipsoid.

It can be shown (Example 11) that the stress represented by a radius vector of the stress ellipsoid acts on the plane parallel to tangent plane to the surface called the stress–director surface, defined by

$$\frac{x^2}{\sigma_1} + \frac{y^2}{\sigma_2} + \frac{z^2}{\sigma_3} = 1 \tag{56}$$

The tangent plane to the stress–director surface is drawn at the point of intersection of the surface with the radius vector. Consequently, Lame's ellipsoid and the stress–director surface together completely define the state of stress at point P.

Example 11:

Show that Lame's ellipsoid and the stress–director surface together completely define the state of stress at a point.

Solution:

If $\sigma_1 = \sigma_2 = \sigma_3$ are the principal axes at a point P, the equation of the ellipsoid is given by

$$\frac{x^2}{\sigma_1^2}+\frac{y^2}{\sigma_2^2}+\frac{z^2}{\sigma_3^2}=1$$

The stress–director surface has the equation

$$\frac{x^2}{\sigma_1}+\frac{y^2}{\sigma_2}+\frac{z^2}{\sigma_3}=1$$

It is known form analytical geometry that for a surface defined by $F(x, y, z) = 0$, the normal to the tangent at a point (x_0, y_0, z_0) has direction cosines proportional to $\frac{\partial F}{\partial x}, \frac{\partial F}{\partial y}, \frac{\partial F}{\partial z}$, evaluated at (x_0, y_0, z_0). Hence, at a point (x_0, y_0, z_0) on the stress ellipsoid, if m is the normal to the tangent.

$$m_x = \alpha\frac{x_0}{\sigma_1}, \quad m_y = \alpha\frac{x_0}{\sigma_2}. \quad m_z = \alpha\frac{x_0}{\sigma_3}$$

Consider a plane through P with normal parallel to m. On this plane, the resultant stress vector will be $\overset{n}{T}$ with components given by

$$\overset{m}{T}_x = \sigma_1 m_x; \quad \overset{m}{T}_y = \sigma_2 m_y; \quad \overset{m}{T}_z = \sigma_3 m_z$$

Substituting for m_x, m_y and m_z

$$\overset{m}{T}_x = \alpha x_0, \quad \overset{m}{T}_y = \alpha y_0, \quad \overset{m}{T}_z = \alpha z_0$$

i.e. the components of stress on the plane with normal m are proportional to the coordinates (x_0, y_0, z_0). Hence the stress–director surface has the following property.

$L(x_0, y_0, z_0)$ is a point on the stress–director surface. m is the normal to the tangent plane at L. On a plane through P with normal m, the resultant stress vector is $\overset{m}{T}$ with components proportional to x_0, y_0 and z_0. This means that the components of PL are proporti nal to $\overset{m}{T}_x$, $\overset{m}{T}_y$ and $\overset{m}{T}_z$.

PQ being an extension of PL and equal to $\overset{n}{T}$ in magnitude, the plane having this resultant stress will have m as its normal.

THE PLANE STATE OF STRESS

If in given state of stress there exists a coordinate system *Oxyz* such that for this system

$$\sigma_z = 0, \qquad \tau_{xz} = 0, \qquad \tau_{yz} = 0 \tag{57}$$

then the state is said to have a 'plane state of stress' parallel to the *xy* plane. This state is also generally known as a two–dimensional state of stress. All the foregoing discussions can be applied and the equations reduce to simpler forms as a result of Eq. (57). The state of stress is shown in Fig. 20.

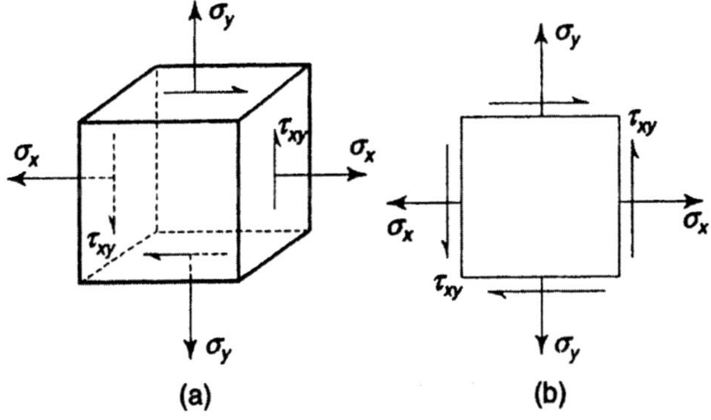

Fig. 20

Consider a plane with the normal lying in the *xy* plane, If n_x, n_y and n_z are the direction cosines of the normal, we have $n_x = \cos\theta$, $n_y = \sin\theta$ and $n_z = 0$ (Fig. 21) From Eq. (9)

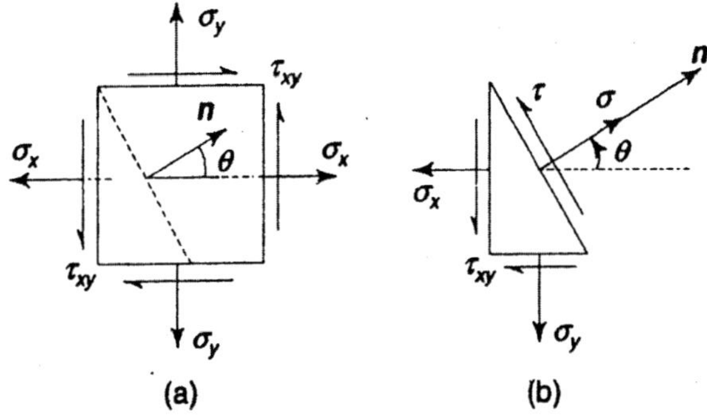

Fig. 21

$$\overset{n}{T}_x = \sigma_x \cos\theta + \tau_{xy} \sin\theta$$

$$\overset{n}{T}_y = \sigma_y \sin\theta + \tau_{xy} \cos\theta \tag{58}$$

$$\overset{n}{T}_z = 0$$

The normal and shear stress components on this plane are from Eqs. (11) and (12)

$$\sigma = \sigma_x \cos^2\theta + \sigma_y \sin^2\theta + 2\tau_{xy} \sin\theta \cos\theta$$

$$= \frac{\sigma_x + \sigma_y}{2} + \frac{\sigma_x - \sigma_y}{2} \cos 2\theta + \tau_{xy} \sin 2\theta \tag{59}$$

and $$\tau^2 = \overset{n}{T}_x^2 + \overset{n}{T}_y^2 - \sigma^2$$

or $$\tau = \frac{\sigma_x - \sigma_y}{2} \sin 2\theta + \tau_{xy} \cos 2\theta \tag{60}$$

The principal stresses are given by Eq. (1.29) as

$$\sigma_1, \sigma_2 = \frac{\sigma_x + \sigma_y}{2} \pm \left[\left(\frac{\sigma_x - \sigma_y}{2}\right)^2 + \tau_{xy}^2\right]^{1/2} \tag{61}$$

$$\sigma_3 = 0$$

The principal planes are given by

(i) the z plane on which $\sigma_3 = \sigma_z = 0$ and

(ii) two planes with normal in the xy plane such that

$$\tan 2\phi = \frac{2\tau_{xy}}{\sigma_x - \sigma_y} \tag{62}$$

The above equation gives two planes at right angles to each other.

If the principal stresses σ_1, σ_2 and σ_3 are arranged such that $\sigma_1 \geq \sigma_2 \geq \sigma_3$, the maximum shear stress at the point will be

$$\tau_{\max} = \frac{\sigma_1 - \sigma_3}{2} \tag{63a}$$

In the xy plane, the maximum shear stress will be

$$\tau_{\max} = \frac{1}{2}(\sigma_1 - \sigma_2) \tag{63b}$$

and from Eq. (61)

$$\tau_{\max} = \left[\left(\frac{\sigma_x - \sigma_y}{2}\right)^2 + \tau_{xy}^2\right]^{1/2} \tag{64}$$

DIFFERENTIAL EQUATIONS OF EQUILIBRIUM

So far, attention has been focussed on the state of stress at a point. In general, the state of stress in a body varies from point to point. One of the fundamental problems in a book of this kind is the determination of the state of stress at every point or at any desired point in a body. One of the important sets of equations used in the analyses of such problems deals with the conditions to be satisfied by the stress components when they vary from point to point. These conditions will be established when the body (and therefore, every part of it) is in equilibrium. We isolate a small element of the body and derive the equations of equilibrium from its free–body diagram (Fig. 22).

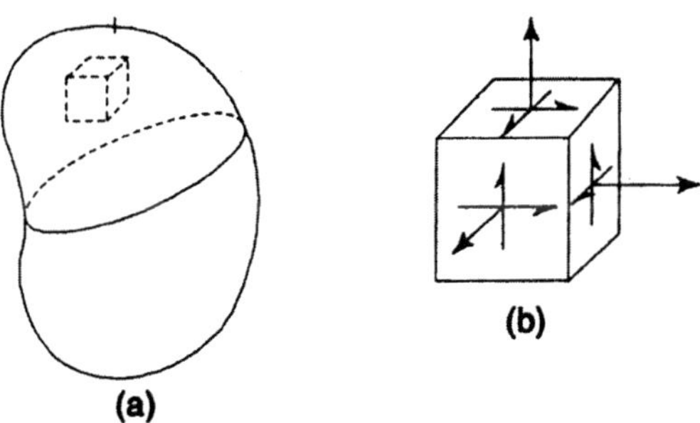

Fig. 22

Consider a small rectangular element with sides Δx, $\Delta\psi$, and Δz isolated from its parent body. Since in the limit, we are going to make Δx, $\Delta\psi$, and Δz tend to zero, we shall deal with average values of the stress components on each face. These stress components are shown in Fig. 23.

The faces are marked as 1, 2, 3 etc. On the left hand face, i.e. face No. 1, the average stress components are σ_x, τ_{xy} and τ_{xz}. On the right hand face, i.e. face No. 2, the average stress components are

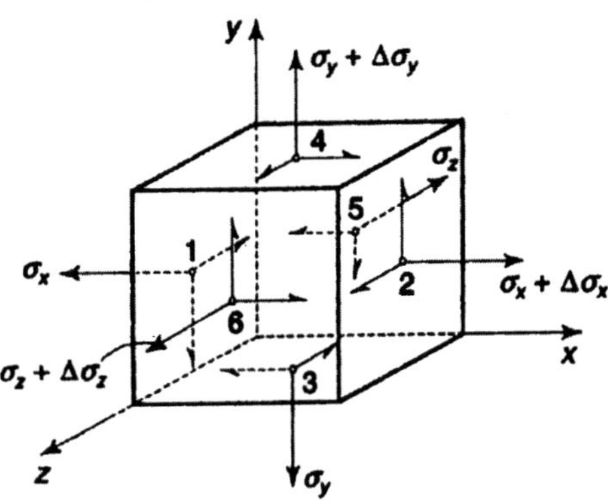

Fig. 23

$$\sigma_x + \frac{\partial \sigma_x}{\partial x}\Delta x, \quad \tau_{xy} + \frac{\partial \tau_{xy}}{\partial x}\Delta x, \quad \tau_{xz} + \frac{\partial \tau_{xz}}{\partial x}\Delta x$$

This is because the right hand face is Δx distance away from the left hand face. Following a similar procedure, the stress components on the six faces of the element are as follows:

Face 1 σ_x, τ_{xy}, τ_{xz}

Face 2 $\sigma_x + \frac{\partial \sigma_x}{\partial x}\Delta x, \quad \tau_{xy} + \frac{\partial \tau_{xy}}{\partial x}\Delta x, \quad \tau_{xz} + \frac{\partial \tau_{xz}}{\partial x}\Delta x$

Face 3 σ_y, τ_{yx}, τ_{yz}

Face 4 $\sigma_y + \frac{\partial \sigma_y}{\partial y}\Delta y, \quad \tau_{yx} + \frac{\partial \tau_{yx}}{\partial y}\Delta y, \quad \tau_{yz} + \frac{\partial \tau_{yz}}{\partial y}\Delta y$

Face 5 σ_z, τ_{zx}, τ_{zy}

Face 6 $\sigma_z + \frac{\partial \sigma_z}{\partial z}\Delta z, \quad \tau_{zx} + \frac{\partial \tau_{zx}}{\partial z}\Delta z, \quad \tau_{zy} + \frac{\partial \tau_{zy}}{\partial z}\Delta z$

Let the body force components per unit volume in the x, y and z directions be γ_x, γ_y and γ_z. For equilibrium in x direction

$$\left(\sigma_x + \frac{\partial \sigma_x}{\partial x}\Delta x\right)\Delta y\Delta z - \sigma_x \Delta y\Delta z + \left(\tau_{yx} + \frac{\partial \tau_{yx}}{\partial y}\Delta y\right)\Delta z\Delta x - \tau_{yx}\Delta z\Delta x +$$
$$\left(\tau_{zx} + \frac{\partial \tau_{zx}}{\partial z}\Delta z\right)\Delta x\Delta y - \tau_{zx}\Delta x\Delta y + \gamma_x \Delta x\Delta y\Delta z = 0$$

Cancelling terms, dividing by Δx, Δy, Δz and going to the limit, we get

$$\frac{\partial \sigma_x}{\partial x} + \frac{\partial \tau_{yx}}{\partial y} + \frac{\partial \tau_{zx}}{\partial z} + \gamma_x = 0$$

Similarly, equating forces in the y and z directions respectively to zero, we get two more equations. On the basis of the fact that the cross shears are equal i.e. $\tau_{xy} = \tau_{yx}$, $\tau_{yz} = \tau_{zy}$, $\tau_{xz} = \tau_{zx}$, we obtain the three differential equations of equilibrium as

$$\frac{\partial \sigma_x}{\partial x} + \frac{\partial \tau_{xy}}{\partial y} + \frac{\partial \tau_{zx}}{\partial z} + \gamma_x = 0$$
$$\frac{\partial \sigma_y}{\partial y} + \frac{\partial \tau_{xy}}{\partial x} + \frac{\partial \tau_{yz}}{\partial z} + \gamma_y = 0 \qquad (65)$$
$$\frac{\partial \sigma_z}{\partial z} + \frac{\partial \tau_{xz}}{\partial x} + \frac{\partial \tau_{yz}}{\partial y} + \gamma_z = 0$$

Equations (65) must be satisfied at all points throughout the volume of the body. It must be recalled that the moment equilibrium conditions established the equality of cross shears in Sec. 1.8.

EQUILIBRIUM EQUATIONS FOR PLANE STRESS STATE

The plane stress has already been defined. If there exists a plane stress state in the xy plane, then $\sigma_z = \tau_{zx} = \tau_{zy} = \gamma_z = 0$ and σ_x, σ_y, τ_{xz}, γ_x and γ_y exist. The differential equations of equilibrium become

$$\frac{\partial \sigma_x}{\partial x} + \frac{\partial \tau_{xy}}{\partial y} + \gamma_x = 0$$

$$\frac{\partial \sigma_y}{\partial y} + \frac{\partial \tau_{xy}}{\partial x} + \gamma_y = 0 \qquad (1.66)$$

Example 12:

The pressure of water on face OB is also shown. With the axes OX and Oy, the stresses at any point (x, y) are given by (γ = specific weight of water and ρ = specific weight of dam material)

$$\sigma_x = \square\ \gamma y$$

$$\sigma_x = \left(\frac{\rho}{\tan\beta} - \frac{2\gamma}{\tan^3\beta}\right)x + \left(\frac{\gamma}{\tan^2\beta} - \rho\right)y$$

$$\tau_{xt} = \tau_{yx} = -\frac{\gamma}{\tan^2\beta}x$$

$$\tau_{yz} = 0, \qquad \tau_{zx} = 0, \qquad \sigma_z = 0$$

Solution:

Check if these stress components satisfy the differential equations of equilibrium. Also, verify if the boundary conditions are satisfied on face *OB*.

The equations of equilibrium are

$$\frac{\partial \sigma_x}{\partial x} + \frac{\partial \tau_{xy}}{\partial y} + \gamma_x = 0$$

and
$$\frac{\partial \sigma_y}{\partial y} + \frac{\partial \tau_{xy}}{\partial x} + \gamma_y = 0$$

Substituting and noting that $\gamma_x = 0$ and $\gamma_y = \rho$, the first equation is satisfied. For the second equation also

$$\frac{\gamma}{\tan^2\beta} - \rho - \frac{\gamma}{\tan^2\beta} + \rho = 0$$

On face *OB*, at any *y* the stress components are $\sigma_x = -\gamma_y$ and $\tau_{xy} = 0$. Hence the boundary conditions are also satisfied.

Example 13:

Consider a function ϕ (x, y), which is called the stress function. If the values of σ_x, σ_y and τ_{xy} are as given below, show that these satisfy the differential equations of equilibrium in the absence of body forces.

$$\sigma_x = \frac{\partial^2 \phi}{\partial y^2}, \quad \sigma_y = \frac{\partial^2 \phi}{\partial x^2}, \quad \tau_{xy} = -\frac{\partial^2 \phi}{\partial x \partial y}$$

Solution:

Substituting in the differential equations of equilibrium

$$\frac{\partial^3 \phi}{\partial y^2 \partial x} - \frac{\partial^3 \phi}{\partial y^2 \partial y} = 0$$

$$\frac{\partial^3 \phi}{\partial x^2 \partial y} - \frac{\partial^3 \phi}{\partial y^2 \partial y} = 0$$

Example 14:

Consider the rectangular beam according to the elementary theory of bending, the 'fibre stress' in the elastic range due to bending is given by

$$\sigma_x = -\frac{My}{I} = -\frac{12My}{bh^3}$$

Solution:

where M is the bending moment which is a function of x. Assume that $\sigma_z = \tau_{zx} = \tau_{zy} = 0$ and also that $\tau_{xy} = 0$ at the top and bottom and further, that $\sigma_z = 0$ at the bottom. Using the differential equations of equilibrium, determine τ_{xy} and σ_y. Compare these with the values given in the elementary strength of materials.

From Eq. (65)

$$\frac{\partial \sigma_x}{\partial x} + \frac{\partial \tau_{xy}}{\partial y} + \frac{\partial \tau_{xz}}{\partial z} = 0$$

Since $\tau_{xz} = 0$ and M is a function of x

$$-\frac{12y}{bh^3}\frac{\partial M}{\partial x} + \frac{\partial \tau_{xy}}{\partial y} = 0$$

or
$$\frac{\partial \tau_{xy}}{\partial y} = \frac{12y}{bh^3}\frac{\partial M}{\partial x} y$$

Integrating
$$\tau_{xy} = \frac{6}{bh^3}\frac{\partial M}{\partial x} y^2 + c_1 f(x) - c_2$$

where $f(x)$ is a function of x alone and c_1, c_2 are constants. It is given that

$$\tau_{xy} = 0 \text{ at } y = \pm\frac{h}{2}$$

$$\therefore \qquad \frac{6}{bh^3}\frac{h^2}{4}\frac{\partial M}{\partial x} = -c_1 f(x) - c_2$$

or $$c_1 f(x) + c_2 = \frac{3}{2bh}\frac{\partial M}{\partial x}$$

$$\therefore \qquad \tau_{xy} = \frac{3}{2bh}\frac{\partial M}{\partial x}\left(\frac{4y^2}{h^2} - 1\right)$$

Form elementary strength of materials, we have

$$\tau_{xy} = \frac{V}{Ib}\int_y^{h/2} y'\,dA$$

where $V = -\dfrac{\partial M}{\partial x}$ is the shear force. Simplifying the above expression

$$\tau_{xy} = -\frac{\partial M}{\partial x}\frac{12}{b^2 h^3}\left(\frac{h^2}{4} - y^2\right)\frac{b}{2}$$

or $$\tau_{xy} = \frac{3}{2bh}\frac{\partial M}{\partial x}\left(\frac{4y^2}{h^3} - 1\right)$$

i.e. the same as the expression obtained above.

Form the next equilibrium equation, i.e. from

$$\frac{\partial \sigma_y}{\partial y} + \frac{\partial \tau_{xy}}{\partial x} + \frac{\partial \tau_{yz}}{\partial z} = 0$$

we get $$\frac{\partial \sigma_y}{\partial y} = \frac{3}{2bh}\left(\frac{4y^2}{h^2} - 1\right)\frac{\partial^2 M}{\partial x^2}$$

$$\therefore \qquad \sigma_y = -\frac{3}{2bh}\frac{\partial^2 M}{\partial x^2}\left(\frac{4y^3}{3h^2} - y\right) + c_3 F(x) + c_4$$

where $F(x)$ is a function of x alone. It is given that $\sigma_y = 0$ at $y = -\dfrac{h}{2}$.

Hence, $$c_3 F(x) + c_4 = \frac{3}{2bh}\frac{\partial^2 M}{\partial x^2}\frac{h}{3}$$

$$= \frac{1}{2b}\frac{\partial^2 M}{\partial x^2}$$

Substituting

$$\sigma_y = -\frac{3}{2bh}\frac{\partial^2 M}{\partial x^2}\left(\frac{4y^3}{3h^2} - y - \frac{h}{3}\right)$$

At $y = + h/2$, the value of σ_y is

$$\sigma_y = -\frac{1}{b}\frac{\partial^2 M}{\partial x^2} = \frac{w}{b}$$

where w is the intensity of loading. Since b is the width of the beam, the stress will be w/b as obtained above.

BOUNDARY CONDITIONS

Equation (66) must be satisfied throughout the volume of the body. When the stresses vary over the plate (i.e. the body having the plane stress state), the stress components σ_x, σ_y and τ_{xy} must be consistent with the externally applied forces at a boundary point.

Consider the two–dimensional body shown in Fig. 24. At a boundary point P, the outward drawn normal is n. Let F_x and F_y be the components of the surface forces per unit area at this point.

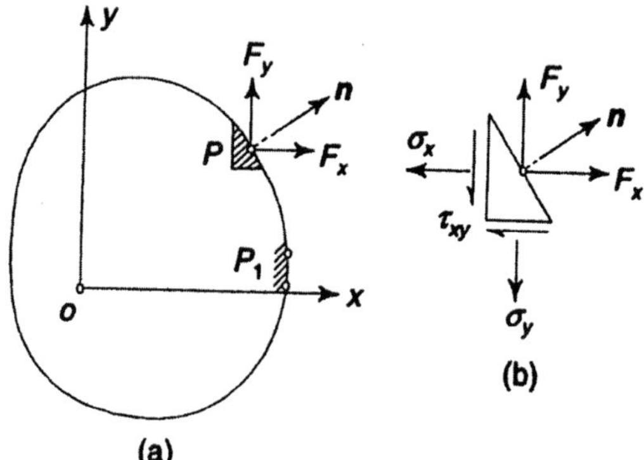

Fig 24

F_x and F_y must be continuations of the stresses σ_x, σ_y and τ_{xy} at the boundary. Hence, using Cauchy's equations

$$\overset{n}{T}_x = F_x = \sigma_x n_x + \tau_{xy} n_x$$

$$\overset{n}{T}_y = F_y = \sigma_y n_y + \tau_{xy} n_y$$

If the boundary of the plate happens to be parallel to y axis, as at point P_1, the boundary conditions become

$$F_x = \sigma_x \quad \text{and} \quad F_y = \tau_{xy}$$

EQUATIONS OF EQUILIBRIUM IN CYLINDRICAL COORDINATES

Till this section, we have been using a rectangular or the Cartesian frame of reference for analyses. Such a frame of reference is useful it the body under analysis happens to possess rectangular or straight boundaries. Numerous problems exist where the bodies under discussion possess radial symmetry; For example, a thick cylinder subjected to internal or external pressure. For the analysis of such problems, it is more convenient to use polar or cylindrical coordinates. In this section, we shall develop some equations in cylindrical coordinates.

$$\sigma_r,\ \sigma_\theta,\ \sigma_z,\ \tau_\theta,\ \tau_{\theta z} \text{ and } \tau_{zr}$$

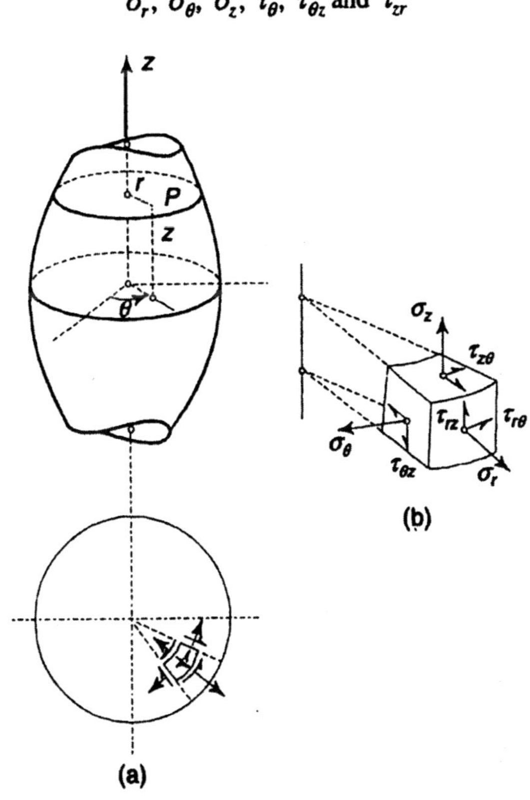

Fig. 25

Consider an axisymmetric body as shown in Fig. 25(a). The axis of the body is usually taken as the z axis. The two other coordinates are r and θ, where θ is measured counter–clockwise. The rectangular stress components at a point P (r, θ, z) are

$$\sigma_r,\ \sigma_\theta\ \sigma_z,\ \tau_\theta\ \tau_{\theta z} \text{ and } \tau_{zr}$$

These are shown acting on the faces of a radial element at point P in Fig. 26(b). σ_r, σ_θ and σ_z are called the radial, circumferential and axial stresses respectively. If the stresses vary from point to point, one can derive the appropriate differential equations of equilibrium. For this purpose, consider a cylindrical element having a radial length Δr with an included angle $\Delta\theta$ and a height Δz, isolated from the body. The free–body diagram of the element is shown in Fig. 26(b). Since the element is very small, we work with the average stresses acting on each face.

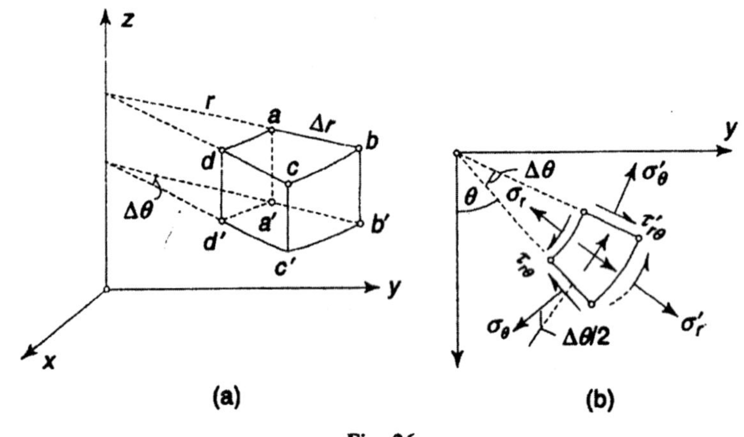

Fig. 26

The area of the $aa'\,d'$ is $r\,\Delta\theta\,\Delta z$ and the area of face $aa'\,c'\,c$ is $(r + \Delta r)\,\Delta\theta\,\Delta z$. The areas of faces $dcc'd'$ and $abb'e'$ are each equal to $\Delta r\,\Delta z$.

The faces $abcd$ and $a'b'c'd'$ have each an area $\left(r+\dfrac{\Delta r}{2}\right)\Delta\theta\,\Delta r$. The average stresses on these faces (which are assumed to be acting at the mid point of each face) are

On face $aa'\,d'\,d$

normal stress σ_r

tangential stresses τ_{rz} and $\tau_{r\theta}$

On face $bb'c'c$

normal stress $\sigma_r + \dfrac{\partial \sigma_r}{\partial r}\Delta r$

tangential stresses $\tau_{rz} + \frac{\partial \tau_{rz}}{\partial r} \Delta r$ and $\tau_{r\theta} + \frac{\partial \tau_{\theta r}}{\partial r} \Delta r$

The changes are because the face *bb'c'c* is Δr distance away from the face *aa' d' d*.

On face *dcc'd*

normal stress σ_θ

tangential stresses $\tau_{r\theta}$ and $\tau_{\theta z}$

On face *abb'a*

normal stress $\sigma_\theta + \frac{\partial \sigma_\theta}{\partial \theta} \Delta \theta$

tangential stresses $\tau_{r\theta} + \frac{\partial \tau_{r\theta}}{\partial \theta} \Delta \theta$ and $\tau_{\theta z} + \frac{\partial \tau_{\theta z}}{\partial \theta} \Delta \theta$

The changes in the above components are because the face *abb'a* is separated by an angle $\Delta\theta$ from the face *dccc'd'*.

On face *a'b'c'd'* -

normal stress θ_z

tangential stresses τ_{rz} and $\tau_{\theta z}$

On face *abcd*

normal stress $\sigma_z + \frac{\partial \sigma_z}{\partial z} \Delta z$

tangential stresses $\tau_{rz} + \frac{\partial \tau_{rz}}{\partial z} \Delta z$ and $\tau_{\theta z} + \frac{\partial \tau_{\theta z}}{\partial \theta} \Delta z$

Let γ_r, γ_θ and γ_z be the body force components per unit volume. If the element is in equilibrium, the sum of forces in r, θ and z directions must vanish individually, Equating the forces in r direction to zero,

$$\left(\sigma_r + \frac{\partial \sigma_r}{\partial r} \Delta r\right)(r + \Delta r)\Delta\theta\, \Delta z + \left(\tau_{rz} + \frac{\partial \tau_{rz}}{\partial z} \Delta z\right)\left(r + \frac{\Delta r}{2}\right)\Delta\theta\, \Delta r$$

$$-\sigma_r r \Delta\theta\, \Delta z - \tau_{rz}\left(r + \frac{\Delta r}{2}\right)\Delta\theta\, \Delta r - \sigma_\theta \sin\frac{\Delta\theta}{2}\Delta r\, \Delta z$$

$$-\tau_{r\theta} \cos\frac{\Delta\theta}{2}\Delta r\, \Delta z - \left(\sigma_\theta + \frac{\partial \sigma_\theta}{\partial \theta}\Delta\theta\right)\sin\frac{\Delta\theta}{2}\Delta r\, \Delta z$$

$$+\left(\tau_{r\theta} + \frac{\partial \tau_{r\theta}}{\partial \theta}\Delta\theta\right)\cos\frac{\Delta\theta}{2}\Delta r\, \Delta z + \gamma_r\left(r + \frac{\Delta r}{2}\right)\Delta\theta\, \Delta r\, \Delta z = 0$$

Cancelling terms, dividing by $\Delta\theta\,\Delta r\,\Delta z$ and going to the limit with $\Delta\theta$, Δr and Δz, all tending to zero

$$r\frac{\partial\sigma_r}{\partial r}+r\frac{\partial\tau_{rz}}{\partial z}+\frac{\partial\tau_{r\theta}}{\partial\theta}+\sigma_r-\sigma_\theta+r\gamma_r=0$$

or
$$\frac{\partial\sigma_r}{\partial r}+\frac{\partial\tau_{rz}}{\partial z}+\frac{1}{r}\frac{\partial\tau_{r\theta}}{\partial\theta}+\frac{\sigma_r-\sigma_\theta}{r}+\gamma_z=0 \qquad (1.67)$$

Similarly, for equilibrium in z and θ directions, we get

$$\frac{\partial\tau_{rz}}{\partial r}+\frac{\partial\sigma_z}{\partial z}+\frac{1}{r}\frac{\partial\tau_{\theta z}}{\partial\theta}+\frac{\tau_{rz}}{\partial}+\gamma_z=0 \qquad (1.68)$$

and
$$\frac{\partial\tau_{r\theta}}{\partial r}+\frac{\partial\tau_{\theta z}}{\partial z}+\frac{1}{r}\frac{\partial\sigma_\theta}{\partial\theta}+\frac{2\tau_{r\theta}}{r}+\gamma_\theta=0 \qquad (1.69)$$

Equations (67) – (69) are the differential equations of equilibrium expressed in polar coordinates.

AXISYMMETRIC CASE AND PLANE STRESS CASE

If an axisymmetric body is loaded symmetrically, the stress components do not depend on θ. Since the deformations are symmetric, $\tau_{r\theta}$ and $\tau_{\theta z}$ do not exist and consequently the above set of equations in the absence of body forces are reduced to

$$\frac{\partial\sigma_r}{\partial r}+\frac{\partial\tau_{rz}}{\partial z}+\frac{\sigma_r-\sigma_\theta}{r}=0$$

$$\frac{\partial\tau_{rz}}{\partial r}+\frac{\partial\sigma_z}{\partial z}+\frac{\tau_{rz}}{r}=0$$

A sphere under diametral compression or a cone under a load at the apex are examples to which the above set of equations can be applied.

If the state of stress is two–dimensional in nature, i.e. plane stress state, then only σ_r, σ_θ $\tau_{r\theta}\gamma_r$ and γ_θ exist. The other stress components vanish. These non–vanishing stress components depend only on θ and r are independent of z in the absence of body forces. The equations of equilibrium reduce to

$$\frac{\partial\sigma_r}{\partial r}+\frac{1}{r}\frac{\partial\tau_{r\theta}}{\partial\theta}+\frac{\sigma_r-\sigma_\theta}{r}=0 \qquad (70)$$

$$\frac{\partial\tau_{r\theta}}{\partial r}+\frac{1}{r}\frac{\partial\sigma_\theta}{\partial\theta}+\frac{2\tau_{r\theta}}{r}=0$$

Example 15:

Consider a function ϕ (r, θ), which is called the stress function. If the values of σ_r, σ_θ and $\tau_{r\theta}$ are as given below, show that in the absence of body forces, these satisfy the differential equations of equilibrium.

$$\sigma_r = \frac{1}{r}\frac{\partial\phi}{\partial r} + \frac{1}{r^2}\frac{\partial^2\phi}{\partial\theta^2}$$

$$\sigma_\theta \frac{\partial^2\phi}{\partial r^2}$$

$$\tau_{r\theta} = -\frac{1}{r}\frac{\partial^2\phi}{\partial r\partial\theta}\frac{1}{r^2}\frac{\partial\phi}{\partial r}$$

Solution:

The equations of equilibrium are

$$\frac{\partial\sigma_r}{\partial r} + \frac{1}{r}\frac{\partial\tau_{r\theta}}{\partial\theta} + \frac{\sigma_r - \sigma_\theta}{r} = 0$$

$$\frac{\partial\tau_{r\theta}}{\partial r} + \frac{1}{r}\frac{\partial\tau_\theta}{\partial\theta} + \frac{2\tau_{r\theta}}{r} = 0$$

Substituting the stress function in the first equation of equilibrium,

$$-\frac{1}{r^2}\frac{\partial\phi}{\partial r} + \frac{1}{r}\frac{\partial^2\phi}{\partial r^2} - \frac{2}{r^3}\frac{\partial^2\phi}{\partial\theta^2} + \frac{1}{r^2}\frac{\partial^3\phi}{\partial\theta^2\,\partial r} + \frac{1}{r}\left(-\frac{1}{r}\frac{\partial^3\phi}{\partial\theta^2\,\partial r} + \frac{1}{r^2}\frac{\partial^2\phi}{\partial\theta^2}\right)$$

$$+\frac{1}{r^2}\frac{\partial\phi}{\partial r} + \frac{2}{r^3}\frac{\partial^2\phi}{\partial\theta^2} - \frac{1}{r}\frac{\partial^2\phi}{\partial r^2} = 0$$

Hence, the first equation is satisfied. Similarly, it can easily be verified that the second condition also holds good.

DEFLECTIONS OF THICK CURVED BARS

In Chapter 5, the problems of thin rings and thin curved members were analyzed, using energy methods. In this section, we shall discuss a few problems involving thick rings. The energy method will be used.

In the straight part of the U-ring, across any section, there is a tangential force P and a moment $(P_x - M_0)$. In the curved part of the member, there will be a tangential force V, a normal force N and a bending moment M. Their value are

$$V = P \cos\theta$$

$$N = P \sin\theta$$

$$M = M_0 - (d + \rho_0 \text{ sing } \theta)\, P$$

To calculate the strain energy stored we proceed as follows:

(i) In the straight part of the member: Owing to the shear force V, the strain energy stored in a small length Δs is

$$\Delta U_V = \frac{\alpha V^2 \Delta s}{2AG} \tag{1}$$

where a is a numerical factor depending on the shape of the cross section, A is the area of the section and G is the shear modulus.

Because of the bending moment M, the energy stored is

$$\Delta U_M = \frac{M^2 \Delta s}{2EI} \tag{2}$$

where I is the moment of inertia about the neutral axis, which for a straight beam passes through the centroid of the section.

In general, the strain energy due to V is small as compared to that due to M.

(ii) In the curved part of the member: Owing to the shear force V, the strain energy stored in a small sectoral element, enclosing an angle $\Delta\phi$, is

$$\Delta U_V = \frac{\alpha V^2 \Delta s}{2AG} \tag{3}$$

If r_0 is the radius of curvature of the centroidal fibre, $\Delta s = \rho_0 \ \Delta\phi$.

Because of the normal force N, which is assumed to be acting at the centroid of the cross-section,

$$\Delta U_N = \frac{N^2 \Delta s}{2AE} \tag{4}$$

Owing to bending moment M, the energy stored is equal to the work done. It $\delta\Delta\phi$ is the change in the angle due to bending.

$$\Delta U_M = \frac{1}{2} M(\delta \ \Delta\phi)$$

$$\delta\Delta\phi = \Delta\phi \ r_0 \left(\frac{1}{r} - \frac{1}{r_0} \right)$$

Substituting for the right-hand part in the above equation

$$\delta\Delta\phi = \Delta\phi \frac{M}{AeE}$$

Hence, $$\Delta U_M = \frac{M^2 \Delta\phi}{2AeE}$$

Putting $\Delta\phi = \frac{\Delta s}{\rho_0}$

$$\Delta U_M = \frac{M^2 \Delta s}{2AeE\rho_0} \tag{5}$$

If N is applied first and then M, Owing to the rotation of the section, the centroid C moves through a distance $\varepsilon_0 \Delta s$, where ε_0 is the strain at C and consequently, the force N does additional work equal to

$$\Delta U_{MN} = N\, \varepsilon_0\, \Delta s$$

ε_0 from Eq. (35) is

$$\varepsilon_0 = \frac{\sigma_x}{E} = -\frac{M}{AeE}\frac{y_0}{(r_0 - y_0)}$$

In the above equation, M is positive, according to the convention. y_0 is the distance of the centroidal fibre from the neutral axis and is equal to $-e$. Also $\rho_0 = +\,e$. With these,

$$\varepsilon_0 = +\frac{M}{A\rho_0 E}$$

Hence the work done by N is

$$\Delta_{MN} = \frac{MN\, \Delta s}{A\rho_0 E} \tag{6}$$

The same result is obtained if M is applied first and then N. This is according to the principal of superposition, which is valid for small deformations. The normal force N acting across the section produces uniform strain ε_n; since the lengths of the fibres are different, face AB will not shift parallel to itself. The extension of the fibre at b will be $\varepsilon_n r_1 \Delta\phi$. The angle enclosed between AB and $A'B'$ is therefore

$$\delta\theta = \frac{\varepsilon_n \Delta\phi (r_2 - r_1)}{(r_2 - y_1)} \varepsilon_n \Delta\phi$$

Owing to this rotation of $A'B'$, the moment M does work equal to

$$\Delta U_{MN} = M\, \varepsilon_n\, \Delta\phi$$

Since $\varepsilon_n = \frac{N}{AE}$

$$\Delta U_{NM} = \frac{MN}{AE}\Delta\phi$$

$$= \frac{MN\, \Delta s}{AE\rho_0}$$

For a straight beam, the work done by N when M is applied is zero since then section rotates about the neutral axis which passes through the centroid. This is also confirmed in the above expression where $\rho_0 = \infty$ for a straight beam and therefore $\Delta U_{MN} = 0$. Combining all the energies detailed above, the total strain energy is.

$$U = \int_x (\Delta U_V + \Delta U_N + \Delta U_M + \Delta U_{MN})$$

$$= \int_x \left(\frac{\alpha V^2}{2AG} + \frac{N^2}{2AE} + \frac{M^2}{2AeE\rho_0} + \frac{MN}{AE\rho_0} \right) ds \tag{7}$$

For the straight part of the beam, the last expression will be zero and the third expression (which becomes indeterminate since $e = 0$ and $\rho_0 = \infty$) is replaced by $M^2/2EI$. With the strain energy calculated as above and using Castigliano's theorem, one can solve for the unknown - either the deflection or the indeterminate reaction. We shall illustrate this through an example.

Example 1:

A ring with a rectangular section is subjected to diametral compression. Determine the bending moment and stress at point A of the inner radius across a section θ r_1 and r_2 are the inner and external radii respectively.

Solution:

We observe that the deformation of the ring will be symmetrical about the horizontal and vertical axes. Consequently, there will be no changes in the slopes of the vertical and horizontal faces of the ring. We can, therefore, consider only a quadrant of the circle for the analysis. M_0 is the unknown internal moment. Its value can be determined from the condition that the change in the slope of this section is zero. We shall use Castigliano's theorem to determine this moment.

Across any section ϕ, the moment is

$$M = M_0 - \frac{P}{2}\rho_0(1 - \cos\phi)$$

In addition, there is a normal force N and a shear force V. Their value are

$$N = \frac{P}{2}\rho_0 \cos\phi \quad \text{and} \quad V = -\frac{P}{2}\sin\phi$$

The total strain energy for the quadrant from Eq. (49) is

$$U = \int_0^{\pi/2} \frac{\alpha P^2 \sin^2\phi}{8AG}\rho_0\, d\phi + \int_0^{\pi/2} \frac{P^2\cos^2\phi}{8AE}\rho_0\, d\phi +$$

$$\int_0^{\pi/2} \frac{\left[M_0 - \frac{P}{2}\rho_0(1-\cos\phi\right]^2}{2AeE} d\phi -$$

$$\int_0^{\pi/2} \frac{\left[M_0 - \frac{P}{2}\rho_0(1-\cos\phi\right] P\cos\phi}{2AE} d\phi \tag{8}$$

$$= \left(\frac{\alpha P^2}{8AG} + \frac{P^2}{8AE}\right)\frac{\pi}{4}\rho_0 +$$

$$\frac{1}{2AeE}\left[M_0^2\frac{\pi}{4} + \frac{P^2}{4}\rho_0^2\left(\frac{\pi}{2}+\frac{\pi}{4}-2\right) - M_0\rho_0\, P\left(\frac{\pi}{2}-1\right)\right] -$$

$$\frac{P}{2AE}\left(M_0 - \frac{P\rho_0}{2} + \frac{P\rho_0}{2}\frac{\pi}{4}\right) \tag{9}$$

In the above expression, M_0 is still an unknown quantity. As the change in slope at the section where M is applied is zero,

$$\frac{\partial U}{\partial M_0} = \frac{1}{2AeE}\left[M_0\pi - \rho_0 P\left(\frac{\pi}{2}-1\right)\right] - \frac{P}{2AE} = 0$$

$$\therefore \quad M_0 = \frac{P\rho_0}{2}\left(1 - \frac{2}{\pi} + \frac{2e}{\pi\rho_0}\right)$$

If we ignore the initial curvature of the member while calculating the strain energy, then

$$U^* = \int_0^{\pi/2} \frac{\alpha P^2 \sin^2\phi}{8AG}\rho_0\, d\phi + \int_0^{\pi/2} \frac{P^2\cos^2\phi}{8AE}\rho_0\, d\phi +$$

$$\int_0^{\pi/2} \frac{\left[M_0 - \frac{P}{2}\rho_0(1-\cos\phi)\right]^2}{2EI} d\phi$$

and $$\frac{\partial U^*}{\partial M_0} = \frac{1}{EI}\int_0^{\pi/2}\left[M_0 - \frac{P}{2}\rho_0(1-\cos\phi)\right]\rho_0\, d\phi = 0$$

i.e. $$M_0\frac{\pi}{2} - \frac{P}{2}\rho_0\frac{\pi}{2} + \frac{P}{2}\rho_0 = 0$$

$$\therefore \quad M_0 = \frac{P\rho_0}{2}\left(\cos\theta + \frac{2e}{\pi\rho_0} - \frac{2}{\pi}\right)$$

i.e. same as given in Eq. (10) with $e \to 0$ and $\rho_0 \to \infty$. i.e. that of a thin ring.

With the value of M_0 known, the bending moment at any section θ is obtained as

$$M_0 = M_0 - \frac{P}{2}\rho_0(1-\cos\theta)$$

$$= \frac{P\rho_0}{2}\left(\cos\theta + \frac{2e}{\pi\rho_0} - \frac{2}{\pi}\right)$$

The normal stress at A can be calculated and adding additional stress due to the normal force N.

$$\sigma_A = -\frac{M}{Ae}\cdot\frac{y}{(r_0-y)} + \frac{N}{A}$$

$$= -\frac{P\rho_0}{2Ae}\left(\cos\theta + \frac{2e}{\pi\rho_0} - \frac{2}{\pi}\right)\frac{y}{r_0-y} - \frac{P\cos\theta}{2A}$$

For point A, from Eqs. (38) and (39b)

$$y = \frac{h}{2} - e, \qquad r_0 = \frac{r_2 - r_1}{\log(r_2/r_1)}, \qquad e = \rho_0 - \frac{r_2 - r_1}{\log(r_2/r_1)} = \rho_0 - r_0$$

$$\sigma_A = \frac{P}{2A}\left\{\frac{\rho_0(\pi\cos\theta - 2) + 2e}{\pi e}\frac{(h-2e)}{(2\rho_0 - h)} + \cos\theta\right\}$$

Example 2:

A circular ring of rectangular section, shown in Fig. 31, is subjected to diametral compression. Determine the change in the vertical diameter.

Solution:

From Eq. (9), the total energy for the complete ring is

$$U = 4\rho_0\left\{\frac{\alpha P^2\pi}{32AG} + \frac{\pi P^2}{32AE} + \frac{1}{2AeE\rho_0}\left[\frac{\pi M_0^2}{2} + \frac{\rho_0^2 P^2}{4}\left(\frac{3\pi}{4} - 2\right) - M_0\,\rho_0\,P\left(\frac{\pi}{2} - 1\right)\right] - \frac{P}{2A\rho_0 E}\left[M_0 + \frac{P\rho_0}{2}\left(\frac{\pi}{4} - 1\right)\right]\right\}$$

where $\quad M_0 = \dfrac{\rho_0 P}{2}\left(1 - \dfrac{2}{\pi} + \dfrac{2e}{\pi\rho_0}\right)$

$$\delta_V = \frac{\partial U}{\partial P}$$

Using the above expression for U (remembering that M is also a function of P), and simplifying

$$\delta_V = 4P\rho_0\left\{\frac{\alpha\pi}{16AG}+\frac{1}{2AE}\left(\frac{2}{\pi}-\frac{\pi}{8}-\frac{2e}{\pi\rho_0}\right)+\frac{\rho_0^2}{2AEe\rho_0}\left(\frac{\pi}{8}-\frac{1}{\pi}+\frac{e^2}{\pi\rho_0^2}\right)\right\}$$

If e is small compared to ρ_0, then

$$\delta_V \approx \frac{\alpha\pi\,\rho_0 P}{4AG}+\frac{2P\rho_0}{AE}\left(\frac{2}{\pi}-\frac{\pi}{8}\right)+\frac{2P\rho_0^3}{AEe\rho_0}\left(\frac{\pi}{8}-\frac{1}{\pi}\right)$$

$$= \frac{\alpha\pi\,P\rho_0}{4AG}+0.488\frac{P\rho_0}{AE}+0.15\frac{P\rho_0^2}{AEe}$$

If we assume that the ring is thin and the effect of the strain energies due to the direct force and shear force are negligible, then the change in the vertical diameter is obtained as

$$\delta_V = \frac{P\rho_0^3}{EI}\left(\frac{\pi}{4}-\frac{2}{\pi}\right)$$

This can be seen from Eq. (35). When ρ_0 is large compared to y and $e \to 0$, $Ae\rho_0$ becomes equal to I according to flexure formula.

Example 3:

Determine the maximum tensile and maximum compressive stresses across the Sec. AA of the member loaded. Load P = 2000 kgf (19620 N).

Solution:

For the section ρ_0 = 11 cm, h = 6 cm, b = 4 cm.

$$\therefore \qquad \log\frac{\rho_0+h/2}{\rho_0-h/2} = \log\frac{7}{4} = 0.5596$$

From Eqs (38) and (39)

$$r_0 = \frac{6}{0.5596} = 10.73, \qquad e = 11-10.73 = 0.27$$

From Eq. (35), owing to bending moment M

$$\sigma_x' = \frac{M}{Ae}\frac{y}{(r_0-y)}$$

$$= \frac{M}{24\times 0.27}\frac{y}{(10.73-y)}$$

For the problem

$$M = p\,(a + a + h/2) = 19p$$

At C, $\quad y = -(e + h/2) = -3.27$

and, at D, $y=\frac{h}{2}-e=2.73$

Hence, $(\sigma_x^{'})c=\frac{19P}{24\times0.27}\times\frac{(-3.27)}{(10.73+3.27)}=0.6848P$

and $(\sigma_x^{'})_D=-\frac{19P}{24\times0.27}\times\frac{2.73}{(10.73-3.27)}=-1.001P$

The stress due to direct loading is

$$\sigma_x^{\cdot}=-\frac{P}{A}=-\frac{P}{24}=-0.0417\ P$$

Hence the combined stresses are

$(\sigma_x)_C$ = (0.6848 – 0.0417) P

= 0.6431P = 129 kgf/cm² (12642 kPa)

and

$(\sigma_x)_D$ = (– 1.001 – 0.0417) P

= – 1.0427 P = – 209 kgf/cm² (20482 kPa).

Example 4:

Determine the stress at point D of a hook having a trapezoidal section with the following dimensions: b_1 = 4 cm, b_2 = 1 cm, r_1 = 3 cm, b_2 = 10 cm, h = 7 cm, *force* P = 3000 kgf (29400 N).

Solution:

For the section

$$\int\frac{dA}{u}=[1+10(4-1)7]\log\frac{10}{3}-(4-1)$$

= 3.363 cm

$$A=\frac{1}{2}(b_1+b_2)h=\frac{35}{2}=17.5\ \text{cm}^2$$

∴ r_0 = A/3.363 = 17.5/3.363 = 5.204 cm

$$\rho_0=3+\frac{(b_1+2b_2)h}{3(b_1+b_2)}=3+\frac{14}{5}=5.80\ \text{cm}$$

∴ $e=\rho_0-r_0=0.596$

The moment across section D is

M = – 3000 ρ_0 = – 17,400 kgf cm (1705 Nm)

The normal stress due to bending is therefore

$$(\sigma_x^{'})_D = -\frac{M}{Ae}\frac{y}{r_0 - y}$$

$$= +\frac{17,400}{17.5 \times 0.596} \times \frac{2.204}{5.204 - 2.24}$$

$=1226$ kgf/cm^2 (120,148 kPa)

The normal stress due to axial loading is

$$(\sigma_x^{''})_D = \frac{3000}{A} = \frac{3000}{17.5} = 171 \text{ kgf/cm}^2$$

The total normal stress is therefore,

$$(\sigma_x)_D = 1397 \text{ kgf/cm}^2 \text{ or } 136{,}907 \text{ kPa}$$

REGARDING EULER–BERNOULLI HYPOTHESIS

We were able to solve the flexure problem because of the nature of the cross-section which remained plane after bending. It is natural to question how far this assumption is valid.

In order to determine the actual deformation of an initially plane section of a beam subjected to a general loading, we will have to use the methods of the theory of elasticity. Since this is beyond the scope of this book, we shall discuss here the condition necessary for a plane section to remain plane. We have from Hooke's law

$$\begin{aligned} \varepsilon_x &= \frac{1}{E}\left[\sigma_x - V(\sigma_y + \sigma_z)\right] \\ \varepsilon_y &= \frac{1}{E}\left[\sigma_y - V(\sigma_z + \sigma_x)\right] \\ \varepsilon_z &= \frac{1}{E}\left[\sigma_z - V(\sigma_x + \sigma_y)\right] \end{aligned} \tag{1}$$

Solving the above equations for the stress σ_x we get

$$\sigma_x = \frac{vE}{(1+v)(1+2v)}(\varepsilon_x + \varepsilon_y + \varepsilon_z) + \frac{E}{1+v}\varepsilon_x$$

or from Eq. (15)

$$\sigma_x = \lambda J_1 + 2G\varepsilon_x \tag{2}$$

where λ is a constant and G is the shear modulus. According to the Euler-Bernoulli hypothesis, we have

$$\sigma_x = \sigma_z = 0$$

Hence,

$$\sigma_x = E\varepsilon_x = E\frac{\partial u_x}{\partial x} \tag{3}$$

Differentiating,

$$\frac{\partial \sigma_x}{\partial x} = E\frac{\partial^2 u_x}{\partial x^2} \tag{4}$$

From equilibrium equation and stress-strain relations

$$\frac{\partial \sigma_x}{\partial x} = -\frac{\partial \tau_{xy}}{\partial y} - \frac{\partial \tau_{xz}}{\partial z}$$

$$= -G\frac{\partial}{\partial y}\left(\frac{\partial u_x}{\partial y} + \frac{\partial u_y}{\partial x}\right) - G\frac{\partial}{\partial z}\left(\frac{\partial u_x}{\partial z} + \frac{\partial u_z}{\partial x}\right)$$

$$= -G\left(\frac{\partial^2 u_x}{\partial y^2} + \frac{\partial^2 u_x}{\partial z^2}\right) - G\frac{\partial}{\partial x}\left(\frac{\partial u_y}{\partial y} + \frac{\partial u_z}{\partial z}\right)$$

$$= -G\left(\frac{\partial^2 u_x}{\partial y^2} + \frac{\partial^2 u_x}{\partial z^2}\right) - G\frac{\partial}{\partial x}(\varepsilon_y + \varepsilon_z) \tag{5}$$

Since $\sigma_y = \sigma_z = 0$, from Eq. (1),

$$\varepsilon_y = \varepsilon_z = -\frac{V}{E}\sigma_x$$

Hence, Eq. (4) becomes

$$\frac{\partial \sigma_x}{\partial x} = -G\left(\frac{\partial^2 u_x}{\partial y^2} + \frac{\partial^2 u_x}{\partial z^2}\right) + \frac{2VG}{E}\frac{\partial \sigma_x}{\partial x}$$

i.e.
$$\frac{\partial \sigma_x}{\partial x}\left(1 - \frac{2\nu G}{E}\right) = -G\left(\frac{\partial^2 u_x}{\partial y^2} + \frac{\partial^2 u_x}{\partial z^2}\right)$$

or
$$\frac{\partial \sigma_x}{\partial x} = -\frac{GE}{E - 2\nu G}\left(\frac{\partial^2 u_x}{\partial y^2} + \frac{\partial^2 u_x}{\partial z^2}\right) \tag{6}$$

Substituting in Eq. (3),

$$E\frac{\partial^2 u_x}{\partial x^2} + \frac{E}{E - 2\nu G}\left(\frac{\partial^2 u_x}{\partial y^2} + \frac{\partial^2 u_x}{\partial z^2}\right) = 0$$

i.e.
$$(E - 2\nu G)\frac{\partial^2 u_x}{\partial x^2} + G\left(\frac{\partial^2 u_x}{\partial y^2} + \frac{\partial^2 u_x}{\partial z^2}\right) = 0$$

or
$$A\frac{\partial^2 u_x}{\partial x^2} + G\left(\frac{\partial^2 u_x}{\partial y^2} + \frac{\partial^2 u_x}{\partial z^2}\right) = 0 \tag{7}$$

where A is a constant. From flexure formula and Eq. (3)

$$\sigma_x = \frac{My}{I_z} = E\frac{\partial u_x}{\partial x} \tag{8}$$

In the above equation, M is a function of x only and y is the distance measured from the neutral axis; I_z is the moment of inertia about the neutral axis which is taken as the z axis. Then

$$E\frac{\partial^2 u_x}{\partial z^2} = \frac{y}{I_z}\frac{\partial M}{\partial x}$$

Integrating Eq. (8)

$$Eu_x = \frac{y}{I_z}\int M\,dx + \phi(y,z)$$

where ϕ is a function of y and z only. Differentiating the above expression

$$E\frac{\partial^2 u_x}{\partial y^2} = \frac{\partial^2 \phi(y,z)}{\partial y^2}$$

and $$E\frac{\partial^2 u_x}{\partial z^2} = \frac{\partial^2 \phi(y,z)}{\partial z^2}$$

Substituting,

$$\frac{Ay}{EI_z}\frac{\partial M(x)}{\partial x} + \frac{G}{E}\left[\frac{\partial^2 \phi(y,z)}{\partial y^2} + \frac{\partial^2 \phi(y,z)}{\partial z^2}\right] = 0$$

or $$K_1\frac{\partial M(x)}{\partial x} = K_2\left[\frac{\partial^2 \phi(y,z)}{\partial y^2} + \frac{\partial^2 \phi(y,z)}{\partial z^2}\right]$$

The left-hand side quantity is a function of x alone or a constant and the right-hand side quantity is a function of y and z alone or a constant. Hence, both these quantities must be equal to a constant, i.e.

$$\frac{\partial M(x)}{\partial x} = a \text{ constant}$$

or $\quad M(x) = K_3x + K_5$

This means that $M(x)$ can only be due to a concentrated load or a pure moment.

In other words, the Euler-Bernoulli hypothesis that $\sigma_x = \dfrac{M_y}{I_z}$

(which is equivalent to plane sections remaining plane) will be valid only in those cases where the bending moment is a constant or varies linearly along the axis of the beam.

EXERCISES

1. The state of stress at a point is characterised by the matrix shown. Determine T_{11} such that there is at least one plane passing through the point in such a way that the resultant stress on that plane is zero. Determine the direction cosines of the normal to that plane.

$$[\tau_{ij}] = \begin{bmatrix} T_{11} & 2 & 1 \\ 2 & 0 & 2 \\ 1 & 2 & 0 \end{bmatrix}$$

$$\left[Ans.\ T_{11} = 2; n_x = \pm\frac{2}{3}; n_y = \pm\frac{1}{3}; n_z = \pm\frac{2}{3} \right]$$

2. If the rectangular components of stress at a point are as in the matrix below, determine the unit normal of a plane parallel to the z axis. i.e. $n_z = 0$, on which the resultant stress vector is tangential to the plane

$$[\tau_{ij}] = \begin{bmatrix} a & 0 & d \\ 0 & b & e \\ d & e & c \end{bmatrix}$$

$$\left[Ans.\ n_x = \pm\left(\frac{b}{b-a}\right)^{1/2};\ n_y = \pm\left(\frac{a}{a-b}\right)^{1/2};\ n_z = 0 \right]$$

3. Determine the principal stresses and their axes for the states of stress characterised by the following stress matrices (units are 1000 KPa).

(i) $$[\tau_{ij}] = \begin{bmatrix} 18 & 0 & 24 \\ 0 & 150 & 0 \\ 24 & 0 & 32 \end{bmatrix}$$

$$\begin{bmatrix} Ans.\ \sigma_1 = 50, n_x = 0.6, n_y = 0, n_z = 0.8 \\ \sigma_2 = 0, n_x = 0.8, n_y = 0, n_z = 0.6 \\ \sigma_3 = -50, n_x = n_z = 0, n_y = 1 \end{bmatrix}$$

(*ii*) $$[\tau_{ij}] = \begin{bmatrix} 3 & -10 & 0 \\ -10 & 0 & 30 \\ 0 & 30 & -27 \end{bmatrix}$$

$$\begin{bmatrix} Ans.\ \sigma_1 = 23, n_x = 0.394, n_y = 0.788, n_z = 0.473 \\ \sigma_2 = 0, n_x = 0.912, n_y = 0.274, n_z = 0.304 \\ \sigma_3 = -47, n_x = 0.914, n_y = 0.188, n_z = 0.288 \end{bmatrix}$$

4. The state of stress at a point is characterised by the components

$$\sigma_x = 12.31, \quad \sigma_y = 8.96, \quad \sigma_z = 4.34$$

$$\tau_{xy} = 4.20, \quad \tau_{xy} = 4.20, \quad \sigma_z = 0.84$$

Find the values of the principal stresses and their directions

$$\begin{bmatrix} Ans.\ \sigma_1 = 16.41, n_x = 0.709, n_y = 0.627, n_z = 0.322 \\ \sigma_2 = 0.65, n_x = 0.153, n_y = 0.583, n_z = 0.798 \\ \sigma_3 = 0.65, n_x = 0.153, n_y = 0.583, n_z = 0.798 \end{bmatrix}$$

5. For problem 8, determine the principal shears and the associated normal stresses.

$$\begin{bmatrix} Ans.\ \tau_3 = 3.94, \sigma_n = 12.48 \\ \tau_2 = 7.88, \sigma_n = 8.53 \\ \tau_1 = 3.95, \sigma_n = 4.52 \end{bmatrix}$$

6. For the state of stress at point characterised by the components (in 1000 KPa)

$$\sigma_x = 12, \quad \sigma_y = 4, \quad \sigma_z = 10, \quad \tau_{xy} = 3, \quad \tau_{yz} = \tau_{xz} = 0$$

determine the principal stresses and their directions.

$$\begin{bmatrix} Ans.\ \sigma_1 = 13; 18^\circ \text{ with } x \text{ axis }; n_z = 0 \\ \sigma_2 = 10; n_x = 0; n_y = 0; n_z = 1 \\ \sigma_3 = 3; -72^\circ \text{ with } x \text{ axis}; n_z = 0 \end{bmatrix}$$

7. Let $\sigma_x = -5c$, $\sigma_y = c$, $\sigma_z = c$, $\tau_{xy} = -c$, $\tau_{yz} = \tau_{xz} = 0$, where $c = 1000$ kPa. Determine the principal stresses, stress deviators, principal axes, greatest shearing stress and octahedral stresses.

$$\begin{bmatrix} Ans.\ \sigma_1 = (-2+\sqrt{10})c; n_z = 0 \text{ and } \theta = 9.2^\circ \text{ with } y \text{ axis} \\ \sigma_2 = c, n_x = n_y = 0; n_z = 1 \\ \sigma_3 = (-2-\sqrt{10})c; n_z = 0 \text{ and } \theta = 9.2^\circ \text{ with } x \text{ axis} \\ \tau_{max} = \sqrt{10}c; \sigma'_x = -4c; \sigma'_y = 2c; \sigma'_z = 2c \\ \sigma_{oct} = -c; \tau_{oct} = \frac{\sqrt{78}}{3}c \end{bmatrix}$$

8. A solid shaft of diameter $d = \sqrt{10}$ cm is subjected to a tensile force $P = 10{,}000$ N and a torque $T = 50000$ N cm. At point A on the surface,

determine the principal stresses, the octahedral shearing stress and the maximum shearing stress.

$$\left[\begin{aligned} Ans.\ \sigma_{1.2} &= \frac{2000}{\pi}\left(1 \pm \sqrt{13/5}\right) Pa \\ \tau_{max} &= \frac{2000}{\pi}\sqrt{\frac{13}{5}} Pa \\ \tau_{oct} &= \frac{4000}{\pi}\frac{\sqrt{22}}{2} Pa \end{aligned}\right]$$

9. In the above problem, if stringers C and D are made of magnesium alloy and stringers A and B of stainless steel, what will be the bending stresses in stringers A and D?

$$E_{st\ st} = 2 \times 10^6 \text{ kgf/cm}^2 \ (196 \times 10^6 \text{ kPa})$$

$$E_{mg\ alloy} = 0.4 \times 10^6 \text{ kgf/cm}^2 \ (39.2 \times 10^6 \text{ kPa})$$

Hint: Assume once again that sections that are plane before bending remain plane after bending. Hence, to produce the same strain, the stress will be proportional to E. Convert all the stringer areas into equivalent areas of one material. For example, the areas of stringers C and D in equivalent steel will be

$$A_C' = A_C \times \frac{E_{mag}}{E_{st}}, \quad \text{and} \quad A_D' = A_D \times \frac{E_{mag}}{E_{st}}$$

The areas of A and B remain unaltered. Solve the problem in the usual manner, using all equivalent steel stringers. Determine the stresses $(\sigma_x)_A'$ and $(\sigma_x)_D'$. Calculate the forces $F_A = (\sigma_x)'_A A'_A = (\sigma_x)'_A A_A$ and $F_D = (\sigma_x)'_D A'_D = (\sigma_x)'_D A_D$. Now, using the original areas calculate the stress as

$$(\sigma_x)'_A = (\sigma_x)'_A A'_A / A_A = (\sigma_x)'_A$$

$$(\sigma_x)'_D = (\sigma_x)'_D A'_D / A_D$$

$$\left[\begin{aligned} Ans.\ (\sigma_x)_A &= -48 \text{ kgf/cm}^2 \ (-47072 \text{ kPa}) \\ (\sigma_x)_D &= 425.6 \text{ kgf/cm}^2 \ (41737 \text{ kPa}) \end{aligned}\right]$$

10. Determine the ratio of the numerical value of σ_{max} and σ_{min} for a curved bar of rectangular cross-section in pure bending if $\rho_0 = 5$ cm and $h = r_2 - r_1 = 4$ cm. [*Ans.* 1.76]

11. For the ring shown in Fig. 31 determine the changes in the horizontal diameter,

    Hint: Apply two horizontal fictitious forces Q along the diameter. Calculate the total strain energy, Apply Castigliano's theorem.

$$\left[Ans.\ \delta_H = \frac{P\rho_0}{A}\left\{ -\frac{\alpha}{2G} + \frac{1}{E}\left(\frac{4}{\pi} - \frac{1}{2}\right) - \frac{1}{Ee\rho_0}\left[2e^2 - \rho_0^2\left(\frac{2}{\pi} - \frac{1}{2}\right)\right]\right\}\right]$$

12. Solve the previous problem if the bar is made of circular cross-section.

[*Ans.* 1.89]

13. Determine the dimensions b_1 and b_3 of an I-section shown in Fig. 6.25, to make σ_{max} and σ_{min} numerically equal in pure bending. The other dimensions are $r_1 = 3$ cm; $r_3 = 4$ cm; $r_4 = 6$ cm; $r_2 = 7$ cm; b_1 =1cm; and $b_1 + b_3 = 5$ cm.

[*Ans.* $b_1 = 3.67$ cm, $b_3 = 1.33$ cm]